QA
9
.C483

**Symbolic Logic and
Mechanical Theorem Proving**

Symbolic Logic and Mechanical Theorem Proving

CHIN-LIANG CHANG

RICHARD CHAR-TUNG LEE

National Institutes of Health
Bethesda, Maryland

ACADEMIC PRESS
A SUBSIDIARY OF HARCOURT BRACE JOVANOVICH, PUBLISHERS
New York   London   Toronto   Sydney   San Francisco

ACADEMIC PRESS, INC.
111 Fifth Avenue, New York, New York 10003

United Kingdom Edition published by
ACADEMIC PRESS, INC. (LONDON) LTD.
24/28 Oval Road, London NW1

LIBRARY OF CONGRESS CATALOG CARD NUMBER: 72-88358

AMS (MOS) 1970 Subject Classifications: 02B10, 02B05; 02-01,
02-04; 02G99; 68A40, 68A45

PRINTED IN THE UNITED STATES OF AMERICA

83 84 85   9 8 7 6 5

TO OUR WIVES

Contents

8. The Equality Relation

9. Some Proof Procedures Based on Herbrand's Theorem

10. Program Analysis

11. Deductive Question Answering, Problem Solving, and Program Synthesis

Preface

Artificial intelligence is a relatively new field of recent attraction to many researchers concerned with programming computers to perform tasks that require "intelligence." Mechanical theorem proving is an important subject in artificial intelligence.

It has long been man's ambition to find a general decision procedure to prove theorems. This desire clearly dates back to Leibniz (1646–1716); it was revived by Peano around the turn of the century and by Hilbert's school in the 1920s. A very important theorem was proved by Herbrand in 1930; he proposed a mechanical method to prove theorems. Unfortunately, his method was very difficult to apply since it was extremely time consuming to carry out by hand. With the invention of digital computers, logicians regained interest in mechanical theorem proving. In 1960, Herbrand's procedure was implemented by Gilmore on a digital computer, followed shortly by a more efficient procedure proposed by Davis and Putnam.

A major breakthrough in mechanical theorem proving was made by J. A. Robinson in 1965; he developed a single inference rule, the resolution principle, which was shown to be highly efficient and very easily implemented on computers. Since then, many improvements of the resolution principle have been made. Mechanical theorem proving has been applied to many areas, such as program analysis, program synthesis, deductive question-answering systems, problem-solving systems, and robot technology. The number of applications keeps growing.

There has also been an increasing awareness of the importance of symbolic logic itself for researchers in computer science. Some background in the first-order logic is now a necessary tool for reading papers in many journals related to computer science. However, most of the available books in symbolic logic are not computer-science oriented; they are almost all for mathematicians and philosophers.

Our goal was to write a book containing an introduction to symbolic logic and a thorough discussion of mechanical theorem proving and its applications. The book consists of three major parts. Chapters 2 and 3 constitute an introduction to symbolic logic, Chapters 4-9 introduce several techniques in mechanical theorem proving, and Chapters 10 and 11 show how theorem proving can be applied to various areas such as question answering, problem solving, program analysis, and program synthesis.

The book can serve several purposes. It can be used for a senior or graduate course on theorem proving. If used for a senior course, the advanced chapters, 6-11, may be neglected. For such a course, the reader needs no background in symbolic logic, only a basic knowledge of elementary set theory. On the other hand, if used as a graduate course in which a background in logic may be assumed, then Chapters 2 and 3 can be skipped. Since we adopted a purely model-theoretic approach to the first-order logic, this book is quite different from other logic books, which have a long tradition of presenting the first-order logic syntactically (axiomatically). Furthermore, this book emphasizes efficient computer implementations of proof techniques which, usually, have been of no concern to logicians. In some universities, mechanical theorem proving is taught as a part of artificial intelligence, not as a separate course. In such a case, this book can be used as a supplementary textbook in artificial intelligence to provide the reader with background in both mechanical theorem proving and application areas.

Since mechanical theorem proving is such a fast-growing field, it is inevitable that some significant results in this field had to be omitted from the book. However, we have tried to include in the Bibliography every published paper related to this field. The Bibliography is divided into three parts. The first part is a collection of material concerning artificial intelligence in general. The second part covers symbolic logic and mechanical theorem proving. The third part consists of material related to the applications of theorem-proving techniques.

Acknowledgments

It is impossible to name all the people who have helped the authors organize the material and clarify our views. Miss D. Koniver and Dr. L. Hodes of the National Institutes of Health, Professor J. Minker of the University of Maryland, Professor Z. Manna of Stanford University, Dr. S. K. Chang of IBM, Yorktown Heights, Dr. R. Waldinger of the Stanford Research Institute, Professor R. Anderson of the University of Houston, Mr. J. Yochelson, Mr. B. Crissey, and Mr. J. Hu of the Johns Hopkins University have all read a part or all of our manuscript. Although not responsible for any error of our own, they have contributed tremendously to the improvement of this book. We also would like to thank Dr. J. Slagle, who taught both of us artificial intelligence at the University of California, Berkeley. It was he who provided us with an intellectual and pleasant working atmosphere which made this work possible. Finally, our deepest thanks go to our loving wives for their patient and understanding support during the writing of this book.

**Symbolic Logic and
Mechanical Theorem Proving**

Chapter 1

Introduction

1.1 ARTIFICIAL INTELLIGENCE, SYMBOLIC LOGIC, AND THEOREM PROVING

Ever since the birth of the first modern computer, computer technology has developed at a fantastic speed. Today we see computers being used not only to solve computationally difficult problems, such as carrying out a fast Fourier transform or inverting a matrix of high dimensionality, but also being asked to perform tasks that would be called intelligent if done by human beings. Some such tasks are writing programs, answering questions, and proving theorems. Artificial intelligence is a field in computer science that is concerned with performing such tasks [Ernst and Newell, 1969; Feigenbaum and Feldman, 1963; Nilsson, 1971; Slagle, 1971].

The second half of the 1960s has been phenomenal in the field of artificial intelligence for the increase of interest in mechanical theorem proving. The widespread intensive interest in mechanical theorem proving is caused not only by the growing awareness that the ability to make logical deductions is an integral part of human intelligence, but is perhaps more a result of the status of mechanical theorem-proving techniques in the late 1960s. The foundation of mechanical theorem proving was developed by Herbrand in 1930. His method was impractical to apply until the invention of the digital computer. It was not until the landmark paper by J. A. Robinson in 1965, together with the development of the resolution principle, that major steps were taken to achieve realistic computer-implemented theorem provers. Since 1965, many refinements of the resolution principle have been made.

Parallel to the progress in improving mechanical theorem-proving techniques was the progress in applying theorem-proving techniques to various problems in artificial intelligence. These were first applied to deductive question answering and later to problem solving, program synthesis, program analysis, and many others.

There are many points of view from which we can study symbolic logic. Traditionally, it has been studied from philosophical and mathematical orientations. In this book, we are interested in the *applications* of symbolic logic to solving intellectually difficult problems. That is, we want to use symbolic logic to represent problems and to obtain their solutions.

We shall now present some very simple examples to demonstrate how symbolic logic can be used to represent problems. Since we have not yet formally discussed symbolic logic, the reader should rely on his intuition to follow the text at this moment.

Let us consider a simple example. Assume that we have the following facts:

F_1:   If it is hot and humid, then it will rain.
F_2:   If it is humid, then it is hot.
F_3:   It is humid now.

The question is: Will it rain?

The above facts are given in English. We shall use symbols to represent them. Let P, Q, and R represent "It is hot," "It is humid," and "It will rain," respectively. We also need some logical symbols. In this case, we shall use $\wedge$ to represent "and" and $\rightarrow$ to represent "imply." Then the above three facts are represented as:

F_1: $P \wedge Q \rightarrow R$

F_2: $Q \rightarrow P$

F_3: Q.

Thus we have translated English sentences into logical formulas. The reader will see later that whenever F_1, F_2, and F_3 are true, the formula

F_4: R

is true. Therefore, we say that F_4 *logically follows from* F_1, F_2, and F_3. That is, it will rain.

Let us consider another example. We have the following facts:

F_1:   Confucius is a man.
F_2:   Every man is mortal.

To represent F_1 and F_2, we need a new concept, called a predicate. We may let $P(x)$ and $Q(x)$ represent "x is a man" and "x is mortal," respectively. We also use $(\forall x)$ to represent "for all x." Then the above facts are represented by

F_1: P(Confucius)
F_2: $(\forall x)(P(x) \rightarrow Q(x))$.

Again, the reader will see later that from F_1 and F_2, we can *logically deduce*

F_3: Q(Confucius)

which means that Confucius is mortal.

In the above two examples, we essentially had to prove that a formula *logically follows from* other formulas. We shall call a statement that a formula logically follows from formulas a *theorem*. A demonstration that a theorem is true, that is that a formula logically follows from other formulas, will be called a *proof* of the theorem. The *problem of mechanical theorem proving* is to consider mechanical methods for finding proofs of theorems.

There are many problems that can be conveniently transformed into theorem-proving problems; we shall list some of them.

1. In a *question-answering system*, facts can be represented by logical formulas. Then, to answer a question from the facts, we prove that a formula corresponding to the answer is derivable from the formulas representing the facts.

2. In the *program-analysis problem*, we can describe the execution of a program by a formula A, and the condition that the program will terminate by another formula B. Then, verifying that the program will terminate is equivalent to proving that formula B follows from formula A.

3. In the *graph-isomorphism problem*, we want to know whether a graph is isomorphic to a subgraph of another graph. This problem is not merely an interesting problem in mathematics; it is also a practical problem. For example, the structure of an organic compound can be described by a graph. Therefore, testing whether a substructure of the structure of an organic compound is the structure of another organic compound is a graph-isomorphism problem. For this problem, we can describe graphs by formulas. Thus, the problem may be stated as proving that the formula representing a graph follows from the formula representing another graph.

4. In the *state-transformation problem*, there are a collection of states and a collection of (state) operators. When one operator is applied to a state, a new state will be obtained. Starting with an initial state, one tries to find a sequence of operators that will transform the initial state into a desired state. In this case, we can describe states and the state transition rules by logical formulas. Thus, transforming the initial state into the desired state may be

regarded as verifying that the formula representing the desired state follows from the formula representing both the states and the state transition rules.

Since many problems can be formulated as theorem-proving problems, theorem proving is a very important field in artificial intelligence science. Thanks to the relentless efforts of many researchers, great progress has been made in using computers to prove theorems. In this book, both the theory and applications of mechanical theorem proving will be considered.

1.2   MATHEMATICAL BACKGROUND

In this section, we provide some basic mathematical concepts that will be used in this book. All of these concepts can be easily found in elementary mathematical textbooks [Lightstone, 1964; Pfeiffer, 1964; Stoll, 1961]. The following basic definitions of set theory are used.

A *set* is a collection of elements (members). A set that contains no elements is called the *empty set* $\emptyset$. Let A and B be two sets. $x \in A$ is used to denote that x is a member of A, or that x belongs to A.

The set A is *identical* to the set B, denoted $A = B$, if and only if both A and B have the same elements.

The set A is a *subset* of the set B, denoted $A \subseteq B$, if and only if each element of A is an element of B.

The set A is a *proper subset* of the set B, denoted $A \subset B$, if and only if $A \subseteq B$ and $A \neq B$.

The *union* of two sets A and B, denoted by $A \cup B$, is the set consisting of all elements that belong to A or to B.

The *intersection* of two sets A and B, denoted by $A \cap B$, is the set consisting of all elements that belong both to A and to B.

The *difference* of two sets A and B, denoted by $A - B$, is the set consisting of all elements that belong to A but do not belong to B.

We now define relations and functions.

An *ordered pair* of elements is denoted by (x, y), where x is called the first *coordinate* and y is called the second *coordinate*.

A *relation* is a set of ordered pairs. For example, the equality relation is a set of ordered pairs, each of which has the first coordinate equal to its second coordinate. The *domain* of a relation R is the set of all first coordinates of elements of R, and its *range* is the set of all second coordinates.

A *function* is a relation such that no two distinct elements have the same first coordinate. If f is a function and x is an element of its domain, then $f(x)$ denotes the second coordinate of the unique element of f whose first coordinate is x. $f(x)$ is called the *value* of f at x, and we say that f assigns the value $f(x)$ to x.

Throughout this book, we use many conventional symbols. For example,

$>$ means "greater than"; $\geqslant$, "greater than or equal to"; $<$, "less than"; $\leqslant$, "less than or equal to"; $=$, "equal to"; $\neq$, "not equal to"; $\triangleq$, "defined as"; and so on. The equality sign "$=$" will be used for many purposes. It may be used to mean "is defined by," "is identical to," "is equivalent to," or "is equal to." This will not cause the reader any confusion since he can always determine its exact meaning from the text.

REFERENCES

Ernst, G. W., and A. Newell (1969): "GPS: a Case Study in Generality and Problem Solving," Academic Press, New York.

Feigenbaum, E., and J. Feldman, eds. (1963): "Computers and Thought," McGraw-Hill, New York.

Lightstone, A. H. (1964): "The Axiomatic Method, an Introduction to Mathematical Logic," Prentice Hall, Englewood Cliffs, New Jersey.

Nilsson, N. J. (1971): "Problem Solving Methods in Artificial Intelligence," McGraw-Hill, New York.

Pfeiffer, P. E. (1964): "Sets, Events and Switching," McGraw-Hill, New York.

Slagle, J. R. (1971): "Artificial Intelligence, the Heuristic Programming Approach," McGraw-Hill, New York.

Stoll, R. (1961): "Sets, Logic and Axiomatic Theories," Freeman, San Francisco.

Stoll, R. (1963): "Set Theory and Logic," Freeman, San Francisco.

Chapter 2

The Propositional Logic

2.1 INTRODUCTION

Symbolic logic considers languages whose essential purpose is to symbolize reasoning encountered not only in mathematics but also in daily life. In this chapter, we shall first study the simplest symbolic logic—the propositional logic (or the propositional calculus). In the next chapter, we shall consider a more general one—the first-order logic (or the first-order predicate calculus).

In the propositional logic, we are interested in declarative sentences that can be either *true* or *false*, but not both. Any such declarative sentence is called a proposition. More formally, a *proposition* is a declarative sentence that is either true or false, but not both. Examples of propositions are: "Snow is white," "Sugar is a hydrocarbon," "Smith has a Ph.D. degree." The "true" or "false" assigned to a proposition is called the *truth value* of the proposition. Customarily, we represent "true" by T and "false" by F. Furthermore, for convenience, we shall use an uppercase symbol or a string of uppercase symbols to denote a proposition. For instance, we may denote the propositions above as follows:

$$P \triangleq \text{Snow is white,}$$
$$Q \triangleq \text{Sugar is a hydrocarbon,}$$
$$R \triangleq \text{Smith has a Ph.D. degree.}$$

The symbols, such as P, Q, and R, that are used to denote propositions are called *atomic formulus*, or *atoms*.

From propositions, we can build *compound* propositions by using *logical connectives*. Examples of compound propositions are: "Snow is white *and* the sky is clear" and "*If* John is not at home, *then* Mary is at home." The logical connectives in the above two compund propositions are "and" and "if ··· then." In the propositional logic, we shall use five logical connectives: $\sim$ (*not*), $\wedge$ (*and*), $\vee$ (*or*), $\rightarrow$ (*if ··· then*), and $\leftrightarrow$ (*if and only if*). These five logical connectives can be used to build compound propositions from propositions. More generally, they can be used to construct more complicated compound propositions from compound propositions by applying them repeatedly. For example, let us represent "Humidity is high" by P, "Temperature is high" by Q, and "One feels comfortable" by C. Then the sentence "If the humidity is high and the temperature is high, then one does not feel comfortable" may be represented by $((P \wedge Q) \rightarrow (\sim C))$. Therefore, we see that a compound proposition can express a rather complicated idea. In the propositional logic, an expression that represents a proposition, such as P, or a compound proposition, such as $((P \wedge Q) \rightarrow (\sim C))$, is called a well-formed formula.

Definition   *Well-formed formulas*, or *formulas* for short, in the propositional logic are defined recursively as follows:

1. An atom is a formula.
2. If G is a formula, then $(\sim G)$ is a formula.
3. If G and H are formulas, then $(G \wedge H)$, $(G \vee H)$, $(G \rightarrow H)$, and $(G \leftrightarrow H)$ are formulas.
4. All formulas are generated by applying the above rules.

It is not difficult to see that expressions such as $(P \rightarrow)$ and $(P \vee)$ are not formulas. Throughout the book, when no confusion is possible, some pairs of parentheses may be dropped. For example, $P \vee Q$ and $P \rightarrow Q$ are the formulas $(P \vee Q)$ and $(P \rightarrow Q)$, respectively. We can further omit the use of parentheses by assigning decreasing ranks to the propositional connectives as follows:

$$\leftrightarrow, \quad \rightarrow, \quad \wedge, \quad \vee, \quad \sim,$$

and requiring that the connective with greater rank always reaches further. Thus $P \rightarrow Q \wedge R$ will mean $(P \rightarrow (Q \wedge R))$, and $P \rightarrow Q \wedge \sim R \vee S$ will mean $(P \rightarrow (Q \wedge ((\sim R) \vee S)))$.

Let G and H be two formulas. Then the truth values of the formulas $(\sim G)$, $(G \wedge H)$, $(G \vee H)$, $(G \rightarrow H)$, and $(G \leftrightarrow H)$ are related to the truth values of G and H in the following way:

1. $\sim G$ is true when G is false, and is false when G is true. $\sim G$ is called the *negation* of G.

2. $(G \wedge H)$ is true if G and H are both true; otherwise, $(G \wedge H)$ is false. $(G \wedge H)$ is called the *conjunction* of G and H.

3. $(G \vee H)$ is true if at least one of G and H is true; otherwise, $(G \vee H)$ is false. $(G \vee H)$ is called the *disjunction* of G and H.

4. $(G \rightarrow H)$ is false if G is true and H is false; otherwise, $(G \rightarrow H)$ is true. $(G \rightarrow H)$ is read as "*If G, then H,*" or "*G implies H.*"

5. $(G \leftrightarrow H)$ is true whenever G and H have the same truth values; otherwise $(G \leftrightarrow H)$ is false. $(G \leftrightarrow H)$ is read as "*G if and only if H.*"

The above relations can be represented conveniently as in Table 2.1. Based on this table, we shall describe ways to evaluate the truth values of a formula in terms of the truth values of atoms occurring in the formula.

TABLE 2.1

G	H	$\sim G$	$(G \wedge H)$	$(G \vee H)$	$(G \rightarrow H)$	$(G \leftrightarrow H)$
T	T	F	T	T	T	T
T	F	F	F	T	F	F
F	T	T	F	T	T	F
F	F	T	F	F	T	T

2.2 INTERPRETATIONS OF FORMULAS IN THE PROPOSITIONAL LOGIC

Suppose that P and Q are two atoms and that the truth values of P and Q are T and F, respectively. Then, according to the second row of Table 2.1 with P and Q substituted for G and H, respectively, we find that the truth values of $(\sim P)$, $(P \wedge Q)$, $(P \vee Q)$, $(P \rightarrow Q)$, and $(P \leftrightarrow Q)$ are F, F, T, F, and F, respectively. Similarly, the truth value of any formula can be evaluated in terms of truth values of atoms.

Example 2.1

Consider the formula

$$G \triangleq (P \wedge Q) \rightarrow (R \leftrightarrow (\sim S)).$$

The atoms in this formula are P, Q, R, and S. Suppose the truth values of P, Q, R, and S are T, F, T, and T, respectively. Then $(P \wedge Q)$ is F since Q is false; $(\sim S)$ is F since S is T; $(R \leftrightarrow (\sim S))$ is F since R is T and $(\sim S)$ is F; and $(P \wedge Q) \rightarrow (R \leftrightarrow (\sim S))$ is T since $(P \wedge Q)$ is F and $(R \leftrightarrow (\sim S))$ is F. Therefore, the formula G is T if P, Q, R, and S are assigned T, F, T, and T, respectively. The assignment of the truth values $\{T, F, T, T\}$ to $\{P, Q, R, S\}$, respectively, will be called an *interpretation* of the formula G. Since each one of P, Q, R, and S can be assigned either T or F, there are $2^4 = 16$ interpretations of the formula G. In Table 2.2, we give the truth values of the formula G under all these 16 interpretations.

TABLE 2.2

TRUTH TABLE OF $(P \wedge Q) \to (R \leftrightarrow (\sim S))$

P	Q	R	S	~S	(P ∧ Q)	(R ↔ (~ S))	(P ∧ Q)→(R ↔ (~ S))
T	T	T	T	F	T	F	F
T	T	T	F	T	T	T	T
T	T	F	T	F	T	T	T
T	T	F	F	T	T	F	F
T	F	T	T	F	F	F	T
T	F	T	F	T	F	T	T
T	F	F	T	F	F	T	T
T	F	F	F	T	F	F	T
F	T	T	T	F	F	F	T
F	T	T	F	T	F	T	T
F	T	F	T	F	F	F	T
F	T	F	F	T	F	T	T
F	F	T	T	F	F	F	T
F	F	T	F	T	F	T	T
F	F	F	T	F	F	T	T
F	F	F	F	T	F	F	T

A table, such as Table 2.2, that displays the truth values of a formula G for all possible assignments of truth values to atoms occurring in G is called a *truth table* of G.

We now give a formal definition of an interpretation of a propositional formula.

Definition   Given a propositional formula G, let $A_1, A_2, ..., A_n$ be the atoms occurring in the formula G. Then an *interpretation* of G is an assignment of truth values to $A_1, ..., A_n$ in which every A_i is assigned either T or F, but not both.

Definition   A formula G is said to be *true under (or in) an interpretation* if and only if G is evaluated to T in the interpretation; otherwise, G is said to be *false under the interpretation*.

If there are n distinct atoms in a formula, then there will be 2^n distinct interpretations for the formula. Sometimes, if $A_1, ..., A_n$ are all atoms occurring in a formula, it may be more convenient to represent an interpretation by a set $\{m_1, \ldots, m_n\}$, where m_i is either A_i or $\sim A_i$. For example, the set $\{P, \sim Q, \sim R, S\}$ represents an interpretation in which P, Q, R, and S are, respectively, assigned T, F, F, and T. That is, if an atom A is in a set that represents an interpretation, then A is assigned T; while if the negation of the atom A is in the set, then A is assigned F. This convention will be kept throughout the book.

2.3   VALIDITY AND INCONSISTENCY IN THE PROPOSITIONAL LOGIC

In this section, we shall consider formulas that are true under all their possible interpretations and formulas that are false under all their possible interpretations.

Example 2.2

Let us consider the formula

$$G \triangleq ((P \rightarrow Q) \wedge P) \rightarrow Q.$$

The atoms in this formula are P and Q. Hence, the formula G has $2^2 = 4$ interpretations. The truth values of G under all its four interpretations are given in Table 2.3. We see that the formula G is true under all its interpretations. This formula will be called a *valid formula* (or a *tautology*).

TABLE 2.3
TRUTH TABLE OF $((P \rightarrow Q) \wedge P) \rightarrow Q$

P	Q	$(P \rightarrow Q)$	$(P \rightarrow Q) \wedge P$	$((P \rightarrow Q) \wedge P) \rightarrow Q$
T	T	T	T	T
T	F	F	F	T
F	T	T	F	T
F	F	T	F	T

Example 2.3

Consider the formula

$$G \triangleq (P \rightarrow Q) \wedge (P \wedge \sim Q).$$

The truth table of G is given in Table 2.4. We see that G is false under all its interpretations. This formula will be called an *inconsistent formula* (or a *contradiction*).

TABLE 2.4

TRUTH TABLE OF $(P \rightarrow Q) \wedge (P \wedge \sim Q)$

P	Q	$\sim Q$	$(P \rightarrow Q)$	$(P \wedge \sim Q)$	$(P \rightarrow Q) \wedge (P \wedge \sim Q)$
T	T	F	T	F	F
T	F	T	F	T	F
F	T	F	T	F	F
F	F	T	T	F	F

We now give formal definitions for validity and inconsistency.

Definition   A formula is said to be *valid* if and only if it is true under all its interpretations. A formula is said to be *invalid* if and only if it is not valid.

Definition   A formula is said to be *inconsistent* (or *unsatisfiable*) if and only if it is false under all its interpretations. A formula is said to be *consistent* (or *satisfiable*) if and only if it is not inconsistent.

By the above definitions, the following observations are obvious:

1. A formula is valid if and only if its negation is inconsistent.
2. A formula is inconsistent if and only if its negation is valid.
3. A formula is invalid if and only if there is at least one interpretation under which the formula is false.
4. A formula is consistent if and only if there is at least one interpretation under which the formula is true.
5. If a formula is valid, then it is consistent, but not vice versa.
6. If a formula is inconsistent, then it is invalid, but not vice versa.

Example 2.4

By using the truth table techniques, the reader should be able to establish the following:

    a. $(P \wedge \sim P)$ is inconsistent; therefore also invalid.
    b. $(P \vee \sim P)$ is valid; therefore also consistent.
    c. $(P \rightarrow \sim P)$ is invalid, yet it is consistent.

If a formula F is true under an interpretation I, then we say that I *satisfies* F, or F *is satisfied* by I. On the other hand, if a formula F is false under an interpretation I, then we say that I *falsifies* F or F *is falsified* by I. For example, the formula $(P \wedge (\sim Q))$ is satisfied by the interpretation $\{P, \sim Q\}$, but is falsified by the interpretation $\{P, Q\}$. When an interpretation I satisfies a formula F, I is also called a *model* of F.

It will be shown later that the proof of the validity or inconsistency of a formula is a very important problem. In the propositional logic, since the number of interpretations of a formula is finite, one can always decide whether or not a formula in the propositional logic is valid (inconsistent) by exhaustively examining all of its possible interpretations.

2.4   NORMAL FORMS IN THE PROPOSITIONAL LOGIC

As will be clear later, it is often necessary to transform a formula from one form to another, especially to a "normal form." This is accomplished by replacing a formula in the given formula by a formula "equivalent" to it and repeating this process until the desired form is obtained. By "equivalent," we mean the following:

Definition   Two formulas F and G are said to be *equivalent* (or F is *equivalent* to G), denoted $F = G$, if and only if the truth values of F and G are the same under every interpretation of F and G.

Example 2.5

We can verify that $(P \rightarrow Q)$ is equivalent to $(\sim P \vee Q)$ by examining the truth table, Table 2.5.

TABLE 2.5
TRUTH TABLE OF $(P \rightarrow Q)$ AND
$(\sim P \vee Q)$

P	Q	$(P \rightarrow Q)$	$(\sim P \vee Q)$
T	T	T	T
T	F	F	F
F	T	T	T
F	F	T	T

We need an adequate supply of equivalent formulas in order to carry out the transformation of formulas. In the propositional logic, let ■ denote the formula that is always true and □ the formula that is always false. Then we have some useful pairs of equivalent formulas given in Table 2.6, where F, G, and H are all formulas. For simplicity, we shall call each of them a "law."

The laws in Table 2.6 can be verified by using truth tables. Laws (2.3a), (2.3b) are often called *commutative* laws; (2.4a), (2.4b) *associative* laws; (2.5a), (2.5b), *distributive* laws; and (2.10a), (2.10b), *De Morgan's* laws.

Because of the associative laws, the parentheses in $(F \vee G) \vee H$ or $F \vee (G \vee H)$ can be dropped. That is, we can write $F \vee G \vee H$ for $(F \vee G) \vee H$ and $F \vee (G \vee H)$. More generally, we can write $F_1 \vee F_2 \vee \cdots \vee F_n$ without

TABLE 2.6

(2.1)	$F \leftrightarrow G = (F \to G) \wedge (G \to F)$		
(2.2)	$F \to G = \sim F \vee G$		
(2.3)	(a) $F \vee G = G \vee F$;	(b)	$F \wedge G = G \wedge F$
(2.4)	(a) $(F \vee G) \vee H = F \vee (G \vee H)$;	(b)	$(F \wedge G) \wedge H = F \wedge (G \wedge H)$
(2.5)	(a) $F \vee (G \wedge H) = (F \vee G) \wedge (F \vee H)$;	(b)	$F \wedge (G \vee H) = (F \wedge G) \vee (F \wedge H)$
(2.6)	(a) $F \vee \square = F$;	(b)	$F \wedge \blacksquare = F$
(2.7)	(a) $F \vee \blacksquare = \blacksquare$;	(b)	$F \wedge \square = \square$
(2.8)	(a) $F \vee \sim F = \blacksquare$;	(b)	$F \wedge \sim F = \square$
(2.9)	$\sim(\sim F) = F$		
(2.10)	(a) $\sim(F \vee G) = \sim F \wedge \sim G$;	(b)	$\sim(F \wedge G) = \sim F \vee \sim G$

any ambiguity, where $F_1, F_2, ..., F_n$ are formulas. $F_1 \vee F_2 \vee \cdots \vee F_n$ is true whenever at least one of the F_i, $1 \leqslant i \leqslant n$, is true; otherwise, it is false. $F_1 \vee F_2 \vee \cdots \vee F_n$ is called the *disjunction* of $F_1, ..., F_n$. Similarly, we can write $F_1 \wedge F_2 \wedge \cdots \wedge F_n$, which is true if all $F_1, F_2, ..., F_n$ are true, and which is false otherwise. $F_1 \wedge F_2 \wedge \cdots \wedge F_n$ is called the *conjunction* of $F_1, ..., F_n$.

Note that the order in which the F_i appear in a disjunction or a conjunction is immaterial.

For example,

$$F_1 \vee F_2 \vee F_3 = F_1 \vee F_3 \vee F_2 = F_2 \vee F_1 \vee F_3 = F_3 \vee F_2 \vee F_1$$
$$= F_2 \vee F_3 \vee F_1 = F_3 \vee F_1 \vee F_2.$$

We now define normal forms as follows.

Definition   A *literal* is an atom or the negation of an atom.

Definition   A formula F is said to be in a *conjunctive normal form* if and only if F has the form $F \triangleq F_1 \wedge \cdots \wedge F_n, n \geqslant 1$, where each of $F_1, ..., F_n$ is a disjunction of literals.

Example 2.6

Let P, Q, and R be atoms. Then $F \triangleq (P \vee \sim Q \vee R) \wedge (\sim P \vee Q)$ is a formula in a conjunctive normal form. For this formula, $F_1 = (P \vee \sim Q \vee R)$ and $F_2 = (\sim P \vee Q)$. Clearly, F_1 is a disjunction of the literals P, $\sim Q$, and R, and F_2 is a disjunction of the literals $\sim P$ and Q.

Definition   A formula F is said to be in a *disjunctive normal form* if and only if F has the form of $F \triangleq F_1 \vee \cdots \vee F_n$, $n \geqslant 1$, where each of $F_1, ..., F_n$ is a conjunction of literals.

Example 2.7

Let P, Q, and R be atoms. Then $F \triangleq (\sim P \wedge Q) \vee (P \wedge \sim Q \wedge \sim R)$ is a formula in a disjunctive normal form. For this formula, $F_1 = (\sim P \wedge Q)$

and $F_2 = (P \wedge \sim Q \wedge \sim R)$. Clearly, F_1 is a conjunction of the literals $\sim P$ and Q, and F_2 is a conjunction of the literals P, $\sim Q$, and $\sim R$.

Any formula can be transformed into a normal form. This is accomplished easily by using the laws given in Table 2.6. The following is an *outline of a transformation procedure*:

Step 1   Use the laws

(2.1) $F \leftrightarrow G = (F \rightarrow G) \wedge (G \rightarrow F)$

(2.2) $F \rightarrow G = \sim F \vee G$

to eliminate the logical connectives $\leftrightarrow$ and $\rightarrow$.

Step 2   Repeatedly use the law

(2.9) $\sim(\sim F) = F$

and De Morgan's laws

(2.10a) $\sim(F \vee G) = \sim F \wedge \sim G$

(2.10b) $\sim(F \wedge G) = \sim F \vee \sim G$

to bring the negation signs immediately before atoms.

Step 3   Repeatedly use the distributive laws

(2.5a) $F \vee (G \wedge H) = (F \vee G) \wedge (F \vee H)$

(2.5b) $F \wedge (G \vee H) = (F \wedge G) \vee (F \wedge H)$

and the other laws to obtain a normal form.

Example 2.8

Obtain a disjunctive normal form for the formula $(P \vee \sim Q) \rightarrow R$.
Since

$$(P \vee \sim Q) \rightarrow R = \sim(P \vee \sim Q) \vee R \qquad \text{by (2.2)}$$
$$= (\sim P \wedge \sim(\sim Q)) \vee R \qquad \text{by (2.10a)}$$
$$= (\sim P \wedge Q) \vee R \qquad \text{by (2.9),}$$

a disjunctive normal form of $(P \vee \sim Q) \rightarrow R$ is $(\sim P \wedge Q) \vee R$.

Example 2.9

Obtain a conjunctive normal form for the formula $(P \wedge (Q \rightarrow R)) \rightarrow S$.

$(P \wedge (Q \rightarrow R)) \rightarrow S$

$$= (P \wedge (\sim Q \vee R)) \rightarrow S \qquad \text{by (2.2)}$$

$$= \sim(P \wedge (\sim Q \vee R)) \vee S \qquad \text{by (2.2)}$$

$$= (\sim P \vee \sim(\sim Q \vee R)) \vee S \qquad \text{by (2.10b)}$$

$$= (\sim P \vee (\sim(\sim Q) \wedge \sim R)) \vee S \qquad \text{by (2.10a)}$$

$$= (\sim P \vee (Q \wedge \sim R)) \vee S \qquad \text{by (2.9)}$$

$$= ((\sim P \vee Q) \wedge (\sim P \vee \sim R)) \vee S \qquad \text{by (2.5a)}$$

$$= S \vee ((\sim P \vee Q) \wedge (\sim P \vee \sim R)) \qquad \text{by (2.3a)}$$

$$= (S \vee (\sim P \vee Q)) \wedge (S \vee (\sim P \vee \sim R)) \qquad \text{by (2.5a)}$$

$$= (S \vee \sim P \vee Q) \wedge (S \vee \sim P \vee \sim R) \qquad \text{by (2.4a)}.$$

A conjunctive normal form of $(P \wedge (Q \rightarrow R)) \rightarrow S$ is $(S \vee \sim P \vee Q) \wedge (S \vee \sim P \vee \sim R)$.

2.5 LOGICAL CONSEQUENCES

In mathematics as well as in daily life, we often have to decide whether one statement follows from some other statements. This leads to the concept of "logical consequence." Before giving a formal definition of logical consequence, let us first consider the following example.

Example 2.10

Suppose the stock prices go down if the prime interest rate goes up. Suppose also that most people are unhappy when stock prices go down. Assume that the prime interest rate does go up. Show that you can conclude that most people are unhappy.

To show the above conclusion, let us denote the statements as follows:

$$P \triangleq \text{Prime interest rate goes up,}$$

$$S \triangleq \text{Stock prices go down,}$$

$$U \triangleq \text{Most people are unhappy.}$$

There are four statements in this example. They are

(1) If the prime interest rate goes up, stock prices go down.
(2) If stock prices go down, most people are unhappy.

(3) The prime interest rate goes up.
(4) Most people are unhappy.

These statements are first symbolized as

(1') $P \to S$

(2') $S \to U$

(3') P

(4') U.

We now show that (4') is true whenever (1') $\wedge$ (2') $\wedge$ (3') is true.
Let us first transform $((P \to S) \wedge (S \to U) \wedge P)$ (representing (1') $\wedge$ (2') $\wedge$ (3')) into a normal form:

$$((P \to S) \wedge (S \to U) \wedge P) = ((\sim P \vee S) \wedge (\sim S \vee U) \wedge P) \qquad \text{by (2.2)}$$

$$= (P \wedge (\sim P \vee S) \wedge (\sim S \vee U)) \qquad \text{by (2.3b)}$$

$$= (((P \wedge \sim P) \vee (P \wedge S)) \wedge (\sim S \vee U)) \qquad \text{by (2.5b)}$$

$$= ((\square \vee (P \wedge S)) \wedge (\sim S \vee U)) \qquad \text{by (2.8b)}$$

$$= (P \wedge S) \wedge (\sim S \vee U) \qquad \text{by (2.6a)}$$

$$= (P \wedge S \wedge \sim S) \vee (P \wedge S \wedge U) \qquad \text{by (2.5b)}$$

$$= (P \wedge \square) \vee (P \wedge S \wedge U) \qquad \text{by (2.8b)}$$

$$= \square \vee (P \wedge S \wedge U) \qquad \text{by (2.7b)}$$

$$= P \wedge S \wedge U \qquad \text{by (2.6a).}$$

Therefore, if $((P \to S) \wedge (S \to U) \wedge P)$ is true, then $(P \wedge S \wedge U)$ is true. Since $(P \wedge S \wedge U)$ is true only if P, S, and U are all true, we conclude that U is true.

Because U is true whenever $(P \to S)$, $(S \to U)$, and P are true, in logic, U is called a *logical consequence* of $(P \to S)$, $(S \to U)$, and P. More formally, we define a logical consequence as follows.

Definition   Given formulas $F_1, F_2, ..., F_n$ and a formula G, G is said to be a *logical consequence* of $F_1, ..., F_n$ (or G *logically follows from* $F_1, ..., F_n$) if and only if for any interpretation I in which $F_1 \wedge F_2 \wedge \cdots \wedge F_n$ is true, G is also true. $F_1, F_2, ..., F_n$ are called *axioms* (or *postulates, premises*) of G.

Theorem 2.1   Given formulas $F_1, ..., F_n$ and a formula G, G is a logical consequence of $F_1, ..., F_n$ if and only if the formula $((F_1 \wedge \cdots \wedge F_n) \to G)$ is valid.

Proof ($\Rightarrow$)   Suppose G is a logical consequence of $F_1, ..., F_n$. Let I be an arbitrary interpretation. If $F_1, ..., F_n$ are true in I, then, by the definition of logical consequence, G is true in I. Hence, $((F_1 \wedge \cdots \wedge F_n) \to G)$ is true in I. On the other hand, if $F_1, \ldots, F_n$ are false in I, then $((F_1 \wedge \cdots \wedge F_n) \to G)$ is true in I. Thus, we have shown that $((F_1 \wedge \cdots \wedge F_n) \to G)$ is true under any interpretation. That is, $((F_1 \wedge \cdots \wedge F_n) \to G)$ is a valid formula.

($\Leftarrow$) Suppose $((F_1 \wedge \cdots \wedge F_n) \to G)$ is a valid formula. For any interpretation I, if $(F_1 \wedge \cdots \wedge F_n)$ is true in I, G must be true in I. Therefore, G is a logical consequence of $F_1, \ldots, F_n$. Q.E.D.

Theorem 2.2 Given formulas $F_1, \ldots, F_n$ and a formula G, G is a logical consequence of $F_1, \ldots, F_n$ if and only if the formula $(F_1 \wedge \cdots \wedge F_n \wedge \sim G)$ is inconsistent.

Proof By Theorem 2.1, G is a logical consequence of $F_1, \ldots, F_n$ if and only if the formula $((F_1 \wedge \cdots \wedge F_n) \to G)$ is valid. Hence, G is a logical consequence of $F_1, \ldots, F_n$ if and only if the negation of $((F_1 \wedge \cdots \wedge F_n) \to G)$ is inconsistent. Since

$$\sim((F_1 \wedge \cdots \wedge F_n) \to G) = \sim(\sim(F_1 \wedge \cdots \wedge F_n) \vee G)$$
$$= (\sim(\sim(F_1 \wedge \cdots \wedge F_n)) \wedge \sim G)$$
$$= (F_1 \wedge \cdots \wedge F_n) \wedge \sim G$$
$$= F_1 \wedge \cdots \wedge F_n \wedge \sim G,$$

we conclude that Theorem 2.2 is true.

Theorems 2.1 and 2.2 are very important. They show that proving that a particular formula is a logical consequence of a finite set of formulas is equivalent to proving that a certain related formula is valid or inconsistent.

If G is a logical consequence of $F_1, \ldots, F_n$, the formula $((F_1 \wedge \cdots \wedge F_n) \to G)$ is called a *theorem*, and G is also called the *conclusion of the theorem*. In mathematics as well as in other fields, many problems can be formulated as problems of proving theorems. This will be elaborated in the next section. Before going to the next section, let us now consider one simple example to show how we can utilize Theorems 2.1 and 2.2.

Example 2.11

Consider the formulas

$$F_1 \triangleq (P \to Q), \qquad F_2 \triangleq \sim Q, \qquad G \triangleq \sim P.$$

Show that G is a logical consequence of F_1 and F_2.

Method 1 We can use the truth-table technique to show that G is true in every model of $(P \to Q) \wedge \sim Q$. From Table 2.7, one notes that there is only one model for $(P \to Q) \wedge \sim Q$, namely, $\{\sim P, \sim Q\}$. $\sim P$ is certainly true in

TABLE 2.7

TRUTH TABLE OF $(P \rightarrow Q) \wedge \sim Q$ AND $\sim P$

P	Q	$P \rightarrow Q$	$\sim Q$	$(P \rightarrow Q) \wedge \sim Q$	$\sim P$
T	T	T	F	F	F
T	F	F	T	F	F
F	T	T	F	F	T
F	F	T	T	T	T

this model. Thus, by the definintion of logical consequence, we conclude that $\sim P$ is a logical consequence of $(P \rightarrow Q)$ and $\sim Q$.

Method 2   We can use Theorem 2.1. This can be done simply by extending the truth table in Table 2.7 or by evaluating the formula $((P \rightarrow Q) \wedge \sim Q) \rightarrow \sim P$.

Table 2.8 shows that $((P \rightarrow Q) \wedge \sim Q) \rightarrow \sim P$ is true in every interpretation. Therefore $((P \rightarrow Q) \wedge \sim Q) \rightarrow \sim P$ is valid and, according to Theorem 2.1, $\sim P$ is a logical consequence of $(P \rightarrow Q)$ and $\sim Q$.

TABLE 2.8

TRUTH TABLE OF $((P \rightarrow Q) \wedge \sim Q) \rightarrow \sim P$

P	Q	$P \rightarrow Q$	$\sim Q$	$(P \rightarrow Q) \wedge \sim Q$	$\sim P$	$((P \rightarrow Q) \wedge \sim Q) \rightarrow \sim P$
T	T	T	F	F	F	T
T	F	F	T	F	F	T
F	T	T	F	F	T	T
F	F	T	T	T	T	T

We can also prove the validity of $((P \rightarrow Q \wedge \sim Q) \rightarrow \sim P$ by transforming it into a conjunctive normal form.

$$
\begin{aligned}
((P \rightarrow Q) \wedge \sim Q) \rightarrow \sim P &= \sim ((P \rightarrow Q) \wedge \sim Q) \vee \sim P && \text{by (2.2)} \\
&= \sim ((\sim P \vee Q) \wedge \sim Q) \vee \sim P && \text{by (2.2)} \\
&= \sim ((\sim P \wedge \sim Q) \vee (Q \wedge \sim Q)) \vee \sim P && \text{by (2.5b)} \\
&= \sim ((\sim P \wedge \sim Q) \vee \square) \vee \sim P && \text{by (2.8b)} \\
&= \sim ((\sim P \wedge \sim Q)) \vee \sim P && \text{by (2.6a)} \\
&= (P \vee Q) \vee \sim P && \text{by (2.10b)} \\
&= (Q \vee P) \vee \sim P && \text{by (2.3a)}
\end{aligned}
$$

$$= Q \vee (P \vee \sim P) \qquad \text{by (2.4a)}$$
$$= Q \vee \blacksquare \qquad \text{by (2.8a)}$$
$$= \blacksquare \qquad \text{by (2.7a).}$$

Thus, $((P \to Q) \wedge \sim Q) \to \sim P$ is valid.

Method 3 We can use Theorem 2.2. In this case, we prove that

$$((P \to Q) \wedge \sim Q) \wedge (\sim(\sim P)) = (P \to Q) \wedge \sim Q \wedge P$$

is inconsistent.

Again, as in Method 2, we can use the truth table technique to show that $(P \to Q) \wedge \sim Q \wedge P$ is false in every interpretation. From Table 2.9, we conclude that $(P \to Q) \wedge \sim Q \wedge P$ is inconsistent and, according to Theorem 2.2, $\sim P$ is a logical consequence of $(P \to Q)$ and $\sim Q$.

TABLE 2.9
TRUTH TABLE OF $(P \to Q) \wedge \sim Q \wedge P$

P	Q	$P \to Q$	$\sim Q$	$(P \to Q) \wedge \sim Q \wedge P$
T	T	T	F	F
T	F	F	T	F
F	T	T	F	F
F	F	T	T	F

We can also prove the inconsistency of $(P \to Q) \wedge \sim Q \wedge P$ by transforming it into a disjunctive normal form:

$$(P \to Q) \wedge \sim Q \wedge P = (\sim P \vee Q) \wedge \sim Q \wedge P \qquad \text{by (2.2)}$$
$$= (\sim P \wedge \sim Q \wedge P) \vee (Q \wedge \sim Q \wedge P) \qquad \text{by (2.5b)}$$
$$= \square \vee \square \qquad \text{by (2.8b)}$$
$$= \square \qquad \text{by (2.6a).}$$

Thus, $(P \to Q) \wedge \sim Q \wedge P$ is inconsistent.

2.6 APPLICATIONS OF THE PROPOSITIONAL LOGIC

Having discussed the various concepts in the previous sections, we now consider applications of the propositional logic. They are best illustrated by examples.

Example 2.12

Given that if the congress refuses to enact new laws, then the strike will not be over unless it lasts more than one year and the president of the firm resigns, will the strike not be over if the congress refuses to act and the strike just starts?

We first transform statements into symbols:

P: The congress refuses to act,
Q: The strike is over,
R: The president of the firm resigns,
S: The strike lasts more than one year.

Then the facts given in the example can be represented by formulas:

F_1: $(P \to (\sim Q \vee (R \wedge S)))$ $\triangleq$ If the congress refuses to enact new laws, then the strike will not be over unless it lasts more than one year and the president of the firm resigns,

F_2: P $\triangleq$ The congress refuses to act,

F_3: $\sim S$ $\triangleq$ The strike just starts.

From the facts F_1, F_2, and F_3, can we conclude that the strike will not be over? That is, can we show that $\sim Q$ is a logical consequence of F_1, F_2, and F_3? By Theorem 2.1, this is equivalent to showing that

TABLE 2.10

TRUTH TABLE OF $(F_1 \wedge F_2 \wedge F_3) \to \sim Q$, WHERE $F_1 \triangleq (P \to (\sim Q \vee (R \wedge S)))$, $F_2 \triangleq P, F_3 \triangleq \sim S$.

P	Q	R	S	F_1	F_2	F_3	$\sim Q$	$(F_1 \wedge F_2 \wedge F_3) \to \sim Q$
T	T	T	T	T	T	F	F	T
T	T	T	F	F	T	T	F	T
T	T	F	T	F	T	F	F	T
T	T	F	F	F	T	T	F	T
T	F	T	T	T	T	F	T	T
T	F	T	F	T	T	T	T	T
T	F	F	T	T	T	F	T	T
T	F	F	F	T	T	T	T	T
F	T	T	T	T	F	F	F	T
F	T	T	F	T	F	T	F	T
F	T	F	T	T	F	F	F	T
F	T	F	F	T	F	T	F	T
F	F	T	T	T	F	F	T	T
F	F	T	F	T	F	T	T	T
F	F	F	T	T	F	F	T	T
F	F	F	F	T	F	T	T	T

$((P \to (\sim Q \vee (R \wedge S))) \wedge P \wedge \sim S) \to \sim Q$ is a valid formula. The truth values of the above formula under all the interpretations are shown in Table 2.10.

From Table 2.10, there is no interpretation under which the formula is false. Hence the formula $((P \to (\sim Q \vee (R \wedge S))) \wedge P \wedge \sim S) \to \sim Q$ is a valid formula. Therefore, $\sim Q$ is a logical consequence of F_1, F_2, and F_3. That is, we can conclude $\sim Q$ from F_1, F_2, and F_3. Hence, the answer is "The strike will not be over."

Example 2.13   (Chemical Synthesis Problem)

Suppose we can perform the following chemical reactions:

$$MgO + H_2 \to Mg + H_2O$$

$$C + O_2 \to CO_2$$

$$CO_2 + H_2O \to H_2CO_3.$$

Suppose we have some quantities of MgO, H_2, O_2, and C. Show that we can make H_2CO_3.

For this problem, we may consider MgO, H_2, O_2, and C as atomic formulas. Then the above chemical reactions can be represented by the following formulas:

A_1: $(MgO \wedge H_2) \to (Mg \wedge H_2O)$

A_2: $(C \wedge O_2) \to CO_2$

A_3: $(CO_2 \wedge H_2O) \to H_2CO_3.$

Since we have MgO, H_2, O_2, and C, these facts can be represented by the following formulas:

A_4: MgO

A_5: H_2

A_6: O_2

A_7: C.

Now, the problem can be regarded as proving that H_2CO_3 is a logical consequence of $A_1, \ldots, A_7$, which, by Theorem 2.2, is true if $(A_1 \wedge \cdots \wedge A_7) \wedge \sim H_2CO_3$ is inconsistent. We prove this by transforming the formula

$$(A_1 \wedge \cdots \wedge A_7 \wedge \sim H_2CO_3)$$

into a disjunctive normal form.

$$(A_1 \wedge \cdots \wedge A_7 \wedge \sim H_2CO_3) = ((MgO \wedge H_2) \rightarrow (Mg \wedge H_2O)) \wedge ((C \wedge O_2)$$
$$\rightarrow CO_2) \wedge ((CO_2 \wedge H_2O) \rightarrow H_2CO_3)$$
$$\wedge MgO \wedge H_2 \wedge O_2 \wedge C \wedge \sim H_2CO_3$$
$$= (\sim MgO \vee \sim H_2 \vee Mg) \wedge (\sim MgO \vee \sim H_2$$
$$\vee H_2O) \wedge (\sim C \vee \sim O_2 \vee CO_2)$$
$$\wedge (\sim CO_2 \vee \sim H_2O \vee H_2CO_3)$$
$$\wedge MgO \wedge H_2 \wedge O_2 \wedge C \wedge \sim H_2CO_3$$
$$= (\sim MgO \vee \sim H_2 \vee Mg) \wedge (\sim MgO \vee \sim H_2$$
$$\vee H_2O) \wedge MgO \wedge H_2 \wedge (\sim C \vee \sim O_2$$
$$\vee CO_2) \wedge C \wedge O_2$$
$$\wedge (\sim CO_2 \vee \sim H_2O \vee H_2CO_3) \wedge \sim H_2CO_3$$
$$= Mg \wedge H_2O \wedge MgO \wedge H_2 \wedge CO_2 \wedge C \wedge O_2$$
$$\wedge (\sim CO_2 \vee \sim H_2O) \wedge \sim H_2CO_3$$
$$= (\sim CO_2 \vee \sim H_2O) \wedge H_2O \wedge CO_2 \wedge Mg$$
$$\wedge MgO \wedge H_2 \wedge C \wedge O_2 \wedge \sim H_2CO_3$$
$$= \square \wedge Mg \wedge MgO \wedge H_2 \wedge C \wedge O_2$$
$$\wedge \sim H_2CO_3$$
$$= \square.$$

Since $\square$ is always false, the formula $(A_1 \wedge \cdots \wedge A_7 \wedge \sim H_2CO_3)$ is inconsistent. Therefore H_2CO_3 is a logical consequence of $A_1, ..., A_7$. That is, we can make H_2CO_3 from MgO, H_2, O_2, and C. The above procedure for proving an inconsistent formula by transforming it into $\square$ is sometimes called the *multiplication method*, because the transformation process is very similar to multiplying out an arithmetic expression. Example 2.13 is only a simple chemical synthesis problem. In reality, there are hundreds of chemical reactions. In order to use computers efficiently, we need an efficient computer program for proving theorems. This will be discussed in detail in Chapters 4–9.

In the above examples, we have shown that the propositional logic can be applied to many problems. The approach is first to symbolize problems by formulas and then to prove that the formulas are valid or inconsistent. We have used the truth-table method and the multiplication method to prove whether a formula is valid (inconsistent). In Chapters 4–9, more efficient methods will be given for proving valid and inconsistent formulas.

REFERENCES

Hilbert, D., and W. Ackermann (1950): "Principles of Mathematical Logic," Chelsea, New York.
Kleene, S. C. (1967): "Mathematical Logic," Wiley, New York.
Mendelson, E. (1964): "Introduction to Mathematical Logic," Van Nostrand-Reinhold, Princeton, New Jersey.
Stoll, R. R. (1961): "Sets, Logic and Axiomatic Theories," Freeman, San Francisco.
Whitehead, A. N., and B. Russell, (1927): "Principia Mathematica," Cambridge Univ. Press, London and New York.

EXERCISES

Section 2.1

1. Symbolize the following statements by formulas.

   (a) A relation is an equivalence relation if and only if it is reflexive, symmetric, and transitive.
   (b) If the humidity is so high, it will rain either this afternoon or this evening.
   (c) Cancer will not be cured unless its cause is determined and a new drug for cancer is found.
   (d) It requires courage and skills to climb that mountain.
   (e) If he is a man who campaigns so hard, he probably will be elected.

2. Let

   $P \triangleq$ He needs a doctor, $Q \triangleq$ He needs a lawyer,

   $R \triangleq$ He has an accident, $S \triangleq$ He is sick,

   $U \triangleq$ He is injured.

   State the following formulas in English.

   (a) $(S \rightarrow P) \wedge (R \rightarrow Q)$ (b) $P \rightarrow (S \vee U)$

   (c) $(P \wedge Q) \rightarrow R$ (d) $(P \wedge Q) \leftrightarrow (S \wedge U)$

   (e) $\sim(S \vee U) \rightarrow \sim P$.

Section 2.2

3. Complete the following truth table (Table 2.11) of the formula

$$(\sim P \vee Q) \wedge (\sim(P \wedge \sim Q)).$$

TABLE 2.11

P	Q	$\sim P$	$\sim Q$	$\sim P \vee Q$	$P \wedge \sim Q$	$\sim(P \wedge \sim Q)$	$(\sim P \vee Q) \wedge (\sim(P \wedge \sim Q))$
T	T						
T	F						
F	T						
F	F						

Section 2.3

4. For each of the following formulas, determine whether it is valid, invalid, inconsistent, consistent, or some combination of these.

(a) $\sim(\sim P) \rightarrow P$ (b) $P \rightarrow (P \wedge Q)$

(c) $\sim(P \vee Q) \vee \sim Q$ (d) $(P \vee Q) \rightarrow P$

(e) $(P \rightarrow Q) \rightarrow (\sim Q \rightarrow \sim P)$ (f) $(P \rightarrow Q) \rightarrow (Q \rightarrow P)$

(g) $P \vee (P \rightarrow Q)$ (h) $(P \wedge (Q \rightarrow P)) \rightarrow P$

(i) $P \vee (Q \rightarrow \sim P)$ (j) $(P \vee \sim Q) \wedge (\sim P \vee Q)$

(k) $\sim P \wedge (\sim(P \rightarrow Q))$ (l) $P \rightarrow \sim P$

(m) $\sim P \rightarrow P$.

5. Consider the following statement:
    If the congress refuses to enact new laws, then the strike will not be over unless it lasts more than one year and the president of the firm resigns, and if either the congress enacts new laws or the strike is not over then the strike lasts more than one year.
    Is the above statement contradictory? Explain.

Section 2.4

6. Transform the following into disjunctive normal forms

(a) $(\sim P \wedge Q) \rightarrow R$ (b) $P \rightarrow ((Q \wedge R) \rightarrow S)$

(c) $\sim(P \vee \sim Q) \wedge (S \rightarrow T)$ (d) $(P \rightarrow Q) \rightarrow R$

(e) $\sim(P \wedge Q) \wedge (P \vee Q)$.

7. Transform the following into conjunctive normal forms

(a) $P \vee (\sim P \wedge Q \wedge R)$ (b) $\sim(P \rightarrow Q) \vee (P \vee Q)$

(c) $\sim(P \rightarrow Q)$ (d) $(P \rightarrow Q) \rightarrow R$

(e) $(\sim P \wedge Q) \vee (P \wedge \sim Q)$.

8. Is it possible to have a formula that is in conjunctive normal form as well as disjunctive normal form? If yes, give an example.

9. Verify each of the following pairs of equivalent formulas by transforming formulas on both sides of the sign $=$ into the same normal form.

(a) $P \wedge P = P$, and $P \vee P = P$

(b) $(P \rightarrow Q) \wedge (P \rightarrow R) = (P \rightarrow (Q \wedge R))$

(c) $(P \rightarrow Q) \rightarrow (P \wedge Q) = (\sim P \rightarrow Q) \wedge (Q \rightarrow P)$

(d) $P \wedge Q \wedge (\sim P \vee \sim Q) = \sim P \wedge \sim Q \wedge (P \vee Q)$

(e) $P \vee (P \rightarrow (P \wedge Q)) = \sim P \vee \sim Q \vee (P \wedge Q)$.

Sections 2.5 and 2.6

10. Prove that $(\sim Q \rightarrow \sim P)$ is a logical consequence of $(P \rightarrow Q)$.

11. If the congress refuses to enact new laws, then the strike will not be over unless it lasts more than one year and the president of the firm resigns. Suppose the congress refuses to act, the strike is over, and the president of the firm does not resign. Has the strike lasted more than one year?

12. Consider the following statements:

   $F_1 \triangleq$ Tom cannot be a good student unless he is smart and his father supports him.

   $F_2 \triangleq$ Tom is a good student only if his father supports him.

   Show that F_2 is a logical consequence of F_1.

13. Show that for the following statements, F_2 is a logical consequence of F_1.

   $F_1 \triangleq$ If the president does not have the appropriate authority or if he does not want to take the responsibility, then neither order will be restored nor will the riots stop spreading unless the rioters become tired of rioting and the local authorities begin to take conciliatory actions.

   $F_2 \triangleq$ If the president does not want to take the responsibility and the rioters are not tired of rioting, then riots will spread.

14. Show that Q is a logical consequence of $(P \rightarrow Q)$ and P. This is related to the so-called *modus ponens* rule.

Chapter 3

The First-Order Logic

3.1 INTRODUCTION

In the propositional logic, the most basic elements are atoms. Through atoms we build up formulas. We then use formulas to express various complex ideas. As discussed in Chapter 2, in this simple logic an atom represents a declarative sentence that can be either true or false, but not both. An atom is treated as a single unit. Its structure and composition are suppressed. However, there are many ideas that cannot be treated in this simple way. For example, consider the following deduction of statements:

> Every man is mortal.
> Since Confucius is a man, he is mortal.

The above reasoning is intuitively correct. However, if we denote

> $P:$ Every man is mortal,
> $Q:$ Confucius is a man,
> $R:$ Confucius is mortal,

then R is not a logical consequence of P and Q within the framework of the propositional logic. This is because the structures of P, Q, and R are not used in the propositional logic. In this chapter, we shall introduce the first-order logic, which has three more logical notions (called *terms*, *predicates*,

26

and *quantifiers*) than does the propositional logic. It will be made clear later that much of everyday and mathematical language can be symbolized by the first-order logic.

Just as in the propositional logic, we first have to define atoms in the first-order logic. Before giving the formal definition of an atom, we first consider some examples.

Suppose we want to represent "x is greater than 3." We first define a predicate $GREATER(x, y)$ to mean "x is greater than y." (Note that a predicate is a relation.) Then the sentence "x is greater than 3" is represented by $GREATER(x, 3)$.

Similarly, we can represent "x loves y" by the predicate $LOVE(x, y)$. Then "John loves Mary" can be represented by $LOVE(\text{John}, \text{Mary})$.

We can also use function symbols in the first-order logic. For example, we can use $plus(x, y)$ to denote "$x + y$" and $father(x)$ to mean the father of x. The sentences "$x + 1$ is greater than x" and "John's father loves John" can be symbolized as $GREATER(plus(x, 1), x)$ and $LOVE(father(\text{John}), \text{John})$, respectively.

In the above examples, $GREATER(x, 3)$, $LOVE(\text{John}, \text{Mary})$, $GREATER$ $(plus(x, 1), x)$, and $LOVE(father(\text{John}), \text{John})$ are all atoms in the first-order logic, where $GREATER$ and $LOVE$ are *predicate symbols*, x is a *variable*, 3, John, and Mary are *individual symbols or constants*, and *father* and *plus* are *function symbols*.

In general, we are allowed to use the following four types of symbols to construct an atom:

i. Individual symbols or constants: These are usually names of objects, such as John, Mary, and 3.

ii. Variable symbols: These are customarily *lowercase* unsubscripted or subscripted letters, $x, y, z, \ldots$.

iii. Function symbols: These are customarily *lowercase* letters $f, g, h, \ldots$ or expressive strings of *lowercase* letters such as *father* and *plus*.

iv. Predicate symbols: These are customarily *uppercase* letters $P, Q, R, \ldots$ or expressive strings of *uppercase* letters such as $GREATER$ and $LOVE$.

Any function or predicate symbol takes a specified number of arguments. If a function symbol f takes n arguments, f is called an *n-place function symbol*. It is noted that an individual symbol or a constant may be considered a function symbol that takes no argument. Similarly, if a predicate symbol P takes n arguments, P is called an *n-place predicate symbol*. For example, *father* is a one-place function symbol, and $GREATER$ and $LOVE$ are two-place predicate symbols.

A function is a mapping that maps a list of constants to a constant. For example, *father* is a function that maps a person named John to a person who is John's father. Therefore, *father*(John) represents a person, even though his name is unknown. We call *father*(John) a *term* in the first-order logic. More formally, we have the following definition.

Definition   *Terms* are defined recursively as follows:

   i. A constant is a term.
   ii. A variable is a term.
   iii. If f is an n-place function symbol, and $t_1, ..., t_n$ are terms, then $f(t_1, ..., t_n)$ is a term.
   iv. All terms are generated by applying the above rules.

Example 3.1

Since x and 1 are both terms and *plus* is a two-place function symbol, $plus(x, 1)$ is a term according to the above definition.

Furthermore, the reader can see that $plus(plus(x, 1), x)$ and *father*(*father* (John)) are also terms; the former denotes $(x + 1) + x$ and the latter denotes the grandfather of John.

A predicate is a mapping that maps a list of constants to T or F. For example, $GREATER$ is a predicate. $GREATER(5, 3)$ is T, but $GREATER(1, 3)$ is F. Having defined terms, we can now formally define an atom in the first-order logic.

Definition   If P is an n-place predicate symbol, and $t_1, ..., t_n$ are terms, then $P(t_1, ..., t_n)$ is an *atom*. No other expressions can be atoms.

Once atoms are defined, we can use the same five logical connectives given in Chapter 2 to build up formulas. Furthermore, since we have introduced variables, we use two special symbols ∀ and ∃ to characterize variables. The symbols ∀ and ∃ are called, respectively, the *universal* and *existential* *quantifiers*. If x is a variable, then $(\forall x)$ is read as "*for all x,*" "*for each x,*" or "*for every x,*" while $(\exists x)$ is read as "*there exists an x,*" "*for some x,*" or "*for at least one x.*" Let us consider a few examples to see how the quantifiers can be used.

Example 3.2

Symbolize the following statements:

(a)   Every rational number is a real number.
(b)   There exists a number that is a prime.
(c)   For every number x, there exists a number y such that $x < y$.

Denote "x is a prime number" by $P(x)$, "x is a rational number" by $Q(x)$, "x is a real number" by $R(x)$, and "x is less than y" by $LESS(x, y)$. Then the above statements can be denoted, respectively, as

(a') $(\forall x)(Q(x) \rightarrow R(x))$

(b') $(\exists x)P(x)$

(c') $(\forall x)(\exists y)LESS(x,y)$.

Each of the expressions (a'), (b'), and (c') is called a formula. Before we give a formal definition of a formula, we have to distinguish between *bound* variables and *free* variables. To do this, we first define the *scope* of a quantifier occurring in a formula as the formula to which the quantifier applies. For example, the scope of both the universal and existential quantifiers in the formula $(\forall x)(\exists y)LESS(x,y)$ is $LESS(x,y)$. The scope of the universal quantifier in the formula $(\forall x)(Q(x) \rightarrow R(x))$ is $(Q(x) \rightarrow R(x))$.

Definition   An occurrence of a variable in a formula is *bound* if and only if the occurrence is within the scope of a quantifier employing the variable, or is the occurrence in that quantifier. An occurrence of a variable in a formula is *free* if and only if this occurrence of the variable is not bound.

Definition   A variable is *free* in a formula if at least one occurrence of it is free in the formula. A variable is *bound* in a formula if at least one occurrence of it is bound.

In the formula $(\forall x)P(x,y)$, since both the occurrences of x are bound, the variable x is bound. However, the variable y is free since the only occurrence of y is free. We note that a variable can be both free and bound in a formula. For example, y is both free and bound in the formula $(\forall x)P(x,y) \wedge (\forall y)Q(y)$.

We now can formally define a formula by using atoms, logical connectives, and quantifiers.

Definition   *Well-formed formulas*, or *formulas* for short, in the first-order logic are defined recursively as follows:

i. An atom is a formula. (Note that "atom" is an abbreviation for an atomic formula.)

ii. If F and G are formulas, then $\sim(F)$, $(F \vee G)$, $(F \wedge G)$, $(F \rightarrow G)$, and $(F \leftrightarrow G)$ are formulas.

iii. If F is a formula and x is a free variable in F, then $(\forall x)F$ and $(\exists x)F$ are formulas.

iv. Formulas are generated only by a finite number of applications of (i), (ii), and (iii).

Throughout this book, parentheses may be omitted by the same conventions established in Chapter 2. We extend these conventions by agreeing that quantifiers have the least ranks. For example, $(\exists x)A \vee B$ stands for $(((\exists x)A) \vee (B))$.

Example 3.3

Translate the statement "Every man is mortal. Confucius is a man. Therefore, Confucius is mortal." into a formula.

Denote "x is a man" by $MAN(x)$, and "x is mortal" by $MORTAL(x)$. Then "every man is mortal" can be represented by

$$(\forall x)(MAN(x) \rightarrow MORTAL(x)),$$

"Confucius is a man" by

$$MAN(\text{Confucius}),$$

and "Confucius is mortal" by

$$MORTAL(\text{Confucius}).$$

The whole statement can now be represented by

$$(\forall x)(MAN(x) \rightarrow MORTAL(x)) \wedge MAN(\text{Confucius}) \rightarrow MORTAL(\text{Confucius}).$$

Example 3.4

The basic axioms of natural numbers are the following:

A_1: For every number, there is one and only one immediate successor.
A_2: There is no number for which 0 is the immediate successor.
A_3: For every number other than 0, there is one and only one immediate predecessor.

Let $f(x)$ and $g(x)$ represent the immediate successor and predecessor of x, respectively. Let "x is equal to y" be denoted by $E(x, y)$. Then the axioms can be represented by the following formulas:

A_1': $(\forall x)(\exists y)(E(y, f(x)) \wedge (\forall z)(E(z, f(x)) \rightarrow E(y, z)))$

A_2': $\sim ((\exists x)E(0, f(x)))$

A_3': $(\forall x)(\sim E(x, 0) \rightarrow ((\exists y)(E(y, g(x)) \wedge (\forall z)(E(z, g(x)) \rightarrow E(y, z)))))$.

3.2 INTERPRETATIONS OF FORMULAS IN THE FIRST-ORDER LOGIC

In the propositional logic, an interpretation is an assignment of truth values to atoms. In the first-order logic, since there are variables involved, we have to do more than that. To define an interpretation for a formula in the first-order logic, we have to specify two things, namely, the domain and an assignment to constants, function symbols, and predicate symbols occurring in the formula. The following is the formal definition of an interpretation of a formula in the first-order logic.

Definition An *interpretation* of a formula F in the first-order logic consists of a nonempty domain D, and an assignment of "values" to each constant, function symbol, and predicate symbol occurring in F as follows:

1. To each constant, we assign an element in D.
2. To each n-place function symbol, we assign a mapping from D^n to D. (Note that $D^n = \{(x_1, \ldots, x_n) \mid x_1 \in D, \ldots, x_n \in D\}$).
3. To each n-place predicate symbol, we assign a mapping from D^n to $\{T, F\}$.

Sometimes, to emphasize the domain D, we speak of an interpretation of the formula *over* D. When we evaluate the truth value of a formula in an interpretation over the domain D, $(\forall x)$ will be interpreted as "for all elements x in D," and $(\exists x)$ as "there is an element in D."

For every interpretation of a formula over a domain D, the formula can be evaluated to T or F according to the following rules:

1. If the truth values of formulas G and H are evaluated, then the truth values of the formulas $\sim G$, $(G \wedge H)$, $(G \vee H)$, $(G \rightarrow H)$, and $(G \leftrightarrow H)$ are evaluated by using Table 2.1 in Chapter 2.
2. $(\forall x)G$ is evaluated to T if the truth value of G is evaluated to T for every d in D; otherwise, it is evaluated to F.
3. $(\exists x)G$ is evaluated to T if the truth value of G is T for at least one d in D; otherwise, it is evaluated to F.

We note that any formula containing free variables cannot be evaluated. In the rest of the book, we shall assume either that formulas do not contain free variables, or that free variables are treated as constants.

Example 3.5

Let us consider the formulas

$$(\forall x)\, P(x) \qquad \text{and} \qquad (\exists x) \sim P(x).$$

Let an interpretation be as follows:

Domain: $D = \{1, 2\}$.

Assignment for P:

$P(1)$	$P(2)$
T	F

It should be easy for the reader to confirm that $(\forall x)P(x)$ is F in this interpretation because $P(x)$ is not T for both $x = 1$ and $x = 2$. On the other hand, since $\sim P(2)$ is T in this interpretation, $(\exists x)\sim P(x)$ is T in this interpretation.

Example 3.6

Consider the formula

$$(\forall x)(\exists y)P(x,y).$$

Let us define an interpretation as follows:

$$D = \{1,2\}$$

$P(1,1)$	$P(1,2)$	$P(2,1)$	$P(2,2)$
T	F	F	T

If $x = 1$, we can see that there is a y, namely 1, such that $P(1,y)$ is T.
If $x = 2$, there is also a y, namely 2, such that $P(2,y)$ is T.
Therefore, in the above interpretation, for every x in D, there is a y such that $P(x,y)$ is T; that is $(\forall x)(\exists y)P(x,y)$ is T in this interpretation.

Example 3.7

Consider the formula

G: $(\forall x)(P(x) \rightarrow Q(f(x),a))$.

There are one constant a, one one-place function symbol f, one one-place predicate symbol P, and one two-place predicate symbol Q in G. The following is an interpretation I of G.

Domain: $D = \{1,2\}$.

Assignment for a:

a
1

Assignment for f:

$f(1)$	$f(2)$
2	1

Assignment for P and Q:

$P(1)$	$P(2)$	$Q(1,1)$	$Q(1,2)$	$Q(2,1)$	$Q(2,2)$
F	T	T	T	F	T

If $x = 1$, then

$$P(x) \rightarrow Q(f(x), a) = P(1) \rightarrow Q(f(1), a)$$
$$= P(1) \rightarrow Q(2, 1)$$
$$= F \rightarrow F = T.$$

If $x = 2$, then

$$P(x) \rightarrow Q(f(x), a) = P(2) \rightarrow Q(f(2), a)$$
$$= P(2) \rightarrow Q(1, 1)$$
$$= T \rightarrow T = T.$$

Since $P(x) \rightarrow Q(f(x), a)$ is true for all elements x in the domain D, the formula $(\forall x)(P(x) \rightarrow Q(f(x), a))$ is true under the interpretation I.

Example 3.8

Evaluate the truth values of the following formulas under the interpretation given in Example 3.7.

(a) $(\exists x)(P(f(x)) \wedge Q(x, f(a)))$,

(b) $(\exists x)(P(x) \wedge Q(x, a))$,

(c) $(\forall x)(\exists y)(P(x) \wedge Q(x, y))$.

For (a):   If $x = 1$,

$$P(f(x)) \wedge Q(x, f(a)) = P(f(1)) \wedge Q(1, f(a))$$
$$= P(2) \wedge Q(1, f(1))$$
$$= P(2) \wedge Q(1, 2)$$
$$= T \wedge T = T.$$

If $x = 2$,

$$P(f(x)) \wedge Q(x, f(a)) = P(f(2)) \wedge Q(2, f(1))$$
$$= P(1) \wedge Q(2, 1)$$
$$= F \wedge F = F.$$

Since there is an element in the domain D, that is, $x = 1$, such that $P(f(x)) \wedge Q(x, f(a))$ is true, the truth value of the formula $(\exists x)(P(f(x)) \wedge Q(x, f(a)))$ is true under the interpretation I.

For (b):   If $x = 1$,

$$P(x) \wedge Q(x,a) = P(1) \wedge Q(1,1)$$
$$= F \wedge T = F.$$

If $x = 2$,

$$P(x) \wedge Q(x,a) = P(2) \wedge Q(2,1)$$
$$= T \wedge F = F.$$

Since there is no element in the domain D such that $P(x) \wedge Q(x,a)$ is true, the formula $(\exists x)(P(x) \wedge Q(x,a))$ is evaluated to be false under the interpretation I.

For (c):   If $x = 1$, then $P(x) = P(1) = F$. Therefore $P(x) \wedge Q(x,y) = F$ for $y = 1$ and $y = 2$. Since there exists an x, that is, $x = 1$, such that $(\exists y)(P(x) \wedge Q(x,y))$ is false, the formula $(\forall x)(\exists y)(P(x) \wedge Q(x,y))$ is false under the interpretation I, that is, the formula is falsified by I.

Once interpretations are defined, all the concepts, such as validity, inconsistency, and logical consequence, defined in Chapter 2 can be defined analogously for formulas of the first-order logic.

Definition   A formula G is *consistent* (*satisfiable*) if and only if there exists an interpretation I such that G is evaluated to T in I. If a formula G is T in an interpretation I, we say that I is a *model* of G and I *satisfies* G.

Definition   A formula G is *inconsistent* (*unsatisfiable*) if and only if there exists no interpretation that satisfies G.

Definition   A formula G is *valid* if and only if every interpretation of G satisfies G.

Definition   A formula G is a *logical consequence* of formulas $F_1, F_2, ..., F_n$ if and only if for every interpretation I, if $F_1 \wedge \cdots \wedge F_n$ is true in I, G is also true in I.

The relations between validity (inconsistency) and logical consequence as stated in Theorems 2.1 and 2.2 are also true for the first-order logic. In fact, the first-order logic can be considered as an extension of the propositional logic. When a formula in the first-order logic contains no variables and quantifiers, it can be treated just as a formula in the propositional logic.

Example 3.9

It is left as an exercise for the reader to prove the following:

(1) $(\forall x)P(x) \wedge (\exists y) \sim P(y)$ is inconsistent.

(2) $(\forall x)P(x) \to (\exists y)P(y)$ is valid.

(3) $P(a) \to \sim ((\exists x)P(x))$ is consistent.

(4) $(\forall x)P(x) \vee ((\exists y) \sim P(y))$ is valid.

Example 3.10

Consider formulas

F_1: $(\forall x)(P(x) \twoheadrightarrow Q(x))$

F_2: $P(a)$.

We will now prove that formula $Q(a)$ is a logical consequence of F_1 and F_2.

Consider any interpretation I that satisfies $(\forall x)(P(x) \to Q(x)) \wedge P(a)$. Certainly in this interpretation, $P(a)$ is T. Assume $Q(a)$ is not T in this interpretation; then $\sim P(a) \vee Q(a)$, that is, $P(a) \to Q(a)$ is F in I. This means that $(\forall x)(P(x) \to Q(x))$ is F in I, which is impossible. Therefore, $Q(a)$ must be T in every interpretation that satisfies

$$(\forall x)(P(x) \to Q(x)) \wedge P(a).$$

This means that $Q(a)$ is a logical consequence of F_1 and F_2.

In the first-order logic, since there are an infinite number of domains, in general, there are an infinite number of interpretations of a formula. Therefore, unlike in the propositional logic, it is not possible to verify a valid or an inconsistent formula by evaluating the formula under all the possible interpretations. In the subsequent chapters, we shall give procedures for verifying inconsistent formulas in the first-order logic.

3.3   PRENEX NORMAL FORMS IN THE FIRST-ORDER LOGIC

In the propositional logic, we have introduced two normal forms—the conjunctive normal form and the disjunctive normal form. In the first-order logic, there is also a normal form called the "prenex normal form." The reason for considering a prenex normal form of a formula is to simplify proof procedures, which will be discussed in the sequel.

Definition   A formula F in the first-order logic is said to be in a *prenex normal form* if and only if the formula F is in the form of

$$(Q_1 x_1) \cdots (Q_n x_n)(M)$$

where every $(Q_i x_i)$, $i = 1, \ldots, n$, is either $(\forall x_i)$ or $(\exists x_i)$, and M is a formula containing no quantifiers. $(Q_1 x_1) \cdots (Q_n x_n)$ is called the *prefix* and M is called the *matrix* of the formula F.

Here are some formulas in prenex normal form:

$$(\forall x)(\forall y)(P(x,y) \land Q(y)), \qquad (\forall x)(\forall y)(\sim P(x,y) \to Q(y)),$$

$$(\forall x)(\forall y)(\exists z)(Q(x,y) \to R(z)).$$

Given a formula, we now consider a method of transforming it into a prenex normal form. This is to be accomplished by first considering some basic pairs of equivalent formulas in the first-order logic. We remember that two formulas F and G are *equivalent*, denoted by $F = G$, if and only if the truth values of F and G are the same under every interpretation. The basic pairs of equivalent formulas given in Table 2.6 in Chapter 2 are still true for the first-order logic. In addition, there are other pairs of equivalent formulas containing quantifiers. We now consider these additional pairs of equivalent formulas. Let F be a formula containing a free variable x. To emphasize that the free variable x is in F, we represent F by $F[x]$. Let G be a formula that does not contain variable x. Then we have the following pairs of equivalent formulas, where Q is either $\forall$ or $\exists$. For simplicity, we shall call each of the following pairs of equivalent formulas a "law."

(3.1a) $(Qx)F[x] \lor G = (Qx)(F[x] \lor G).$

(3.1b) $(Qx)F[x] \land G = (Qx)(F[x] \land G).$

(3.2a) $\sim((\forall x)F[x]) = (\exists x)(\sim F[x]).$

(3.2b) $\sim((\exists x)F[x]) = (\forall x)(\sim F[x]).$

Laws (3.1a) and (3.1b) are obviously true, since G does not contain x and therefore can be brought into the scope of the quantifier Q. Laws (3.2a) and (3.2b) are not difficult to prove. Let I be any arbitrary interpretation over a domain D. If $\sim((\forall x)F[x])$ is true in I, then $(\forall x)F[x]$ is false in I. This means there is an element e in D such that $F[e]$ is false, that is, $\sim F[e]$ is true in I. Therefore, $(\exists x)(\sim F[x])$ is true in I. On the other hand, if $\sim((\forall x)F[x])$ is false in I, then $(\forall x)F[x]$ is true in I. This means that $F[x]$ is true for every element x in D, that is, $\sim F[x]$ is false for every element x in D. Therefore, $(\exists x)(\sim F[x])$ is false in I. Since $\sim((\forall x)F[x])$ and $(\exists x)(\sim F[x])$ always assume the same truth value for every arbitrary interpretation, by definition, $\sim((\forall x)F[x]) = (\exists x)(\sim F[x])$. Hence, law (3.2a) is proved. Similarly, we can prove law (3.2b).

Suppose $F[x]$ and $H[x]$ are two formulas containing x. There are two other laws:

(3.3a) $(\forall x)F[x] \land (\forall x)H[x] = (\forall x)(F[x] \land H[x]).$

(3.3b) $(\exists x)F[x] \lor (\exists x)H[x] = (\exists x)(F[x] \lor H[x]).$

That is, the universal quantifier $\forall$ and the existential quantifier $\exists$ can distribute over $\land$ and $\lor$, respectively.

The proofs of (3.3a) and (3.3b) are not difficult. We leave the proofs to the reader. However, the universal quantifier $\forall$ and the existential quantifier $\exists$ *cannot* distribute over $\vee$ and $\wedge$, respectively. That is,

$$(\forall x)F[x] \vee (\forall x)H[x] \neq (\forall x)(F[x] \vee H[x])$$

and

$$(\exists x)F[x] \wedge (\exists x)H[x] \neq (\exists x)(F[x] \wedge H[x]).$$

For cases like these, we have to do something special. Since every bound variable in a formula can be considered as a dummy variable, every bound variable x can be renamed z, and the formula $(\forall x)H[x]$ becomes $(\forall z)H[z]$; that is, $(\forall x)H[x] = (\forall z)H[z]$. Suppose we choose the variable z that does not appear in $F[x]$. Then,

$$(\forall x)F[x] \vee (\forall x)H[x] = (\forall x)F[x] \vee (\forall z)H[z]$$
$$\text{(by renaming all } x\text{'s occurring in } (\forall x)\,H[x] \text{ as } z)$$

$$= (\forall x)(\forall z)(F[x] \vee H[z])$$
$$\text{(by 3.1a).}$$

Similarly, we can have

$$(\exists x)F[x] \wedge (\exists x)H[x] = (\exists x)F[x] \wedge (\exists z)H[z]$$
$$\text{(by renaming all } x\text{'s occurring in } (\exists x)\,H[x] \text{ as } z)$$

$$= (\exists x)(\exists z)(F[x] \wedge H[z])$$
$$\text{(by 3.1b).}$$

Therefore, for these two cases, we still can bring all the quantifiers to the left of the formula. In general, we have

(3.4a) $(Q_1 x)F[x] \vee (Q_2 x)H[x] = (Q_1 x)(Q_2 z)(F[x] \vee H[z])$

(3.4b) $(Q_3 x)F[x] \wedge (Q_4 x)H[x] = (Q_3 x)(Q_4 z)(F[x] \wedge H[z])$

where Q_1, Q_2, Q_3, and Q_4 are either $\forall$ or $\exists$, and z does not appear in $F[x]$. Of course, if $Q_1 = Q_2 = \exists$ and $Q_3 = Q_4 = \forall$, then we do not have to rename x's in $(Q_2 x)H[x]$ or $(Q_4 x)H[x]$. We can use (3.3) directly.

Using laws (2.1)–(2.10) and laws (3.1)–(3.4), we can always transform a given formula into prenex normal form. The following is an outline of the transforming procedure.

Transforming Formulas into Prenex Normal Form

Step 1   Use the laws

$$F \leftrightarrow G = (F \rightarrow G) \wedge (G \rightarrow F) \tag{2.1}$$

$$F \rightarrow G = \sim F \vee G \tag{2.2}$$

to eliminate the logical connectives $\leftrightarrow$ and $\rightarrow$.

Step 2   Repeatedly use the law

$$\sim(\sim F) = F \tag{2.9}$$

De Morgan's laws

$$\sim(F \vee G) = \sim F \wedge \sim G \tag{2.10a}$$

$$\sim(F \wedge G) = \sim F \vee \sim G \tag{2.10b}$$

and the laws

$$\sim((\forall x)F[x]) = (\exists x)(\sim F[x]) \tag{3.2a}$$

$$\sim((\exists x)F[x]) = (\forall x)(\sim F[x]) \tag{3.2b}$$

to bring the negation signs immediately before atoms.

Step 3   Rename bound variables if necessary.

Step 4   Use the laws

$$(Qx)F[x] \vee G = (Qx)(F[x] \vee G) \tag{3.1a}$$

$$(Qx)F[x] \wedge G = (Qx)(F[x] \wedge G) \tag{3.1b}$$

$$(\forall x)F[x] \wedge (\forall x)H[x] = (\forall x)(F[x] \wedge H[x]) \tag{3.3a}$$

$$(\exists x)F[x] \vee (\exists x)H[x] = (\exists x)(F[x] \vee H[x]) \tag{3.3b}$$

$$(Q_1 x)F[x] \vee (Q_2 x)H[x] = (Q_1 x)(Q_2 z)(F[x] \vee H[z]) \tag{3.4a}$$

$$(Q_3 x)F[x] \wedge (Q_4 x)H[x] = (Q_3 x)(Q_4 z)(F[x] \wedge H[z]) \tag{3.4b}$$

to move the quantifiers to the left of the entire formula to obtain a prenex normal form.

Example 3.11

Transform the formula $(\forall x)P(x) \rightarrow (\exists x)Q(x)$ into prenex normal form.

$$
\begin{aligned}
(\forall x)P(x) \rightarrow (\exists x)Q(x) &= \sim((\forall x)P(x)) \vee (\exists x)Q(x) && \text{by (2.2)} \\
&= (\exists x)(\sim P(x)) \vee (\exists x)Q(x) && \text{by (3.2a)} \\
&= (\exists x)(\sim P(x) \vee Q(x)) && \text{by (3.3b).}
\end{aligned}
$$

Therefore, a prenex normal form of $(\forall x)P(x) \rightarrow (\exists x)Q(x)$ is $(\exists x)(\sim P(x) \vee Q(x))$.

Example 3.12

Obtain a prenex normal form for the formula

$$(\forall x)(\forall y)((\exists z)(P(x,z) \wedge P(y,z)) \rightarrow (\exists u)Q(x,y,u)).$$

$$(\forall x)(\forall y)((\exists z)(P(x,z) \wedge P(y,z)) \rightarrow (\exists u)Q(x,y,u))$$
$$= (\forall x)(\forall y)\big(\sim ((\exists z)(P(x,z) \wedge P(y,z)))$$ by (2.2)
$$\vee (\exists u)Q(x,y,u)\big)$$
$$= (\forall x)(\forall y)((\forall z)(\sim P(x,z) \vee \sim P(y,z))$$ by (3.2b) and (2.10b)
$$\vee (\exists u)Q(x,y,u))$$
$$= (\forall x)(\forall y)(\forall z)(\exists u)(\sim P(x,z)$$ using (3.1a), move the
$$\vee \sim P(y,z) \vee Q(x,y,u))$$ quantifiers to the left.

Therefore, we obtain the last formula as a prenex normal form of the first formula.

3.4   APPLICATIONS OF THE FIRST-ORDER LOGIC

In this section we shall give a few examples to illustrate some applications of the first-order logic to problem solving. As in the propositional logic, the usual approach is first to symbolize problems by formulas and then to prove that the formulas are valid or inconsistent.

Example 3.13

Consider Example 3.3. There are two axioms:

A_1: $(\forall x)(MAN(x) \rightarrow MORTAL(x))$.
A_2: $MAN(\text{Confucius})$.

From A_1 and A_2, show that Confucius is mortal. That is, show that $MORTAL(\text{Confucius})$ is a logical consequence of A_1 and A_2.
   We have

$A_1 \wedge A_2$: $(\forall x)(MAN(x) \rightarrow MORTAL(x)) \wedge MAN(\text{Confucius})$.

If $A_1 \wedge A_2$ is true in an interpretation I, then both A_1 and A_2 are true in I. Since $(MAN(x) \rightarrow MORTAL(x))$ is true for all x, when x is replaced by "Confucius," $(MAN(\text{Confucius}) \rightarrow MORTAL(\text{Confucius}))$ is true in I. That is, $\sim MAN(\text{Confucius}) \vee MORTAL(\text{Confucius})$ is true in I. However, $\sim MAN(\text{Confucius})$ is false in I since $MAN(\text{Confucius})$ is true in I. Hence, $MORTAL(\text{Confucius})$ must be true in I. We have therefore shown that $MORTAL(\text{Confucius})$ is true in I whenever $(A_1 \wedge A_2)$ is true in I. By definition, $MORTAL(\text{Confucius})$ is a logical consequence of A_1 and A_2.

Example 3.14

No used-car dealer buys a used car for his family. Some people who buy used cars for their families are absolutely dishonest. Conclude that some absolutely dishonest people are not used-car dealers.
   We let $U(x)$, $B(x)$, and $D(x)$ denote "x is a used-car dealer," "x buys

a used car for his family," and "x is absolutely dishonest," respectively. Then we have

A_1: $(\forall x)(U(x) \rightarrow \sim B(x))$

A_2: $(\exists x)(B(x) \wedge D(x))$.

We have to show that

A_3: $(\exists x)(D(x) \wedge \sim U(x))$

is a logical consequence of A_1 and A_2.

Assume that A_1 and A_2 are true in an interpretation I over a domain D. Since A_2 is true in I, there is an x in D, say a, such that $B(a) \wedge D(a)$ is true in I. Therefore $B(a)$ is true in I, that is, $\sim B(a)$ is false in I. A_1 can be written in the following form:

A_1: $(\forall x)(\sim U(x) \vee \sim B(x))$.

Since A_1 is true in I and $\sim B(a)$ is false in I, $\sim U(a)$ must be true in I. However, since $B(a) \wedge D(a)$ is true in I, $D(a)$ is true in I. Therefore, $D(a) \wedge \sim U(a)$ is true in I. Thus, A_3, that is, $(\exists x)(D(x) \wedge \sim U(x))$, is true in I. Hence, A_3 is a logical consequence of A_1 and A_2.

Example 3.15

Some patients like all doctors. No patient likes any quack. Therefore, no doctor is a quack. Denote

$P(x)$: x is a patient,

$D(x)$: x is a doctor,

$Q(x)$: x is a quack,

$L(x,y)$: x likes y.

Then the facts and the conclusion may be symbolized as follows:

F_1: $(\exists x)(P(x) \wedge (\forall y)(D(y) \rightarrow L(x,y)))$

F_2: $(\forall x)(P(x) \rightarrow (\forall y)(Q(y) \rightarrow \sim L(x,y)))$

G: $(\forall x)(D(x) \rightarrow \sim Q(x))$.

We now show that G is a logical consequence of F_1 and F_2. Let I be an arbitrary interpretation over a domain D. Suppose F_1 and F_2 are true in I. Since F_1, that is, $(\exists x)(P(x) \wedge (\forall y)(D(y) \rightarrow L(x,y)))$ is true in I, there is some element, say e, in D such that $(P(e) \wedge (\forall y)(D(y) \rightarrow L(e,y)))$ is true in I. That is, both $P(e)$ and $(\forall y)(D(y) \rightarrow L(e,y))$ are true in I. On the other hand, since $(P(x) \rightarrow (\forall y)(Q(y) \rightarrow \sim L(x,y)))$ is true in I for all elements of x in D, certainly $(P(e) \rightarrow (\forall y)(Q(y) \rightarrow \sim L(e,y)))$ is true in I. Since $P(e)$ is true in I, $(\forall y)(Q(y) \rightarrow \sim L(e,y))$ must be true in I. Hence we know that for every element y in D, both $(D(y) \rightarrow L(e,y))$ and $(Q(y) \rightarrow \sim L(e,y))$ are true in I. If $D(y)$ is false in I, then $(D(y) \rightarrow \sim Q(y))$ is true in I. If $D(y)$ is true in I, then $L(e,y)$ must be true in I, since $(D(y) \rightarrow L(e,y))$ is true in I. Therefore, $Q(y)$ must be false in I, since $(Q(y) \rightarrow \sim L(e,y))$ is true in I. Consequently, $(D(y) \rightarrow \sim Q(y))$ is true in I. Therefore, $(D(y) \rightarrow \sim Q(y))$ is true for every y in D; i.e., $(\forall y)(D(y) \rightarrow \sim Q(y))$ is true in I. Hence, we have shown that if F_1 and F_2 are true in I, $(\forall y)(D(y) \rightarrow \sim Q(y))$ is true in I. This shows that G is a logical consequence of F_1 and F_2.

In the above examples, we have shown that the conclusions follow from the given facts. A demonstration that a conclusion follows from axioms is called a *proof*. A procedure for finding proofs is called a *proof procedure*. In Chapters 4–9, we shall give proof procedures that *mechanically* use inference rules to find a proof. These proof procedures are very efficient. The reader will find out that the above three examples are extremely easy to prove if these proof procedures are used.

REFERENCES

Hilbert, D., and W. Ackermann (1950): "Principles of Mathematical Logic," Chelsea, New York.

Kleene, S. C. (1967): "Mathematical Logic," Wiley, New York.

Korfhage, R. R. (1966): "Logic and Algorithms," Wiley, New York.

Mendelson, E., (1964): "Introduction to Mathematical Logic," Van Nostrand-Reinhold, Princeton, New Jersey.

Stoll, R. R. (1961): "Sets, Logic and Axiomatic Theories," Freeman, San Francisco.

Whitehead, A., and B. Russell (1927): "Principia Mathematica," Cambridge Univ. Press, London and New York.

EXERCISES

Section 3.1

1. Let $P(x)$ and $Q(x)$ represent "x is a rational number" and "x is a real number," respectively. Symbolize the following sentences:

    1.1 Every rational number is a real number.
    1.2 Some real numbers are rational numbers.
    1.3 Not every real number is a rational number.

2. Let $C(x)$ mean "x is a used-car dealer," and $H(x)$ mean "x is honest." Translate each of the following into English:

    2.1 $(\exists x)C(x)$

2.2 $(\exists x) H(x)$

2.3 $(\forall x)(C(x) \rightarrow \sim H(x))$

2.4 $(\exists x)(C(x) \wedge H(x))$

2.5 $(\exists x)(H(x) \rightarrow C(x))$.

3. Let $P(x)$, $L(x)$, $R(x,y,z)$, and $E(x,y)$ represent "x is a point," "x is a line," "z passes through x and y," and "$x = y$," respectively. Translate the following:

   For every two points, there is one and only one line passing through both points.

4. An Abelian group is a set A with a binary operator $+$ that has certain properties. Let $P(x,y,z)$ and $E(x,y)$ represent $x+y = z$ and $x = y$, respectively. Express the following axioms for Abelian groups symbolically.

   (a)   For every x and y in A, there exists a z in A such that $x+y = z$ (closure).
   (b)   If $x+y = z$ and $x+y = w$, then $z = w$ (uniqueness).
   (c) $(x+y)+z = x+(y+z)$ (associativity).
   (d) $x+y = y+x$ (symmetry).
   (e)   For every x and y in A, there exists a z such that $x+z = y$ (right solution).

Section 3.2

5. For the following interpretation $(D = \{a,b\})$,

$P(a,a)$	$P(a,b)$	$P(b,a)$	$P(b,b)$
T	F	F	T

   determine the truth value of the following formulas:

   (a) $(\forall x)(\exists y) P(x,y)$ (b) $(\forall x)(\forall y) P(x,y)$

   (c) $(\exists x)(\forall y) P(x,y)$ (d) $(\exists y) \sim P(a,y)$

   (e) $(\forall x)(\forall y)(P(x,y) \rightarrow P(y,x))$ (f) $(\forall x) P(x,x)$.

6. Consider the following formula:

   $A:$ $(\exists x) P(x) \rightarrow (\forall x) P(x)$.

   a. Prove that this formula is always true if the domain D contains only one element.
   b. Let $D = \{a,b\}$. Find an interpretation over D in which A is evaluated to F.

7. Consider the following interpretation:

Domain: $D = \{1,2\}$.

Assignment of constants a and b:

a	b
1	2

Assignment for function f:

$f(1)$	$f(2)$
2	1

Assignment for predicate P:

$P(1,1)$	$P(1,2)$	$P(2,1)$	$P(2,2)$
T	T	F	F

Evaluate the truth value of the following formulas in the above interpretation:

(1) $P(a, f(a)) \wedge P(b, f(b))$

(2) $(\forall x)(\exists y) P(y, x)$

(3) $(\forall x)(\forall y)(P(x, y) \rightarrow P(f(x), f(y)))$.

8. Let F_1 and F_2 be as follows:

F_1: $(\forall x)(P(x) \rightarrow Q(x))$

F_2: $\sim Q(a)$.

Prove that $\sim P(a)$ is a logical consequence of F_1 and F_2.

Section 3.3

9. Transform the following formulas into prenex normal forms:

(1) $(\forall x)(P(x) \rightarrow (\exists y) Q(x, y))$

(2) $(\exists x)(\sim ((\exists y) P(x, y)) \rightarrow ((\exists z) Q(z) \rightarrow R(x)))$

(3) $(\forall x)(\forall y)((\exists z) P(x, y, z) \wedge ((\exists u) Q(x, u) \rightarrow (\exists v) Q(y, v)))$.

Section 3.4

10. Consider the following statements:

F_1:   Every student is honest.

F_2:   John is not honest.

From the above statements, prove that John is not a student.

11. Consider the following premises:

   (1)   Every athelete is strong.
   (2)   Everyone who is both strong and intelligent will succeed in his career.
   (3)   Peter is an athelete.
   (4)   Peter is intelligent.

   Try to conclude that Peter will succeed in his career.

12. Assume that St. Francis is loved by everyone who loves someone. Also assume that no one loves nobody. Deduce that St. Francis is loved by everyone.

Chapter 4

Herbrand's Theorem

4.1 INTRODUCTION

In the previous chapters, we have discussed how to solve problems by proving theorems. In this and the following chapters, we shall consider proof procedures. Actually, finding a general decision procedure to verify the validity (inconsistency) of a formula was considered long ago. It was first tried by Leibniz (1646–1716) and further revived by Peano around the turn of the century and by Hilbert's school in the 1920s. It was not until Church [1936] and Turing [1936] that this was proved impossible. Church and Turing independently showed that there is no general decision procedure to check the validity of formulas of the first-order logic. However, there are proof procedures which can verify that a formula is valid if indeed it is valid. For invalid formulas, these procedures in general will never terminate. In view of the result of Church and Turing, this is the best we can expect to get from a proof procedure.

A very important approach to mechanical theorem proving was given by Herbrand in 1930. By definition, a valid formula is a formula that is true under all interpretations. Herbrand developed an algorithm to find an interpretation that can falsify a given formula. However, if the given formula is indeed valid, no such interpretation can exist and his algorithm will halt after a finite number of trials. Herbrand's method is the basis for most modern automatic proof procedures.

Gilmore [1960] was one of the first persons to implement Herbrand's procedure on a computer. Since a formula is valid if and only if its negation is inconsistent, his program was designed to detect the inconsistency of the negation of the given formula. During the execution of his program, propositional formulas are generated that are periodically tested for inconsistency. If the negation of the given formula is inconsistent, his program will eventually detect this fact. Gilmore's program managed to prove a few simple formulas, but encountered decisive difficulties with most other formulas of the first order logic. Careful studies of his program revealed that his method of testing the inconsistency of a propositional formula is inefficient. Gilmore's method was improved by Davis and Putnam [1960] a few months after his result was published. However, their improvement was still not enough. Many valid formulas of the first-order logic still could not be proved by computers in a reasonable amount of time.

A major breakthrough was made by Robinson [1965a], who introduced the so-called resolution principle. Resolution proof procedure is much more efficient than any earlier procedure. Since the introduction of the resolution principle, several refinements have been suggested in attempts to further increase its efficiency. Some of these refinements are semantic resolution [Slagle, 1967; Meltzer, 1966; Robinson, 1965b; Kowalski and Hayes, 1969], lock resolution [Boyer, 1971], linear resolution [Loveland, 1970a, b; Luckham, 1970; Anderson and Bledsoe, 1970; Yates et al., 1970; Reiter, 1971; Kowalski and Kuehner, 1970], unit resolution [Wos et al., 1964; Chang, 1970a], and set-of-support strategy [Wos et al., 1965]. In this chapter, we shall first prove Herbrand's theorem. The resolution principle and some refinements of the resolution principle will be discussed in subsequent chapters.

4.2 SKOLEM STANDARD FORMS

Herbrand's and the resolution proof procedures that are to be discussed later in this book are actually refutation procedures. That is, instead of proving a formula valid, they prove that the negation of the formula is inconsistent. This is just a matter of convenience. There is no loss of generality in using refutation procedures. Furthermore, these refutation procedures are applied to a "standard form" of a formula. This standard form was introduced by Davis and Putnam [1960], and will be used throughout this book.

Essentially what Davis and Putnam did was to exploit the following ideas:

1. A formula of the first-order logic can be transformed into prenex normal form where the matrix contains no quantifiers and the prefix is a sequence of quantifiers.

2. The matrix, since it does not contain quantifiers, can be transformed into a conjunctive normal form.

3. Without affecting the inconsistency property, the existential quantifiers in the prefix can be eliminated by using Skolem functions.

In Chapter 3, we have already discussed how to transform a formula into a prenex normal form. By the techniques given in Chapter 2, we also know how to transform the matrix into a conjunctive normal form. We now discuss how to eliminate the existential quantifiers.

Let a formula F be already in a prenex normal form $(Q_1 x_1) \cdots (Q_n x_n) M$, where M is in a *conjunctive* normal form. Suppose Q_r is an existential quantifier in the prefix $(Q_1 x_1) \cdots (Q_n x_n)$, $1 \leqslant r \leqslant n$. If no universal quantifier appears before Q_r, we choose a new constant c different from other constants occurring in M, replace all x_r appearing in M by c, and delete $(Q_r x_r)$ from the prefix. If $Q_{s_1}, \ldots, Q_{s_m}$ are all the universal quantifiers appearing before Q_r, $1 \leqslant s_1 < s_2 \cdots < s_m < r$, we choose a new m-place function symbol f different from other function symbols, replace all x_r in M by $f(x_{s_1}, x_{s_2}, \ldots, x_{s_m})$, and delete $(Q_r x_r)$ from the prefix. After the above process is applied to all the existential quantifiers in the prefix, the last formula we obtain is a *Skolem standard form* (*standard form* for short) of the formula F. The constants and functions used to replace the existential variables are called *Skolem* functions.

Example 4.1

Obtain a standard form of the formula

$$(\exists x)(\forall y)(\forall z)(\exists u)(\forall v)(\exists w) P(x, y, z, u, v, w).$$

In the above formula, $(\exists x)$ is preceded by no universal quantifiers, $(\exists u)$ is preceded by $(\forall y)$ and $(\forall z)$, and $(\exists w)$ by $(\forall y)$, $(\forall z)$ and $(\forall v)$. Therefore, we replace the existential variable x by a constant a, u by a two-place function $f(y, z)$, and w by a three-place function $g(y, z, v)$. Thus, we obtain the following standard form of the formula:

$$(\forall y)(\forall z)(\forall v) P(a, y, z, f(y, z), v, g(y, z, v)).$$

Example 4.2

Obtain a standard form of the formula

$$(\forall x)(\exists y)(\exists z)((\sim P(x, y) \wedge Q(x, z)) \vee R(x, y, z)).$$

First, the matrix is transformed into a conjunctive normal form:

$$(\forall x)(\exists y)(\exists z)((\sim P(x, y) \vee R(x, y, z)) \wedge (Q(x, z) \vee R(x, y, z))).$$

Then, since $(\exists y)$ and $(\exists z)$ are both preceded by $(\forall x)$, the existential variables y and z are replaced, respectively, by one-place functions $f(x)$ and $g(x)$.

Thus, we obtain the following standard form of the formula:

$$(\forall x)\big((\sim P(x, f(x)) \vee R(x, f(x), g(x))) \wedge (Q(x, g(x)) \vee R(x, f(x), g(x)))\big).$$

Definition   A *clause* is a finite disjunction of zero or more literals.

When it is convenient, we shall regard a set of literals as synonymous with a clause. For example, $P \vee Q \vee \sim R = \{P, Q, \sim R\}$. A clause consisting of r literals is called an *r-literal* clause. A one-literal clause is called a *unit* clause. When a clause contains no literal, we call it the *empty* clause. Since the empty clause has no literal that can be satisfied by an interpretation, the empty clause is always false. We customarily denote the empty clause by $\square$.

The disjunctions $\sim P(x, f(x)) \vee R(x, f(x), g(x))$ and $Q(x, g(x)) \vee R(x, f(x), g(x))$ in the standard form in Example 4.2 are clauses. *A set S of clauses is regarded as a conjunction of all clauses in S, where every variable in S is considered governed by a universal quantifier.* By this convention, a standard form can be simply represented by a set of clauses. For example, the standard form of Example 4.2 can be represented by the set

$$\{\sim P(x, f(x)) \vee R(x, f(x), g(x)), Q(x, g(x)) \vee R(x, f(x), g(x))\}.$$

As we said in the beginning of this section, we can eliminate existential quantifiers without affecting the inconsistency property. This is shown in the following theorem.

Theorem 4.1   Let S be a set of clauses that represents a standard form of a formula F. Then F is inconsistent if and only if S is inconsistent.

Proof   Without loss of generality, we may assume that F is in a prenex normal form, that is, $F = (Q_1 x_1) \cdots (Q_n x_n) M[x_1, \ldots, x_n]$. (We use $M[x_1, \ldots, x_n]$ to indicate that the matrix M contains variables $x_1, \ldots, x_n$.) Let Q_r be the first existential quantifier. Let $F_1 = (\forall x_1) \cdots (\forall x_{r-1})(Q_{r+1} x_{r+1}) \cdots (Q_n x_n) M[x_1, \ldots, x_{r-1}, f(x_1, \ldots, x_{r-1}), x_{r+1}, \ldots, x_n]$, where f is a Skolem function corresponding to x_r, $1 \leqslant r \leqslant n$. We want to show that F is inconsistent if and only if F_1 is inconsistent. Suppose F is inconsistent. If F_1 is consistent, then there is an interpretation I such that F_1 is true in I. That is, for all $x_1, \ldots, x_{r-1}$, there exists at least one element, which is $f(x_1, \ldots, x_{r-1})$, such that $(Q_{r+1} x_{r+1}) \cdots (Q_n x_n) M[x_1, \ldots, x_{r-1}, f(x_1, \ldots, x_{r-1}), x_{r+1}, \ldots, x_n]$ is true in I. Thus, F is true in I, which contradicts the assumption that F is inconsistent. Therefore F_1 must be inconsistent. On the other hand, suppose that F_1 is inconsistent. If F is consistent, then there is an interpretation I over a domain D such that F is true in I. That is, for all $x_1, \ldots, x_{r-1}$, there exists an element x_r such that $(Q_{r+1} x_{r+1}) \cdots (Q_n x_n) M[x_1, \ldots, x_{r-1}, x_r, x_{r+1}, \ldots, x_n]$ is true in I. Extend the interpretation I to include a function f that maps $(x_1, \ldots, x_{r-1})$ to x_r for all $x_1, \ldots, x_{r-1}$ in D, that is, $f(x_1, \ldots, x_{r-1}) = x_r$. Let this extension of I be denoted by I'. Then, clearly, for all $x_1, ..., x_{r-1}$, $(Q_{r+1} x_{r+1}) \cdots (Q_n x_n) M[x_1, ..., x_{r-1}, f(x_1, ..., x_{r-1}), x_{r+1}, ..., x_n]$ is true in I'. That is, F_1 is true in I', which contradicts the assumption that F_1 is inconsistent. Therefore F must be inconsistent. Assume there are m existential quantifiers in F. Let $F_0 = F$. Let F_k be obtained from F_{k-1} by replacing the first existential quantifier in F_{k-1} by a Skolem function, $k = 1, ..., m$. Clearly, $S = F_m$. Using the same arguments given above, we can show that F_{k-1} is inconsistent if and only if F_k is inconsistent for $k = 1, ..., m$. Therefore, we conclude that F is inconsistent if and only if S is inconsistent.   Q.E.D.

Let S be a standard form of a formula F. If F is inconsistent, then by Theorem 4.1, $F = S$. If F is not inconsistent, we note that, in general, F is not equivalent to S. For example, let $F \triangleq (\exists x) P(x)$ and $S \triangleq P(a)$. Clearly, S is a standard form of F. However, let I be the interpretation defined below:

Domain: $D = \{1, 2\}$.

Assignment for a:

$\overline{a}$
1

Assignment for P:

$P(1)$	$P(2)$
F	T

Then, clearly, F is true in I, but S is false in I. Therefore $F \neq S$.

It is noted that a formula may have more than one standard form. For the sake of simplicity, when we transform a formula F into a standard form S, we should replace existential quantifiers by Skolem functions that are as simple as possible. That is, we should use Skolem functions with the least number of arguments. This means that we should move existential quantifiers to the left as far as possible. Furthermore, if we have $F = F_1 \wedge \cdots \wedge F_n$, we can separately obtain a set S_i of clauses, where each S_i represents a standard form of F_i, $i = 1, ..., n$. Then, let $S = S_1 \cup \cdots \cup S_n$. By arguments similar to those given in the proof of Theorem 4.1, it is not difficult to see that F is inconsistent if and only if S is inconsistent.

Example 4.3

In this example, we shall show how to express the following theorem in a standard form:

If $x \cdot x = e$ for all x in group G, where $\cdot$ is a binary operator and e is the identity in G, then G is commutative.

We shall first symbolize the above theorem together with some basic axioms in group theory and then represent the negation of the above theorem by a set of clauses.

We know that group G satisfies the following four axioms:

A_1: ` $x, y \in G$ implies that $x \cdot y \in G$ (closure property);

A_2: $x, y, z \in G$ implies that $x \cdot (y \cdot z) = (x \cdot y) \cdot z$ (associativity property);

A_3: $x \cdot e = e \cdot x = x$ for all $x \in G$ (identity property);

A_4: for every $x \in G$ there exists an element $x^{-1} \in G$ such that $x \cdot x^{-1} = x^{-1} \cdot x = e$ (inverse property).

Let $P(x, y, z)$ stand for $x \cdot y = z$ and $i(x)$ for x^{-1}. Then the above axioms can be represented by

A_1': $(\forall x)(\forall y)(\exists z) P(x, y, z)$

A_2': $(\forall x)(\forall y)(\forall z)(\forall u)(\forall v)(\forall w)(P(x, y, u) \wedge P(y, z, v) \wedge P(u, z, w) \rightarrow P(x, v, w))$
$\wedge (\forall x)(\forall y)(\forall z)(\forall u)(\forall v)(\forall w)(P(x, y, u) \wedge P(y, z, v) \wedge P(x, v, w) \rightarrow P(u, z, w))$

A_3': $(\forall x) P(x, e, x) \wedge (\forall x) P(e, x, x)$

A_4': $(\forall x) P(x, i(x), e) \wedge (\forall x) P(i(x), x, e)$.

The conclusion of the theorem is

B: If $x \cdot x = e$ for all $x \in G$, then G is commutative, i.e., $u \cdot v = v \cdot u$ for all $u, v \in G$.

B can be represented by

B': $(\forall x) P(x, x, e) \rightarrow ((\forall u)(\forall v)(\forall w)(P(u, v, w) \rightarrow P(v, u, w)))$.

Now, the entire theorem is represented by the formula $F = A_1' \wedge \cdots \wedge A_4' \rightarrow B'$. Thus, $\sim F = A_1' \wedge A_2' \wedge A_3' \wedge A_4' \wedge \sim B'$. To obtain a set S of clauses for $\sim F$, we first obtain a set S_i of clauses for each axiom A_i', $i = 1, 2, 3, 4$ as follows.

S_1: $\{P(x, y, f(x, y))\}$

S_2: $\{\sim P(x, y, u) \vee \sim P(y, z, v) \vee \sim P(u, z, w) \vee P(x, v, w)$,
          $\sim P(x, y, u) \vee \sim P(y, z, v) \vee \sim P(x, v, w) \vee P(u, z, w)\}$

S_3: $\{P(x, e, x), P(e, x, x)\}$.

S_4': $\{P(x, i(x), e), P(i(x), x, e)\}$.

Since

$$\sim B' = \sim\big((\forall x)\, P(x,x,e) \to ((\forall u)(\forall v)(\forall w)(P(u,v,w) \to P(v,u,w)))\big)$$

$$= \sim\big(\sim(\forall x)\, P(x,x,e) \vee ((\forall u)(\forall v)(\forall w)(\sim P(u,v,w) \vee P(v,u,w)))\big)$$

$$= (\forall x)\, P(x,x,e) \wedge \sim((\forall u)(\forall v)(\forall w)(\sim P(u,v,w) \vee P(v,u,w)))$$

$$= (\forall x)\, P(x,x,e) \wedge (\exists u)(\exists v)(\exists w)(P(u,v,w) \wedge \sim P(v,u,w)),$$

a set of clauses for $\sim B'$ is given below.

$T: \quad \{P(x,x,e),$

$\qquad P(a,b,c),$

$\qquad \sim P(b,a,c)\}.$

Thus, the set $S = S_1 \cup S_2 \cup S_3 \cup S_4 \cup T$ is the set consisting of the following clauses:

(1) $P(x,y,f(x,y))$

(2) $\sim P(x,y,u) \vee \sim P(y,z,v) \vee \sim P(u,z,w) \vee P(x,v,w)$

(3) $\sim P(x,y,u) \vee \sim P(y,z,v) \vee \sim P(x,v,w) \vee P(u,z,w)$

(4) $P(x,e,x)$

(5) $P(e,x,x)$

(6) $P(x,i(x),e)$

(7) $P(i(x),x,e)$

(8) $P(x,x,e)$

(9) $P(a,b,c)$

(10) $\sim P(b,a,c).$

In Example 4.3, we have shown how to obtain a set S of clauses for the formula $\sim F$. By Theorems 2.2 and 4.1, we know that F is valid if and only if S is inconsistent. As we said at the beginning of this section, we shall use refutation procedures to prove theorems. Thus, from here on, we shall assume that the input to a refutation procedure is always a set of clauses, such as the set S obtained in the above example. Furthermore, we shall use "unsatisfiable" ("satisfiable"), instead of "inconsistent" ("consistent"), for sets of clauses.

4.3 THE HERBRAND UNIVERSE OF A SET OF CLAUSES

By definition, a set S of clauses is unsatisfiable if and only if it is false under all interpretations over all domains. Since it is inconvenient and impossible to consider all interpretations over all domains, it would be nice if we could fix on one special domain H such that S is unsatisfiable if and only if S is false under all the interpretations over this domain. Fortunately, there does exist such a domain, which is called the *Herbrand universe* of S, defined as follows.

Definition Let H_0 be the set of constants appearing in S. If no constant appears in S, then H_0 is to consist of a single constant, say $H_0 = \{a\}$. For $i = 0, 1, 2, \ldots$, let H_{i+1} be the union of H_i and the set of all terms of the form $f^n(t_1, \ldots, t_n)$ for all n-place functions f^n occurring in S, where t_j, $j = 1, \ldots, n$, are members of the set H_i. Then each H_i is called the *i-level constant set* of S, and H_∞, or $\lim_{i \to \infty} H_i$, is called the *Herbrand universe* of S.

Example 4.4

Let $S = \{P(a), \sim P(x) \vee P(f(x))\}$. Then

$H_0 = \{a\}$

$H_1 = \{a, f(a)\}$

$H_2 = \{a, f(a), f(f(a))\}$

$\vdots$

$H_\infty = \{a, f(a), f(f(a)), f(f(f(a))), \ldots\}.$

Example 4.5

Let $S = \{P(x) \vee Q(x), R(z), T(y) \vee \sim W(y)\}$. Since there is no constant in S, we let $H_0 = \{a\}$. There is no function symbol in S, hence $H = H_0 = H_1 = \cdots = \{a\}$.

Example 4.6

Let $S = \{P(f(x), a, g(y), b)\}$. Then

$H_0 = \{a, b\}$

$H_1 = \{a, b, f(a), f(b), g(a), g(b)\}$

$H_2 = \{a, b, f(a), f(b), g(a), g(b), f(f(a)), f(f(b)), f(g(a)), f(g(b)), g(f(a)),$
$\qquad g(f(b)), g(g(a)), g(g(b))\}$

$\vdots$

In the sequel, by expression we mean a term, a set of terms, an atom, a set of atoms, a literal, a clause, or a set of clauses. When no variable appears in an expression, we sometimes call the expression a *ground expression* to emphasize this fact. Thus we may use a ground term, a ground atom, a ground literal, and a ground clause to mean that no variable occurs in the respective expressions. Furthermore, a *subexpression* of an expression E is an expression that occurs in E.

Definition   Let S be a set of clauses. The set of ground atoms of the form $P^n(t_1, ..., t_n)$ for all n-place predicates P^n occurring in S, where $t_1, ..., t_n$ are elements of the Herbrand universe of S, is called the *atom set*, or the *Herbrand base* of S.

Definition   A *ground instance* of a clause C of a set S of clauses is a clause obtained by replacing variables in C by members of the Herbrand universe of S.

Example 4.7

Let $S = \{P(x), Q(f(y)) \vee R(y)\}$. $C = P(x)$ is a clause in S and $H = \{a, f(a), f(f(a)), ...\}$ is the Herbrand universe of S. Then $P(a)$ and $P(f(f(a)))$ are both ground instances of C.

We now consider interpretations over the Herbrand universe. Let S be a set of clauses. As discussed in Chapter 3, an interpretation over the Herbrand universe of S is an assignment of constants, function symbols, and predicate symbols occurring in S. In the following, we shall define a special interpretation over the Herbrand universe of S, called the H-interpretation of S.

Definition   Let S be a set of clauses; H, the Herbrand universe of S; and I, an interpretation of S over H. I is said to be an *H-interpretation* of S if it satisfies the following conditions:

1. I maps all constants in S to themselves.
2. Let f be an n-place function symbol and $h_1, ..., h_n$ be elements of H. In I, f is assigned a function that maps $(h_1, ..., h_n)$ (an element in H^n) to $f(h_1, ..., h_n)$ (an element in H).

There is no restriction on the assignment to each n-place predicate symbol in S. Let $A = \{A_1, A_2, ..., A_n, ...\}$ be the atom set of S. An H-interpretation I can be conveniently represented by a set

$$I = \{m_1, m_2, ..., m_n, ...\}$$

in which m_j is either A_j or $\sim A_j$ for $j = 1, 2, ...$. The meaning of this set is that if m_j is A_j, then A_j is assigned "true"; otherwise, A_j is assigned "false."

Example 4.8

Consider the set $S = \{P(x) \vee Q(x), R(f(y))\}$. The Herbrand universe H of S is $H = \{a, f(a), f(f(a)), ...\}$. There are three predicate symbols: P, Q, and R. Hence the atom set of S is

$$A = \{P(a), Q(a), R(a), P(f(a)), Q(f(a)), R(f(a)), ...\}.$$

Some H-interpretations for S are as follows:

$$I_1 = \{P(a), Q(a), R(a), P(f(a)), Q(f(a)), R(f(a)), \ldots\}$$

$$I_2 = \{\sim P(a), \sim Q(a), \sim R(a), \sim P(f(a)), \sim Q(f(a)), \sim R(f(a)), \ldots\}$$

$$I_3 = \{P(a), Q(a), \sim R(a), P(f(a)), Q(f(a)), \sim R(f(a)), \ldots\}.$$

An interpretation of a set S of clauses does not necessarily have to be defined over the Herbrand universe of S. Thus an interpretation may not be an H-interpretation. For example, let $S = \{P(x), Q(y, f(y, a))\}$. If the domain is $D = \{1, 2\}$, then the following is an interpretation of S.

$D = \{1, 2\}.$

a	$f(1, 1)$	$f(1, 2)$	$f(2, 1)$	$f(2, 2)$
2	1	2	2	1

$P(1)$	$P(2)$	$Q(1, 1)$	$Q(1, 2)$	$Q(2, 1)$	$Q(2, 2)$
T	F	F	T	F	T

For an interpretation such as the one defined above, we can define an H-interpretation I^* corresponding to I. We use the above example to illustrate this point. First, we find the atom set of S,

$$A = \{P(a), Q(a, a), P(f(a, a)), Q(a, f(a, a)), Q(f(a, a), a), Q(f(a, a), f(a, a)), \ldots\}.$$

Next, we evaluate each member of A by using the above table.

$$P(a) = P(2) = F$$

$$Q(a, a) = Q(2, 2) = T$$

$$P(f(a, a)) = P(f(2, 2)) = P(1) = T$$

$$Q(a, f(a, a)) = Q(2, f(2, 2)) = Q(2, 1) = F$$

$$Q(f(a, a), a) = Q(f(2, 2), 2) = Q(1, 2) = T$$

$$Q(f(a, a), f(a, a)) = Q(f(2, 2), f(2, 2)) = Q(1, 1) = F$$

$$\vdots$$

Therefore, the H-interpretation I^* corresponding to I is

$$I^* = \{\sim P(a), Q(a, a), P(f(a, a)), \sim Q(a, f(a, a)), Q(f(a, a), a),$$

$$\sim Q(f(a, a), f(a, a)), \ldots\}.$$

In case there is no constant in S, the element a that is used to initiate the Herbrand universe of S can be mapped into any element of the domain

D. In this case, if there is more than one element in D, then there is more than one H-interpretation corresponding to I. For example, let $S = \{P(x), Q(y, f(y, z))\}$ and let an interpretation I for S be as follows:

$$D = \{1, 2\}$$

$f(1, 1)$	$f(1, 2)$	$f(2, 1)$	$f(2, 2)$
1	2	2	1

$P(1)$	$P(2)$	$Q(1, 1)$	$Q(1, 2)$	$Q(2, 1)$	$Q(2, 2)$
T	F	F	T	F	T

Then the two H-interpretations corresponding to I are

$$I^* = \{ \sim P(a), Q(a,a), P(f(a, a)), \sim Q(a, f(a, a)), Q(f(a, a), a),$$
$$\sim Q(f(a,a), f(a,a)), \ldots \} \quad \text{if} \quad a = 2,$$
$$I^* = \{ P(a), \sim Q(a,a), P(f(a,a)), \sim Q(a, f(a,a)), \sim Q(f(a,a), a),$$
$$\sim Q(f(a,a), f(a,a)), \ldots \} \quad \text{if} \quad a = 1.$$

We can formalize the concepts mentioned above as follows:

Definition   Given an interpretation I over a domain D, an H-*interpretation* I^* *corresponding to* I is an H-interpretation that satisfies the following condition:

Let $h_1, \ldots, h_n$ be elements of H (the Herbrand universe of S). Let every h_i be mapped to some d_i in D. If $P(d_1, \ldots, d_n)$ is assigned $T(F)$ by I, then $P(h_1, \ldots, h_n)$ is also assigned $T(F)$ in I^*.

In fact, it is not hard to prove the following lemma. The proof is left as an exercise.

Lemma 4.1   If an interpretation I over some domain D satisfies a set S of clauses, then any one of the H-interpretations I^* corresponding to I also satisfies S.

Theorem 4.2   A set S of clauses is unsatisfiable if and only if S is false under all the H-interpretations of S.

Proof ($\Rightarrow$)   The first half of the above theorem is obvious since, by definition, S is unsatisfiable if and only if S is false under all the interpretations over *any* domain.

($\Leftarrow$)   To prove the second half of the above theorem, assume that S is false under all the H-interpretations of S. Suppose S is not unsatisfiable. Then there is an interpretation I over some domain D such that S is true under I. Let I^* be an H-interpretation corresponding to I. According to Lemma 4.1, S is true under I^*. This contradicts the assumption that S is false under all the H-interpretations of S. Therefore, S must be unsatisfiable.

Q.E.D.

Thus we have obtained the objective stated at the beginning of this section. That is, we need consider only interpretations over the Herbrand universe, or more strongly, H-interpretations, for checking whether or not a set of clauses is unsatisfiable. Because of the above theorem, from here on, whenever we mention an interpretation, we mean an H-interpretation.

Let $\emptyset$ denote the empty set. Each of the following observations is obvious. We shall leave their proofs to the reader.

1. A ground instance C' of a clause C is satisfied by an interpretation I if and only if there is a ground literal L' in C' such that L' is also in I, that is, $C' \cap I \neq \emptyset$.

2. A clause C is satisfied by an interpretation I if and only if every ground instance of C is satisfied by I.

3. A clause C is falsified by an interpretation I if and only if there is at least one ground instance C' of C such that C' is not satisfied by I.

4. A set S of clauses is unsatisfiable if and only if for every interpretation I, there is at least one ground instance C' of some clause C in S such that C' is not satisfied by I.

Example 4.9

1. Consider the clause $C = \sim P(x) \vee Q(f(x))$. Let $I_1, I_2,$ and I_3 be defined as follows:

$$I_1 = \{\sim P(a), \sim Q(a), \sim P(f(a)), \sim Q(f(a)), \sim P(f(f(a))), \sim Q(f(f(a))), \ldots\}$$

$$I_2 = \{\ P(a), \quad Q(a), \quad P(f(a)), \quad Q(f(a)), \quad P(f(f(a))), \quad Q(f(f(a))), \ldots\}$$

$$I_3 = \{\ P(a), \sim Q(a), \quad P(f(a)), \sim Q(f(a)), \quad P(f(f(a))), \sim Q(f(f(a))), \ldots\}.$$

The reader should be able to see that C is satisfied by I_1 and I_2, but falsified by I_3.

2. Consider $S = \{P(x), \sim P(a)\}$. There are only two H-interpretations:

$$I_1 = \{P(a)\} \quad \text{and} \quad I_2 = \{\sim P(a)\}.$$

S is falsified by both H-interpretations, and therefore is unsatisfiable.

4.4 SEMANTIC TREES

Having introduced the Herbrand universe, we now consider semantic trees [Robinson, 1968a; Kowalski and Hayes, 1969]. It will be seen in the sequel that finding a proof for a set of clauses is equivalent to generating a semantic tree.

Definition If A is an atom, then the two literals A and $\sim A$ are said to be each other's *complement*, and the set $\{A, \sim A\}$ is called a *complementary pair*.

We note that a clause is a tautology if it contains a complementary pair. In the sequel, when we use "tautology," we shall specifically mean a clause that is a tautology.

Definition Given a set S of clauses, let A be the atom set of S. A *semantic tree* for S is a (downward) tree T, where each link is attached with a finite set of atoms or negations of atoms from A in such a way that:

i. For each node N, there are only finitely many immediate links $L_1, ..., L_n$ from N. Let Q_i be the conjunction of all the literals in the set attached to $L_i, i = 1, ..., n$. Then $Q_1 \vee Q_2 \vee \cdots \vee Q_n$ is a valid propositional formula.

ii. For each node N, let $I(N)$ be the union of all the sets attached to the links of the branch of T down to and including N. Then $I(N)$ does not contain any complementary pair.

Definition Let $A = \{A_1, A_2, ..., A_k, ...\}$ be the atom set of a set S of clauses. A semantic tree for S is said to be *complete* if and only if for every tip node N of the semantic tree, that is, a node that has no links sprouting from it, $I(N)$ contains either A_i or $\sim A_i$ for $i = 1, 2, ...$.

Example 4.10

Let $A = \{P, Q, R\}$ be the atom set of a set S of clauses. Then each one of the two trees in Fig. 4.1 is a complete semantic tree for S. (See p. 58.)

Example 4.11

Consider $S = \{P(x), P(a)\}$. The atom set of S is $\{P(a)\}$. A complete semantic tree for S is shown in Fig. 4.2. (See p. 58.)

Example 4.12

Consider $S = \{P(x), Q(f(x))\}$. The atom set of S is

$$\{P(a), Q(a), P(f(a)), Q(f(a)), P(f(f(a))), Q(f(f(a))), ...\}.$$

Fig. 4.3 shows a semantic tree for S.

It is noted that for each node N in a semantic tree for S, $I(N)$ is a subset of some interpretation for S. For this reason, $I(N)$ will be called a *partial interpretation* for S.

When the atom set A of a set S of clauses is infinite, any complete semantic tree for S will be infinite. As is easily seen, a complete semantic

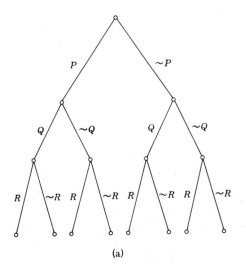

(a)

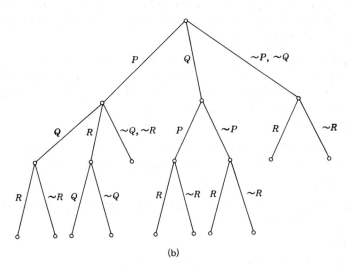

(b)

Figure 4.1

Figure 4.2

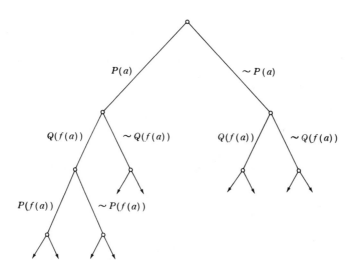

Figure 4.3

tree for S corresponds to an exhaustive survey of all possible interpretations for S. If S is unsatisfiable, then S fails to be true in each of these interpretations. Thus, we may stop expanding nodes from a node N if $I(N)$ falsifies S. This motivates the following definitions.

Definition A node N is a *failure node* if $I(N)$ falsifies some ground instance of a clause in S, but $I(N')$ does not falsify any ground instance of a clause in S for every ancestor node N' of N.

Definition A semantic tree T is said to be *closed* if and only if every branch of T terminates at a failure node.

Definition A node N of a closed semantic tree is called an *inference node* if all the immediate descendant nodes of N are failure nodes.

Example 4.13

Let $S = \{P, Q \vee R, \sim P \vee \sim Q, \sim P \vee \sim R\}$. The atom set of S is $A = \{P, Q, R\}$. Figure 4.4a is a complete semantic tree for S, while Fig. 4.4b is a closed semantic tree for S.

Example 4.14

Consider $S = \{P(x), \sim P(x) \vee Q(f(x)), \sim Q(f(a))\}$. The atom set of S is

$$A = \{P(a), Q(a), P(f(a)), Q(f(a)), \ldots\}.$$

Figure 4.5 shows a closed semantic tree for S.

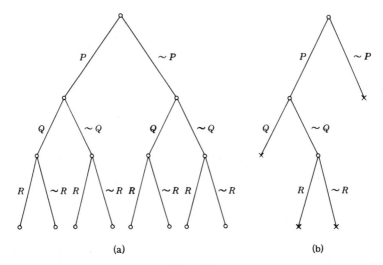

(a)   (b)

Figure 4.4

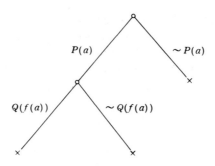

Figure 4.5

4.5 HERBRAND'S THEOREM

Herbrand's theorem is a very important theorem in symbolic logic; it is a base for most modern proof procedures in mechanical theorem proving. Herbrand's theorem is closely related to Theorem 4.2 given in Section 4.3. That is, to test whether a set S of clauses is unsatisfiable, we need consider only interpretations over the Herbrand universe of S. If S is false under all interpretations over the Herbrand universe of S, then we can conclude that S is unsatisfiable. Since there are usually many, possibly an infinite number, of these interpretations, we should organize them in some systematic way.

This can be done by using a semantic tree. We shall give two versions of Herbrand's theorem. The one stated most often in the literature is the second version; however, the first version is useful in this book.

Theorem 4.3   (Herbrand's Theorem, Version I)   A set S of clauses is unsatisfiable if and only if corresponding to every complete semantic tree of S, there is a finite closed semantic tree.

Proof ($\Rightarrow$)   Suppose S is unsatisfiable. Let T be a complete semantic tree for S. For each branch B of T, let I_B be the set of all literals attached to all links of the branch B. Then I_B is an interpretation for S. Since S is unsatisfiable, I_B must falsify a ground instance C' of a clause C in S. However, since C' is finite, there must exist a failure node N_B (which is a finite number of links away from the root node) on the branch B. Since every branch of T has a failure node, there is a closed semantic tree T' for S. Furthermore, since only a finite number of links are connected to each node of T', T' must be finite (that is, the number of nodes in T' is finite), for otherwise, by König's lemma [Knuth, 1968], we could find an infinite branch containing no failure node. Thus we complete the proof of the first half of the theorem.

($\Leftarrow$)   Conversely, if corresponding to every complete semantic tree T for S there is a finite closed semantic tree, then every branch of T contains a failure node. This means that every interpretation falsifies S. Hence S is unsatisfiable. This completes the proof of the second half of the theorem.

Theorem 4.4   (Herbrand's Theorem, Version II).   A set S of clauses is unsatisfiable if and only if there is a finite unsatisfiable set S' of ground instances of clauses of S.

Proof ($\Rightarrow$)   Suppose S is unsatisfiable. Let T be a complete semantic tree for S. Then, by Herbrand's theorem (version I), there is a finite closed semantic tree T' corresponding to T. Let S' be the set of all the ground instances of clauses that are falsified at all the failure nodes of T'. S' is finite since there are a finite number of failure nodes in T'. Since S' is false in every interpretation of S', S' is unsatisfiable.

($\Leftarrow$)   Suppose there is a finite unsatisfiable set S' of ground instances of clauses in S. Since every interpretation I of S contains an interpretation I' of S', if I' falsifies S', then I must also falsify S'. However, S' is falsified by every interpretation I'. Consequently, S' is falsified by every interpretation I of S. Therefore, S is falsified by every interpretation of S. Hence, S is unsatisfiable.   Q.E.D.

Example 4.15

Let $S = \{P(x), \sim P(f(a))\}$. This set S is unsatisfiable. Hence, by Herbrand's theorem, there is a finite unsatisfiable set S' of ground instances of clauses in S. We have found that one of these sets is $S' = \{P(f(a)), \sim P(f(a))\}$.

Example 4.16

Let $S = \{\sim P(x) \vee Q(f(x), x), P(g(b)), \sim Q(y, z)\}$. This set S is unsatisfiable. One of the unsatisfiable sets of ground instances of clauses in S is

$$S' = \{\sim P(g(b)) \vee Q(f(g(b)), g(b)), P(g(b)), \sim Q(f(g(b)), g(b))\}.$$

Example 4.17

Let the set S consist of the following clauses:

$$S = \{\sim P(x, y, u) \vee \sim P(y, z, v) \vee \sim P(x, v, w) \vee P(u, z, w),$$

$$\sim P(x, y, u) \vee \sim P(y, z, v) \vee \sim P(u, z, w) \vee P(x, v, w),$$

$$P(g(x, y), x, y), P(x, h(x, y), y), P(x, y, f(x, y)),$$

$$\sim P(k(x), x, k(x))\}.$$

This set S is also unsatisfiable. However, it is not very easy to find by hand a finite unsatisfiable set S' of ground instances of clauses in S. One way to find such a set S' is to generate a closed semantic tree T' for S. Then the set S' of all the ground instances falsified at all the failure nodes of T' is such a desired set. The following is a desired set S'. The reader may want to check that each ground clause in S' is a ground instance of some clause in S, and that S' is unsatisfiable.

$$S' = \{P(a, h(a, a), a),$$

$$\sim P(k(h(a, a)), h(a, a), k(h(a, a))),$$

$$P(g(a, k(h(a, a))), a, k(h(a, a))),$$

$$\sim P(g(a, k(h(a, a))), a, k(h(a, a))) \vee \sim P(a, h(a, a), a)$$

$$\vee \sim P(g(a, k(h(a, a))), a, k(h(a, a))) \vee P(k(h(a, a)), h(a, a), k(h(a, a)))\}.$$

4.6 IMPLEMENTATION OF HERBRAND'S THEOREM

The second version of Herbrand's theorem suggests a refutation procedure. That is, given an unsatisfiable set S of clauses to prove, if there is a mechanical procedure that can successively generate sets $S_1', \ldots, S_n', \ldots$ of ground instances of clauses in S and can successively test $S_1', S_2', \ldots$ for unsatisfiability, then, as guaranteed by Herbrand's theorem, this procedure can detect a finite N such that S_N' is unsatisfiable.

Gilmore was one of the first men to implement the above idea [Gilmore, 1960]. In 1960, he wrote a computer program that successively generated sets $S_0', S_1', \ldots$, where S_i' is the set of all the ground instances obtained by replacing the variables in S by the constants in the i-level constant set H_i of S. Since each S_i' is a conjunction of ground clauses, one can use any method available in the propositional logic to check its unsatisfiability. Gilmore used the multiplication method. That is, as each S_i' is produced, S_i' is multiplied out into a disjunctive normal form. Any conjunction in the disjunctive normal form containing a complementary pair is removed. Should some S_i' be empty, then S_i' is unsatisfiable and a proof is found.

Example 4.18

Consider

$$S = \{P(x), \sim P(a)\}.$$

$$H_0 = \{a\}$$

$$S_0' = P(a) \wedge \sim P(a) = \square.$$

Thus S is proved to be unsatisfiable.

Example 4.19

Consider

$$S = \{P(a), \sim P(x) \vee Q(f(x)), \sim Q(f(a))\}.$$

$$H_0 = \{a\}.$$

$$S_0' = P(a) \wedge (\sim P(a) \vee Q(f(a))) \wedge \sim Q(f(a))$$
$$= (P(a) \wedge \sim P(a) \wedge \sim Q(f(a))) \vee (P(a) \wedge Q(f(a)) \wedge \sim Q(f(a)))$$
$$= \square \vee \square = \square.$$

Thus S is proved to be unsatisfiable.

The multiplication method used by Gilmore is inefficient. As is easily seen, even for a small set of ten two-literal ground clauses, there are 2^{10} conjunctions. To overcome this inefficiency, Davis and Putnam [1960] introduced a more efficient method for testing the unsatisfiability of a set of ground clauses. We shall now describe their method with some modification.

The Method of Davis and Putnam

Let S be a set of ground clauses. Essentially, the method consists of the following four rules.

    I.   *Tautology Rule*   Delete all the ground clauses from S that are tautologies. The remaining set S' is unsatisfiable if and only if S is.

    II.   *One-Literal Rule*   If there is a unit ground clause L in S, obtain S' from S by deleting those ground clauses in S containing L. If S' is empty,

S is satisfiable. Otherwise, obtain a set S'' from S' by deleting $\sim L$ from S'. S'' is unsatisfiable if and only if S is. Note that if $\sim L$ is a ground unit clause, then the clause becomes $\square$ when $\sim L$ is deleted from the clause.

III.   *Pure-Literal Rule*   A literal L in a ground clause of S is said to be *pure* in S if and only if the literal $\sim L$ does not appear in any ground clause in S. If a literal L is pure in S, delete all the ground clauses containing L. The remaining set S' is unsatisfiable if and only if S is.

IV.   *Splitting Rule*   If the set S can be put into the form

$$(A_1 \vee L) \wedge \cdots \wedge (A_m \vee L) \wedge (B_1 \vee \sim L) \wedge \cdots \wedge (B_n \vee \sim L) \wedge R,$$

where A_i, B_i, and R are free of L and $\sim L$, then obtain the sets $S_1 = A_1 \wedge \cdots \wedge A_m \wedge R$ and $S_2 = B_1 \wedge \cdots \wedge B_n \wedge R$. S is unsatisfiable if and only if $(S_1 \vee S_2)$ is unsatisfiable, that is, both S_1 and S_2 are unsatisfiable.

We can now show that the above rules are sound. That is, if the original set S is unsatisfiable, then the remaining set after one of the rules is applied is still unsatisfiable, and vice versa.

For Rule I   Since a tautology is satisfied by every interpretation, S' is unsatisfiable if and only if S is.

For Rule II   If S' is empty, then all the ground clauses in S contain L. Hence any interpretation containing L can satisfy S. Therefore S is satisfiable. We still have to show that S'' is unsatisfiable if and only if S is unsatisfiable. Suppose S'' is unsatisfiable. If S is satisfiable, then there is a model M of S containing L. For S'', M must satisfy all the clauses which do not contain L. Furthermore, since M falsifies $\sim L$, M must satisfy all the clauses that originally contain $\sim L$. Therefore, M must satisfy S''. This contradicts the assumption that S'' is unsatisfiable. Hence, S must be unsatisfiable. Conversely, suppose S is unsatisfiable. If S'' is satisfiable, then there is a model M'' of S''. Thus any interpretation of S containing M'' and L must be a model of S. This contradicts the assumption that S has no model. Hence S'' must be unsatisfiable. Therefore, S'' is unsatisfiable if and only if S is.

For Rule III   Suppose S' is unsatisfiable. Then S must be unsatisfiable since S' is a subset of S. Conversely, suppose S is unsatisfiable. If S' is satisfiable, then there is a model M of S'. Since neither L nor $\sim L$ is in S', neither L nor $\sim L$ is in M. Thus any interpretation of S that contains M and L is a model of S. This contradicts the assumption that S has no model. Hence S' must be satisfiable. Therefore, S' is unsatisfiable if and only if S is unsatisfiable.

For Rule IV   Suppose S is unsatisfiable. If $(S_1 \vee S_2)$ is satisfiable, then either S_1 or S_2 has a model. If $S_1(S_2)$ has a model M, then any interpretation of S containing $\sim L(L)$ is a model of S. This contradicts the assumption that S has no model. Hence $(S_1 \vee S_2)$ is unsatisfiable. Conversely, suppose $(S_1 \vee S_2)$ is unsatisfiable. If S is satisfiable, S must have a model M. If M contains $\sim L(L)$, M can satisfy $S_1(S_2)$. This contradicts the assumption that $(S_1 \vee S_2)$ is unsatisfiable. Hence S must be unsatisfiable. Therefore, S is unsatisfiable if and only if $(S_1 \vee S_2)$ is.

The above rules are all very important. We shall see in the subsequent chapters that these rules have many extensions. We now give some examples to show how these rules can be used.

Example 4.20

Show that $S = (P \vee Q \vee \sim R) \wedge (P \vee \sim Q) \wedge \sim P \wedge R \wedge U$ is unsatisfiable.

(1) $(P \vee Q \vee \sim R) \wedge (P \vee \sim Q) \wedge \sim P \wedge R \wedge U$

(2) $(Q \vee \sim R) \wedge (\sim Q) \wedge R \wedge U$ Rule II on $\sim P$

(3) $\sim R \wedge R \wedge U$ Rule II on $\sim Q$

(4) $\square \wedge U$ Rule II on $\sim R$.

Since the last formula contains the empty clause $\square$, S is unsatisfiable.

Example 4.21

Show that $S = (P \vee Q) \wedge \sim Q \wedge (\sim P \vee Q \vee \sim R)$ is satisfiable.

(1) $(P \vee Q) \wedge \sim Q \wedge (\sim P \vee Q \vee \sim R)$

(2) $P \wedge (\sim P \vee \sim R)$ Rule II on $\sim Q$

(3) $\sim R$ Rule II on P

(4) $\blacksquare$ Rule II on $\sim R$.

The last set is an empty set. Hence S is satisfiable.

Example 4.22

Show that $S = (P \vee \sim Q) \wedge (\sim P \vee Q) \wedge (Q \vee \sim R) \wedge (\sim Q \vee \sim R)$ is satisfiable.

(1) $(P \vee \sim Q) \wedge (\sim P \vee Q) \wedge (Q \vee \sim R) \wedge (\sim Q \vee \sim R)$

(2) $(\sim Q \wedge (Q \vee \sim R) \wedge (\sim Q \vee \sim R))$

   $\vee (Q \wedge (Q \vee \sim R) \wedge (\sim Q \vee \sim R))$ Rule IV on P

(3) $\sim R \vee \sim R$ Rule II on $\sim Q$ and Q

(4) $\blacksquare \vee \blacksquare$ Rule II on $\sim R$.

Since both of the split sets are satisfiable, S is satisfiable.

Example 4.23

Show that $S = (P \vee Q) \wedge (P \vee \sim Q) \wedge (R \vee Q) \wedge (R \vee \sim Q)$ is satisfiable.

(1) $(P \vee Q) \wedge (P \vee \sim Q) \wedge (R \vee Q) \wedge (R \vee \sim Q)$

(2) $(R \vee Q) \wedge (R \vee \sim Q)$ Rule III on P

(3)   ■   Rule III on R.

Thus, S is satisfiable.

The above method for testing the unsatisfiability (inconsistency) is more efficient than the multiplication method. This method can be applied to any formula in the propositional logic. That is, first transform the given propositional formulas into a conjunctive normal form, and then apply the above four rules on the conjunctive normal form. The reader is encouraged to apply this method to Example 2.13 of Chapter 2.

REFERENCES

Anderson, R., and W. W. Bledsoe (1970): A linear format for resolution with merging and a new technique for establishing completeness, *J. Assoc. Comput. Mach.* **17** 525–534.

Boyer, R. S. (1971): "Locking: A Restriction of Resolution," Ph.D. Thesis, University of Texas at Austin, Texas.

Chang, C. L. (1970a): The unit proof and the input proof in theorem proving, *J. Assoc. Comput. Mach.* **17** 698–707.

Church, A. (1936): An unsolvable problem of number theory, *Amer. J. Math.* **58** 345–363.

Davis, M. (1963): Eliminating the irrelevant from mechanical proofs, *Proc. Symp. Appl. Math.* **15** 15–30.

Davis, M. and H. Putnam (1960): A computing procedure for quantification theory, *J. Assoc. Comput. Mach.* **7** 201–215.

Gilmore, P. C. (1960): A proof method for quantification theory: Its justification and realization, *IBM J. Res. Develop.* 28–35.

Knuth, D. E. (1968): "The Art of Computer Programming," Addison-Wesley, Reading, Massachusetts.

Kowalski, R. and P. Hayes (1969): Semantic trees in automatic theorem proving, *in*: "Machine Intelligence," vol. 4 (B. Meltzer and D. Michie, eds.), American Elsevier, New York, pp. 87–101.

Kowalski, R. and D. Keuhner (1970): Linear resolution with selection function, Metamathematics Unit, Edinburgh University, Scotland.

Loveland, D. W. (1970a): A linear format for resolution, *Proc. IRIA Symp. Automatic Demonstration, Versailles, France, 1968*, Springer-Verlag, New York, pp. 147–162.

Loveland, D. W. (1970b): Some linear Herbrand proof procedures: An analysis, Dept. Computer Science, Carnegie-Mellon University.

Luckham, D. (1970): Refinements in resolution theory, *Proc. IRIA Symp. Automatic Demonstration, Versailles, France, 1968*, Springer-Verlag, New York, pp. 163–190.

Meltzer, B. (1966): Theorem-proving for computers: Some results on resolution and renaming, *Computer J.* **8** 341–343.

Prawitz, D., H. Prawitz, and N. Voghera (1960): A mechanical proof procedure and its realization in an electronic computer, *J. Assoc. Comput. Mach.* **7** 102–128.

Reiter, R. (1971): Two results on ordering for resolution with merging and linear format, *J. Assoc. Comput. Mach.* **18** 630–646.

Robinson, J. A. (1965a): A machine oriented logic based on the resolution principle, *J. Assoc. Comput. Mach.* **12** 23–41.

Robinson, J. A. (1965b): Automatic deduction with hyper-resolution, *Internat. J. Comput. Math.* **1** 227–234.

Robinson, J. A. (1968a): The generalized resolution principle, *in:* "Machine Intelligence," vol. 3, (D. Michie, ed.), American Elsevier, New York, pp. 77–94.

Slagle, J. R. (1967): Automatic theorem proving with renamable and sematic resolution, *J. Assoc. Comput. Mach.* **14** 687–697.

Turing, A. M. (1936): On computable numbers, with an application to the entschiedungsproblem, *Proc. London Math. Soc.* 2, 42, pp. 230–265.

Wang, H. (1960b): Towards mechanical mathematics, *IBM J. Res. Develop.* **4** 224–268.

Wos, L., D. Carson, and G. A. Robinson (1964): The unit preference strategy in theorem proving, *Proc. AFIPS 1964 Fall Joint Computer Conf.* **26** pp. 616–621.

Wos, L., G. A. Robinson, and D. F. Carson (1965): Efficiency and completeness of the set of support strategy in theorem proving, *J. Assoc. Comput. Mach.* **12** 536–541.

Yates, R., B. Raphael, and T. Hart (1970): Resolution graphs, *Artificial Intelligence* **1** 257–290.

EXERCISES

Section 4.2

1. Find a standard form for each of the following formulas:

   (a) $\sim ((\forall x) P(x) \to (\exists y)(\forall z) Q(y, z))$

   (b) $(\forall x)\left(\left(\sim E(x, 0) \to \left((\exists y)\left(E(y, g(x)) \land (\forall z)\left(E(z, g(x)) \to E(y, z)\right)\right)\right)\right)\right)$

   (c) $\sim ((\forall x) P(x) \to (\exists y) P(y))$.

2. Suppose $(\exists x)(\forall y) M[x, y]$ is a prenex normal form of a formula F, where $M[x, y]$ is the matrix that contains only variables x and y. Let f be a function symbol not occurring in $M[x, y]$. Prove that F is valid if and only if $(\exists x) M[x, f(x)]$ is valid.

3. Let S_1 and S_2 be standard forms of formulas F_1 and F_2, respectively. If $S_1 = S_2$, is it true that $F_1 = F_2$? Explain.

4. Consider the following statements:

   F_1: Everyone who saves money earns interst.

   F_2: If there is no interest, then nobody saves money.

   Let $S(x, y), M(x), I(x)$, and $E(x, y)$ represent "x saves y," "x is money," "x is interest," and "x earns y," respectively.

   (1) Symbolize F_1 and F_2.

   (2) Find the standard forms of F_1 and $\sim F_2$.

Section 4.3

5. Let $S = \{P(f(x), a, g(f(x), b))\}$.

    (1) Find H_0 and H_1.
    (2) Find all the ground instances of S over H_0.
    (3) Find all the ground instances of S over H_1.

6. Prove Lemma 4.1 of this chapter.
7. Let F denote a formula. Let S be a standard form of $\sim F$. Find the necessary and sufficient condition for F such that the Herbrand universe of S is finite.
8. Consider the following clause C and interpretation I:

    C: $P(x) \vee Q(x, f(x))$

    I: $\{\sim P(a), \sim P(f(a)), \sim P(f(f(a))), \ldots,$

        $\sim Q(a, a), Q(a, f(a)), \sim Q(a, f(f(a))), \ldots,$

        $\sim Q(f(a), a), Q(f(a), f(a)), \sim Q(f(a), f(f(a))), \ldots\}$.

    Does I satisfy C?

9. Consider the following set S of clauses:

$$S = \begin{cases} P(x) \\ Q(f(y)). \end{cases}$$

    Let an interpretation I be defined as below:

    $I = \{P(a), P(f(a)), P(f(f(a))), \ldots,$

        $Q(a), \sim Q(f(a)), Q(f(f(a))), \ldots\}$.

    Does I satisfy S?

10. Consider $S = \{P(x), \sim P(f(y))\}$.

    1. Give H_0, H_1, H_2, and H_3 of S.
    2. Is it possible to find an interpretation that satisfies S? If yes, give one. If no, why?

Section 4.4

11. Let $S = \{P, \sim P \vee Q, \sim Q\}$. Give a closed semantic tree of S.
12. Consider $S = \{P(x), \sim P(x) \vee Q(x, a), \sim Q(y, a)\}$.

    (a) Give the atom set of S.
    (b) Give a complete semantic tree of S.
    (c) Give a closed semantic tree of S.

Section 4.5

13. Consider $S = \{P(x, a, g(x, b)), \sim P(f(y), z, g(f(a), b))\}$. Find an unsatisfiable set S' of ground instances of clauses in S.

14. Let $S = \{P(x), Q(x, f(x)) \vee \sim P(x), \sim Q(g(y), z)\}$. Find an unsatisfiable set S' of ground instances of clauses in S.

Section 4.6

15. Use the rules suggested by Davis and Putnam to prove that the following formulas are unsatisfiable.

   (1) $(\sim P \vee Q) \wedge \sim Q \wedge P$

   (2) $(P \vee Q) \wedge (R \vee Q) \wedge \sim R \wedge \sim Q$

   (3) $(P \vee Q) \wedge (\sim P \vee Q) \wedge (\sim R \vee \sim Q) \wedge (R \vee \sim Q)$.

16. Use the rules suggested by Davis and Putnam to prove that the following formulas are satisfiable.

   (1) $P \wedge Q \wedge R$

   (2) $(P \vee Q) \wedge (\sim P \vee Q) \wedge R$

   (3) $(P \vee Q) \wedge \sim Q$.

Chapter 5

The Resolution Principle

5.1 INTRODUCTION

In the previous chapters, we have considered Herbrand's theorem. Based upon this theorem, we have given a refutation procedure. However, Herbrand's procedure has one major drawback: It requires the generation of sets S_1', S_2',... of ground instances of clauses. For most cases, this sequence grows exponentially. To see this, let us give one simple example. Consider

$$S = \{P(x, g(x), y, h(x, y), z, k(x, y, z)), \sim P(u, v, e(v), w, f(v, w), x)\}.$$

Since

$$H_0 = \{a\},$$
$$H_1 = \{a, g(a), h(a, a), k(a, a, a), e(a), f(a, a)\}$$
$$\vdots \quad ,$$

the number of elements in S_0', S_1', ... is 2, 1512, ..., respectively. It is interesting to note that the earliest set that is unsatisfiable is S_5'. However, H_5' has of the order of 10^{64} elements. Consequently, S_5' has of the order of 10^{256} elements. With current computer technology, it is impossible even to store S_5' in a computer, not to mention test its unsatisfiability.

In order to avoid the generation of sets of ground instances as required in Herbrand's procedure, we shall introduce in this chapter the resolution principle due to Robinson [1965]. It can be applied directly to any set S of clauses (not necessarily ground clauses) to test the unsatisfiability of S.

The essential idea of the resolution principle is to check whether S contains the empty clause $\square$. If S contains $\square$, then S is unsatisfiable. If S does not contain $\square$, the next thing to check is whether $\square$ can be derived from S. It will be clear later that, by Herbrand's theorem (version I), checking for the presence of $\square$ is equivalent to counting the number of nodes of a closed semantic tree for S. By Theorem 4.3, S is unsatisfiable if and only if there is a finite closed semantic tree T for S. Clearly S contains $\square$ if and only if T consists of only one node—the root node. If S does not contain $\square$, T must contain more than one node. However, if we can reduce the number of nodes in T to one, eventually $\square$ can be forced to appear. This is what the resolution principle does. Indeed, we can view the resolution principle as an inference rule that can be used to generate new clauses from S. If we put these new clauses into S, some nodes of the original T can be forced to become failure nodes. Thus the number of nodes in T can be reduced and the empty clause $\square$ will be eventually derived.

In this chapter, we shall first consider the resolution principle for the propositional logic. Then we shall extend it to the first-order logic.

5.2   THE RESOLUTION PRINCIPLE FOR THE PROPOSITIONAL LOGIC

The resolution principle is essentially an extension of the *one-literal rule* of Davis and Putnam, given in Section 4.6 of Chapter 4. Consider the following clauses:

C_1: P

C_2: $\sim P \vee Q$.

Using the one-literal rule, from C_1 and C_2 we can obtain a clause

C_3: Q.

What the one-literal rule has required us to do is first to examine whether there is a complementary pair of a literal (for example, P) in C_1 and a literal (for example, $\sim P$) in C_2, then to delete this pair from C_1 and C_2 to obtain clause C_3, which is Q.

Extending the above rule and applying it to any pair of clauses (not necessarily unit clauses), we have the following rule, which is called the *resolution principle: For any two clauses C_1 and C_2, if there is a literal L_1 in C_1 that is complementary to a literal L_2 in C_2, then delete L_1 and L_2 from C_1 and C_2, respectively, and construct the disjunction of the remaining clauses. The constructed clause is a resolvent of C_1 and C_2.*

Example 5.1

Consider the following clauses:

C_1: $P \vee R$

C_2: $\sim P \vee Q$.

Clause C_1 has the literal P, which is complementary to $\sim P$ in C_2. Therefore, by deleting P and $\sim P$ from C_1 and C_2, respectively, and constructing the disjunction of the remaining clauses R and Q, we obtain a resolvent $R \vee Q$.

Example 5.2

Consider clauses

C_1: $\sim P \vee Q \vee R$

C_2: $\sim Q \vee S$.

The resolvent of C_1 and C_2 is $\sim P \vee R \vee S$.

Example 5.3

Consider clauses

C_1: $\sim P \vee Q$

C_2: $\sim P \vee R$.

Since there is no literal in C_1 that is complementary to any literal in C_2, there is no resolvent of C_1 and C_2.

An important property of a resolvent is that any resolvent of two clauses C_1 and C_2 is a logical consequence of C_1 and C_2. This is stated in the following theorem.

Theorem 5.1   *Given two clauses C_1 and C_2, a resolvent C of C_1 and C_2 is a logical consequence of C_1 and C_2.*

Proof   Let C_1, C_2, and C be denoted as follows: $C_1 = L \vee C_1'$, $C_2 = \sim L \vee C_2'$, and $C = C_1' \vee C_2'$, where C_1' and C_2' are disjunctions of literals. Suppose C_1 and C_2 are true in an interpretation I. We want to prove that the resolvent C of C_1 and C_2 is also true in I. To prove this, note that either L or $\sim L$ is false in I. Assume L is false in I. Then C_1 must not be a unit clause, for otherwise, C_1 would be false in I. Therefore, C_1' must be true in I. Thus, the resolvent C, that is, $C_1' \vee C_2'$, is true in I. Similarly, we can show that if $\sim L$ is false in I, then C_2' must be true in I. Therefore, $C_1' \vee C_2'$ must be true in I.   Q.E.D.

If we have two unit clauses, then the resolvent of them, if there is one, is the empty clause $\square$. More importantly, if a set S of clauses is unsatisfiable, we can use the resolution principle to generate $\square$ from S. This result will be presented as a theorem in Section 5.6 of this chapter. Meanwhile, we define the notion of deduction.

Definition   Given a set S of clauses, a (resolution) *deduction* of C from S is a finite sequence C_1, C_2, ..., C_k of clauses such that each C_i either is a clause in S or a resolvent of clauses preceding C_i, and $C_k = C$. A deduction of $\square$ from S is called a *refutation*, or a *proof* of S.

We say that a clause C can be *deduced* or *derived* from S if there is a deduction of C from S. We shall now present several examples to show how the resolution principle can be used to prove the unsatisfiability of a set of clauses.

Example 5.4

Consider the set

(1) $\sim P \vee Q$ ⎫

(2) $\sim Q$ ⎬ S.

(3) P ⎭

From (1) and (2), we can obtain a resolvent

(4) $\sim P$.

From (4) and (3), we can obtain a resolvent

(5) $\square$.

Since $\square$ is derived from S by resolution, according to Theorem 5.1 $\square$ is a logical consequence of S. However, $\square$ can only be a logical consequence of an unsatisfiable set of clauses. Hence, S is unsatisfiable.

Example 5.5

For the set

(1) $P \vee Q$ ⎫

(2) $\sim P \vee Q$ ⎬ S

(3) $P \vee \sim Q$

(4) $\sim P \vee \sim Q$ ⎭

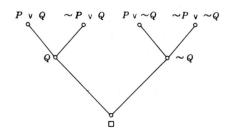

Figure 5.1

we generate the following resolvents

(5) Q from (1) and (2)

(6) $\sim Q$ from (3) and (4)

(7) $\square$ from (5) and (6).

Since $\square$ is derived, S is unsatisfiable. The above deduction can be represented by the tree in Fig. 5.1, called a *deduction tree*.

The resolution principle is a very powerful inference rule. In the sequel, we shall define this principle in the context of the first-order logic. We shall also show that the resolution principle is *complete* in proving the unsatisfiability of a set of clauses. That is, given a set S of clauses, S is unsatisfiable if and only if there is a resolution deduction of the empty clause $\square$ from S. In Section 5.7, we shall apply the resolution principle to several examples so that the reader can appreciate the usefulness of this inference rule.

5.3   SUBSTITUTION AND UNIFICATION

In the last section, we considered the resolution principle for the propositional logic. We shall extend it to the first-order logic in later sections. In Section 5.2, we noted that the most important part of applying the resolution principle is finding a literal in a clause that is complementary to a literal in another clause. For clauses containing no variables, this is very simple. However, for clauses containing variables, it is more complicated. For example, consider the clauses:

C_1: $P(x) \vee Q(x)$

C_2: $\sim P(f(x)) \vee R(x)$.

There is no literal in C_1 that is complementary to any literal in C_2. However, if we substitute $f(a)$ for x in C_1 and a for x in C_2, we obtain

C_1': $P(f(a)) \vee Q(f(a))$

C_2': $\sim P(f(a)) \vee R(a)$.

We know that C_1' and C_2' are ground instances of C_1 and C_2, respectively, and $P(f(a))$ and $\sim P(f(a))$ are complementary to each other. Therefore, from C_1' and C_2', we can obtain a resolvent

C_3': $Q(f(a)) \vee R(a)$.

More generally, if we substitute $f(x)$ for x in C_1, we obtain

C_1^*: $P(f(x)) \vee Q(f(x))$.

Again, C_1^* is an *instance* of C_1. This time the literal $P(f(x))$ in C_1^* is complementary to the literal $\sim P(f(x))$ in C_2. Therefore, we can obtain a resolvent from C_1^* and C_2,

C_3: $Q(f(x)) \vee R(x)$.

C_3' is an instance of clause C_3. By substituting appropriate terms for the variables in C_1 and C_2 as above, we can generate new clauses from C_1 and C_2. Furthermore, clause C_3 is the *most general clause* in the sense that all the other clauses which can be generated by the above process are instances of C_3. C_3 will also be called a resolvent of C_1 and C_2. In the sequel, we shall consider how to generate resolvents from clauses (possibly containing variables). Since obtaining resolvents from clauses frequently requires substitutions for variables, we give the following definitions.

Definition A *substitution* is a finite set of the form $\{t_1/v_1, \ldots, t_n/v_n\}$, where every v_i is a variable, every t_i is a term different from v_i, and no two elements in the set have the same variable after the stroke symbol. When $t_1, \ldots, t_n$ are ground terms, the substitution is called a *ground substitution*. The substitution that consists of no elements is called the *empty substitution* and is denoted by ε. We shall use Greek letters to represent substitutions.

Example 5.6

The following two sets are substitutions:

$$\{f(z)/x, y/z\}, \qquad \{a/x, g(y)/y, f(g(b))/z\}.$$

Definition Let $\theta = \{t_1/v_1, \ldots, t_n/v_n\}$ be a substitution and E be an expression. Then $E\theta$ is an expression obtained from E by replacing simultaneously each occurrence of the variable v_i, $1 \leqslant i \leqslant n$, in E by the term t_i. $E\theta$ is called an *instance* of E. (We note that the definition of an instance is compatible with that of a ground instance of a clause, defined in Chapter 4.)

Example 5.7

Let $\theta = \{a/x, f(b)/y, c/z\}$ and $E = P(x, y, z)$. Then $E\theta = P(a, f(b), c)$.

Definition   Let $\theta = \{t_1/x_1, ..., t_n/x_n\}$ and $\lambda = \{u_1/y_1, ..., u_m/y_m\}$ be two substitutions. Then the *composition* of θ and λ is the substitution, denoted by $\theta \circ \lambda$, that is obtained from the set

$$\{t_1\lambda/x_1, ..., t_n\lambda/x_n, u_1/y_1, ..., u_m/y_m\}$$

by deleting any element $t_j\lambda/x_j$ for which $t_j\lambda = x_j$, and any element u_i/y_i such that y_i is among $\{x_1, x_2, ..., x_n\}$.

Example 5.8

Let

$$\theta = \{t_1/x_1, t_2/x_2\} = \{f(y)/x, z/y\}$$

$$\lambda = \{u_1/y_1, u_2/y_2, u_3/y_3\} = \{a/x, b/y, y/z\}.$$

Then

$$\{t_1\lambda/x_1, t_2\lambda/x_2, u_1/y_1, u_2/y_2, u_3/y_3\} = \{f(b)/x, y/y, a/x, b/y, y/z\}.$$

However, since $t_2\lambda = x_2, t_2\lambda/x_2$, i.e., y/y, should be deleted from the set. In addition, since y_1 and y_2 are among $\{x_1, x_2, x_3\}$, u_1/y_1 and u_2/y_2, that is, a/x and b/y, should be deleted. Thus we obtain

$$\theta \circ \lambda = \{f(b)/x, y/z\}.$$

It should be noted that the composition of substitutions is associative, and that the empty substitution ε is both a left and a right identity. That is, $(\theta \circ \lambda) \circ \mu = \theta \circ (\lambda \circ \mu)$ and $\varepsilon \circ \theta = \theta \circ \varepsilon$ for all θ, λ, and μ.

In the resolution proof procedure, in order to identify a complementary pair of literals, very often we have to unify (match) two or more expressions. That is, we have to find a substitution that can make several expressions identical. Therefore, we now consider the unification of expressions.

Definition   A substitution θ is called a *unifier* for a set $\{E_1, ..., E_k\}$ if and only if $E_1\theta = E_2\theta = \cdots = E_k\theta$. The set $\{E_1, ..., E_k\}$ is said to be *unifiable* if there is a unifier for it.

Definition   A unifier σ for a set $\{E_1, ..., E_k\}$ of expressions is a *most general unifier* if and only if for each unifier θ for the set there is a substitution λ such that $\theta = \sigma \circ \lambda$.

Example 5.9

The set $\{P(a, y), P(x, f(b))\}$ is unifiable since the substitution $\theta = \{a/x, f(b)/y\}$ is a *unifier* for the set.

5.4   UNIFICATION ALGORITHM

In this section, we shall give a unification algorithm for finding a most general unifier for a finite unifiable set of nonempty expressions. When the set is not unifiable, the algorithm will also detect this fact.

Consider $P(a)$ and $P(x)$. These two expressions are not identical: The disagreement is that a occurs in $P(a)$ but x in $P(x)$. In order to unify $P(a)$ and $P(x)$, we first have to find the disagreement, and then try to eliminate it. For $P(a)$ and $P(x)$, the disagreement is $\{a,x\}$. Since x is a variable, x can be replaced by a, and thus the disagreement can be eliminated. This is the idea upon which the unification algorithm is based.

Definition   The disagreement set of a nonempty set W of expressions is obtained by locating the first symbol (counting from the left) at which not all the expressions in W have exactly the same symbol, and then extracting from each expression in W the subexpression that begins with the symbol occupying that position. The set of these respective subexpressions is the *disagreement set* of W.

Example 5.10

If W is $\left\{ P(x, \underline{f(y,z)}), P(x,\underline{a}), P\left(x,\underline{g(h(k(x)))}\right) \right\}$, then the first symbol position at which not all atoms in W are exactly the same is the fifth, since they all have the first four symbols $P(x,$ in common. Thus, the disagreement set consists of the respective subexpressions (underlined terms) that begin in symbol position number five, and it is in fact the set $\{ f(y,z), a, g(h(k(x))) \}$.

Unification Algorithm

Step 1   Set $k = 0$, $W_k = W$, and $\sigma_k = \varepsilon$.

Step 2   If W_k is a singleton, stop; σ_k is a most general unifier for W. Otherwise, find the disagreement set D_k of W_k.

Step 3   If there exist elements v_k and t_k in D_k such that v_k is a variable that does not occur in t_k, go to Step 4. Otherwise, stop; W is not unifiable.

Step 4   Let $\sigma_{k+1} = \sigma_k \{t_k/v_k\}$ and $W_{k+1} = W_k\{t_k/v_k\}$. (Note that $W_{k+1} = W\sigma_{k+1}$.)

Step 5   Set $k = k+1$ and go to Step 2.

Example 5.11

Find a most general unifier for $W = \{P(a, x, f(g(y))), P(z, f(z), f(u))\}$.

1. $\sigma_0 = \varepsilon$ and $W_0 = W$. Since W_0 is not a singleton, σ_0 is not a most general unifier for W.

2. The disagreement set $D_0 = \{a,z\}$. In D_0, there is a variable $v_0 = z$ that does not occur in $t_0 = a$.

3. Let

$$\sigma_1 = \sigma_0 \circ \{t_0/v_0\} = \varepsilon \circ \{a/z\} = \{a/z\},$$

$$W_1 = W_0 \{t_0/v_0\}$$

$$= \{P(a, x, f(g(y))), P(z, f(z), f(u))\} \{a/z\}$$

$$= \{P(a, x, f(g(y))), P(a, f(a), f(u))\}.$$

4. W_1 is not a singleton, hence we find the disagreement set D_1 of W_1:

$$D_1 = \{x, f(a)\}.$$

5. From D_1, we find that $v_1 = x$ and $t_1 = f(a)$.
6. Let

$$\sigma_2 = \sigma_1 \circ \{t_1/v_1\} = \{a/z\} \circ \{f(a)/x\} = \{a/z, f(a)/x\},$$

$$W_2 = W_1 \{t_1/v_1\}$$

$$= \{P(a, x, f(g(y))), P(a, f(a), f(u))\} \{f(a)/x\}$$

$$= \{P(a, f(a), f(g(y))), P(a, f(a), f(u))\}.$$

7. W_2 is not a singleton, hence we find the disagreement set D_2 of W_2: $D_2 = \{g(y), u\}$. From D_2, we find that $v_2 = u$ and $t_2 = g(y)$.
8. Let

$$\sigma_3 = \sigma_2 \circ \{t_2/v_2\} = \{a/z, f(a)/x\} \circ \{g(y)/u\} = \{a/z, f(a)/x, g(y)/u\},$$

$$W_3 = W_2 \{t_2/v_2\}$$

$$= \{P(a, f(a), f(g(y))), P(a, f(a), f(u))\} \{g(y)/u\}$$

$$= \{P(a, f(a), f(g(y))), P(a, f(a), f(g(y)))\}$$

$$= \{P(a, f(a), f(g(y)))\}.$$

9. Since W_3 is a singleton, $\sigma_3 = \{a/z, f(a)/x, g(y)/u\}$ is a most general unifier for W.

Example 5.12

Determine whether or not the set $W = \{Q(f(a), g(x)), Q(y, y)\}$ is unifiable.

1. Let $\sigma_0 = \varepsilon$ and $W_0 = W$.
2. W_0 is not a singleton, hence we find the disagreement set D_0 of W_0: $D_0 = \{f(a), y\}$. From D_0, we know that $v_0 = y$ and $t_0 = f(a)$.
3. Let

$$\sigma_1 = \sigma_0 \circ \{t_0/v_0\} = \varepsilon \circ \{f(a)/y\} = \{f(a)/y\},$$

$$W_1 = W_0 \{t_0/v_0\} = \{Q(f(a), g(x)), Q(y, y)\} \{f(a)/y\}$$

$$= \{Q(f(a), g(x)), Q(f(a), f(a))\}.$$

4. W_1 is not a singleton, hence we find the disagreement set D_1 of W_1:
$D_1 = \{g(x), f(a)\}$.

5. However, no element of D_1 is a variable. Hence the unification algorithm is terminated and we conclude that W is not unifiable.

It is noted that the above unification algorithm will always terminate for any finite nonempty set of expressions, for otherwise there would be generated an infinite sequence $W\sigma_0, W\sigma_1, W\sigma_2, \ldots$ of finite nonempty sets of expressions with the property that each successive set contains one less variable than its predecessor (namely, $W\sigma_k$ contains v_k but $W\sigma_{k+1}$ does not). This is impossible since W contains only finitely many distinct variables.

We indicated before that if W is unifiable, the unification algorithm will always find a most general unifier for W. This is proved in the following theorem.

Theorem 5.2 (Unification Theorem) If W is a finite nonempty unifiable set of expressions, then the unification algorithm will always terminate at Step 2, and the last σ_k is a most general unifier for W.

Proof Since W is unifiable, we let θ be any unifier for W. For $k = 0, 1, \ldots$, we show that there is a substitution λ_k such that $\theta = \sigma_k \circ \lambda_k$ by induction on k. For $k = 0$, we let $\lambda_0 = \theta$. Then $\theta = \sigma_0 \circ \lambda_0$, since $\sigma_0 = \varepsilon$. Suppose $\theta = \sigma_k \circ \lambda_k$ holds for $0 \leqslant k \leqslant n$. If $W\sigma_n$ is a singleton, then the unification algorithm terminates at Step 2. Since $\theta = \sigma_n \circ \lambda_n, \sigma_n$ is a most general unifier for W. If $W\sigma_n$ is not a singleton, then the unification algorithm will find the disagreement set D_n of $W\sigma_n$. Because $\theta = \sigma_n \circ \lambda_n$ is a unifier of W, λ_n must unify D_n. However, since D_n is the disagreement set, there must be a variable in D_n. Let t_n be any other element different from v_n. Then, since λ_n unifies $D_n, v_n \lambda_n = t_n \lambda_n$. Now if v_n occurs in t_n, then $v_n \lambda_n$ occurs in $t_n \lambda_n$. However, this is impossible since v_n and t_n are distinct, and $v_n \lambda_n = t_n \lambda_n$. Therefore, v_n does not occur in t_n. Hence the unification algorithm will not terminate at Step 3, but will go to Step 4 to set $\sigma_{n+1} = \sigma_n \circ \{t_n/v_n\}$. Let $\lambda_{n+1} = \lambda_n - \{t_n\lambda_n/v_n\}$. Then, since v_n does not occur in t_n, $t_n\lambda_{n+1} = t_n(\lambda_n - \{t_n\lambda_n/v_n\}) = t_n\lambda_n$. Thus, we have

$$\{t_n/v_n\} \circ \lambda_{n+1} = \{t_n\lambda_{n+1}/v_n\} \cup \lambda_{n+1}$$

$$= \{t_n\lambda_n/v_n\} \cup \lambda_{n+1}$$

$$= \{t_n\lambda_n/v_n\} \cup (\lambda_n - \{t_n\lambda_n/v_n\})$$

$$= \lambda_n.$$

That is, $\lambda_n = \{t_n/v_n\} \circ \lambda_{n+1}$. Therefore,

$$\theta = \sigma_n \circ \lambda_n = \sigma_n \circ \{t_n/v_n\} \circ \lambda_{n+1} = \sigma_{n+1} \circ \lambda_{n+1}.$$

Hence, for all $k \geqslant 0$, there is a substitution λ_k such that $\theta = \sigma_k \circ \lambda_k$. Since the unification algorithm must terminate, and since it will not terminate at Step 3, it must terminate at Step 2. Furthermore, since $\theta = \sigma_k \circ \lambda_k$ for all k, the last σ_k is a most general unifier for W.   Q.E.D.

5.5 THE RESOLUTION PRINCIPLE FOR THE FIRST-ORDER LOGIC

Having introduced the unification algorithm in the last section, we can now consider the resolution principle for the first-order logic.

Definition   If two or more literals (with the same sign) of a clause C have a most general unifier σ, then $C\sigma$ is called a *factor* of C. If $C\sigma$ is a unit clause, it is called a *unit factor* of C.

Example 5.13

Let $C = P(x) \vee P(f(y)) \vee \sim Q(x)$. Then the first and the second literals (underlined) have a most general unifier $\sigma = \{f(y)/x\}$. Hence, $C\sigma = P(f(y)) \vee \sim Q(f(y))$ is a factor of C.

Definition   Let C_1 and C_2 be two clauses (called *parent clauses*) with no variables in common. Let L_1 and L_2 be two literals in C_1 and C_2, respectively. If L_1 and $\sim L_2$ have a most general unifier σ, then the clause

$$(C_1\sigma - L_1\sigma) \cup (C_2\sigma - L_2\sigma)$$

is called a *binary resolvent* of C_1 and C_2. The literals L_1 and L_2 are called the *literals resolved upon*.

Example 5.14

Let $C_1 = P(x) \vee Q(x)$ and $C_2 = \sim P(a) \vee R(x)$. Since x appears in both C_1 and C_2, we rename the variable in C_2 and let $C_2 = \sim P(a) \vee R(y)$. Choose $L_1 = P(x)$ and $L_2 = \sim P(a)$. Since $\sim L_2 = P(a)$, L_1 and $\sim L_2$ have the most general unifier $\sigma = \{a/x\}$. Therefore,

$$
\begin{aligned}
(C_1\sigma - L_1\sigma) &\cup (C_2\sigma - L_2\sigma) \\
&= (\{P(a), Q(a)\} - \{P(a)\}) \cup (\{\sim P(a), R(y)\} - \{\sim P(a)\}) \\
&= \{Q(a)\} \cup \{R(y)\} = \{Q(a), R(y)\} = Q(a) \vee R(y).
\end{aligned}
$$

Thus $Q(a) \vee R(y)$ is a binary resolvent of C_1 and C_2. $P(x)$ and $\sim P(a)$ are the literals resolved upon.

Definition   A *resolvent* of (parent) clauses C_1 and C_2 is one of the following binary resolvents:

1. a binary resolvent of C_1 and C_2,
2. a binary resolvent of C_1 and a factor of C_2,
3. a binary resolvent of a factor of C_1 and C_2,
4. a binary resolvent of a factor of C_1 and a factor of C_2.

Example 5.15

Let $C_1 = P(x) \vee P(f(y)) \vee R(g(y))$ and $C_2 = \sim P(f(g(a))) \vee Q(b)$. A factor of C_1 is $C_1' = P(f(y)) \vee R(g(y))$. A binary resolvent of C_1' and C_2 is $R(g(g(a))) \vee Q(b)$. Therefore, $R(g(g(a))) \vee Q(b)$ is a resolvent of C_1 and C_2.

The resolution principle, or resolution for short, is an inference rule that generates resolvents from a set of clauses. This rule was introduced in 1965 by Robinson. It is more efficient than the earlier proof procedures such as the direct implementation of Herbrand's theorem used by Gilmore and Davis and Putnam. Furthermore, resolution is *complete*. That is, it will always generate the empty clause □ from an unsatisfiable set of clauses. Before we prove this statement, we first give an example in plane geometry.

Example 5.16

Show that alternate interior angles formed by a diagonal of a trapezoid are equal.

To prove this theorem, we first axiomatize it. Let $T(x,y,u,v)$ mean that $xyuv$ is a trapezoid with upper-left vertex x, upper-right vertex y, lower-right vertex u, and lower-left vertex v; let $P(x,y,u,v)$ mean that the line segment xy is parallel to the line segment uv; and let $E(x,y,z,u,v,w)$ mean that the angle xyz is equal to the angle uvw. Then we have the following axioms:

A_1: $(\forall x)(\forall y)(\forall u)(\forall v)[T(x,y,u,v) \rightarrow P(x,y,u,v)]$ definition of a trapezoid

A_2: $(\forall x)(\forall y)(\forall u)(\forall v)[P(x,y,u,v) \rightarrow E(x,y,v,u,v,y)]$ alternate interior angles of parallel lines are equal

A_3: $T(a,b,c,d)$ given in Fig. 5.2.

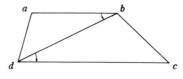

Figure 5.2

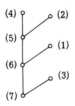

Figure 5.3

From these axioms, we should be able to conclude that $E(a,b,d,c,d,b)$ is true, that is,

$$A_1 \wedge A_2 \wedge A_3 \rightarrow E(a,b,d,c,d,b)$$

is a valid formula. Since we want to prove this by refutation, we negate the conclusion and prove that

$$A_1 \wedge A_2 \wedge A_3 \wedge \sim E(a,b,d,c,d,b)$$

is unsatisfiable. To do this, we transform it into a standard form as follows:

$$S = \{ \sim T(x,y,u,v) \vee P(x,y,u,v), \ \sim P(x,y,u,v) \vee E(x,y,v,u,v,y),$$

$$T(a,b,c,d), \ \sim E(a,b,d,c,d,b)\}.$$

The above standard form S is a set of four clauses. We now show that the set S is unsatisfiable by resolution:

(1)	$\sim T(x,y,u,v) \vee P(x,y,u,v)$	a clause in S
(2)	$\sim P(x,y,u,v) \vee E(x,y,v,u,v,y)$	a clause in S
(3)	$T(a,b,c,d)$	a clause in S
(4)	$\sim E(a,b,d,c,d,b)$	a clause in S
(5)	$\sim P(a,b,c,d)$	a resolvent of (2) and (4)
(6)	$\sim T(a,b,c,d)$	a resolvent of (1) and (5)
(7)	$\square$	a resolvent of (3) and (6).

Since the last clause is the empty clause that is derived from S, we conclude that S is unsatisfiable. The above proof steps can be easily represented by a tree shown in Fig. 5.3.

The tree in Fig. 5.3 is called a deduction tree. That is, a *deduction tree* from a set S of clauses is an (upward) tree, to each initial node of which is attached a clause in S, and to each noninitial node of which is attached a resolvent of clauses attached to its immediate predecessor nodes. We call the deduction tree a *deduction tree of R* if *R* is the clause attached to the root node of the deduction tree. Since a deduction tree is merely the tree representation of a deduction, in the sequel we shall use deduction and deduction tree interchangeably.

5.6 COMPLETENESS OF THE RESOLUTION PRINCIPLE

In Chapter 4, we presented the concept of semantic trees to prove Herbrand's theorem. In this section, we shall use the semantic tree to prove the completeness of the resolution principle. In fact, there is a close relationship between the semantic tree and resolution, as is demonstrated by the following example.

Example 5.17

Consider the following set *S* of clauses:

(1) P

(2) $\sim P \vee Q$

(3) $\sim P \vee \sim Q$.

The atom set of *S* is $\{P, Q\}$. Let *T* be a complete semantic tree as shown in Fig. 5.4a. *T* has a closed semantic tree *T'* shown in Fig. 5.4b. Node (2) in Fig. 5.4b is an inference node. Its two successors, nodes (4) and (5), are failure nodes. The clauses falsified at nodes (4) and (5) are $\sim P \vee \sim Q$ and $\sim P \vee Q$, respectively. As can be seen, these two clauses must have a complementary pair of literals, and therefore can be resolved. Resolving

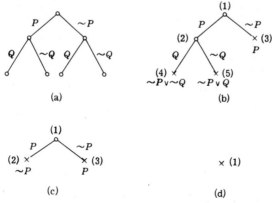

(a)   (b)   (c)   (d)

Figure 5.4   (a) *T*. (b) *T'*. $(\sim P \vee Q)$ and $(\sim P \vee \sim Q)$ can be resolved to give $\sim P$. (c) *T''*. *P* and $\sim P$ can be resolved to give □. (d) *T'''*.

$\sim P \vee \sim Q$ and $\sim P \vee Q$, we obtain $\sim P$. Note that $\sim P$ is falsified by the partial interpretation corresponding to node (2). If we put $\sim P$ into S, then we can have a closed semantic tree T'' for $S \cup \{\sim P\}$, as shown in Fig. 5.4c. In Fig. 5.4c, node (1) is an inference node. This time we can obtain $\square$ by resolving P and $\sim P$. Putting $\square$ into $S \cup \{\sim P\}$, we obtain the closed semantic tree T''' for $S \cup \{\sim P\} \cup \{\square\}$ as shown in Fig. 5.4d. The above "collapsing" of the semantic tree actually corresponds to a resolution proof as follows:

(4) $\sim P$ a resolvent of (2) and (3)

(5) $\square$ a resolvent of (4) and (1).

In the following, we shall use the concept developed above to prove the completeness of the resolution principle. That is, after constructing a closed semantic tree for an unsatisfiable set of clauses, we can gradually force the tree to collapse and at the same time obtain a resolution proof. Before proving the completeness theorem, we prove a lemma called the *lifting lemma*.

Lemma 5.1 (Lifting Lemma) If C_1' and C_2' are instances of C_1 and C_2, respectively, and if C' is a resolvent of C_1' and C_2', then there is a resolvent C of C_1 and C_2 such that C' is an instance of C.

Proof We rename, if necessary, the variables in C_1 and C_2 so that variables in C_1 are all different from those in C_2. Let L_1' and L_2' be the literals resolved upon, and let

$$C' = (C_1'\gamma - L_1'\gamma) \cup (C_2'\gamma - L_2'\gamma),$$

where γ is a most general unifier of L_1' and $\sim L_2'$. Since C_1' and C_2' are instances of C_1 and C_2, respectively, there is a substitution θ such that $C_1' = C_1\theta$ and $C_2' = C_2\theta$. Let $L_i^1, ..., L_i^{r_i}$ be the literals in C_i corresponding to L_i' (i.e., $L_i^1\theta = \cdots = L_i^{r_i}\theta = L_i'$), $i = 1, 2$. If $r_i > 1$, obtain a most general unifier λ_i for $\{L_i^1, ..., L_i^{r_i}\}$ and let $L_i = L_i^1\lambda_i$, $i = 1, 2$. (Note that $L_i^1\lambda_i, ..., L_i^{r_i}\lambda_i$ are the same, since λ_i is a most general unifier.) Then L_i is a literal in the factor $C_i\lambda_i$ of C_i. If $r_i = 1$, let $\lambda_i = \varepsilon$ and $L_i = L_i^1\lambda_i$. Let $\lambda = \lambda_1 \cup \lambda_2$. Thus, clearly L_i' is an instance of L_i. Since L_1' and $\sim L_2'$ are unifiable, L_1 and $\sim L_2$ are unifiable. Let σ be a most general unifier of L_1 and $\sim L_2$. Let

$$C = ((C_1\lambda)\sigma - L_1\sigma) \cup ((C_2\lambda)\sigma - L_2\sigma)$$
$$= ((C_1\lambda)\sigma - (\{L_1^1, ..., L_1^{r_1}\}\lambda)\sigma) \cup ((C_2\lambda)\sigma - (\{L_2^1, ..., L_2^{r_2}\}\lambda)\sigma)$$
$$= (C_1(\lambda \circ \sigma) - \{L_1^1, ..., L_1^{r_1}\}(\lambda \circ \sigma)) \cup (C_2(\lambda \circ \sigma) - \{L_2^1, ..., L_2^{r_2}\}(\lambda \circ \sigma)).$$

C is a resolvent of C_1 and C_2. Clearly, C' is an instance of C since

$$C' = (C_1'\gamma - L_1'\gamma) \cup (C_2'\gamma - L_2'\gamma)$$
$$= ((C_1\theta)\gamma - (\{L_1^1, ..., L_1^{r_1}\}\theta)\gamma) \cup ((C_2\theta)\gamma - (\{L_2^1, ..., L_2^{r_2}\}\theta)\gamma)$$
$$= (C_1(\theta \circ \gamma) - \{L_1^1, ..., L_1^{r_1}\}(\theta \circ \gamma)) \cup (C_2(\theta \circ \gamma) - \{L_2^1, ..., L_2^{r_2}\}(\theta \circ \gamma))$$

and $\lambda \circ \sigma$ is more general than $\theta \circ \gamma$. Thus we complete the proof of this lemma.

Theorem 5.3   (Completeness of the Resolution Principle)   A set S of clauses is unsatisfiable if and only if there is a deduction of the empty clause $\square$ from S.

Proof $(\Rightarrow)$ Suppose S is unsatisfiable. Let $A = \{A_1, A_2, A_3, ...\}$ be the atom set of S. Let T be a complete semantic tree as shown in Fig. 5.5. By Herbrand's theorem (version I), T has a finite closed semantic tree T'. If T' consists of only one (root) node, then $\square$ must be in S, for no other clause can be falsified at the root of a semantic tree. In this case, the theorem is obviously true. Assume T' consists of more than one node. Then T' has at least one inference node, for, if it did not, then every node would have at least one nonfailure descendent. We could then find an infinite branch through T', violating the fact that T' is a finite closed semantic tree. Let N be an inference node in T'. Let N_1 and N_2 be the failure nodes immediately below N. Let

$$I(N) = \{m_1, m_2, ..., m_n\},$$
$$I(N_1) = \{m_1, m_2, ..., m_n, m_{n+1}\},$$
$$I(N_2) = \{m_1, m_2, ..., m_n, \sim m_{n+1}\}.$$

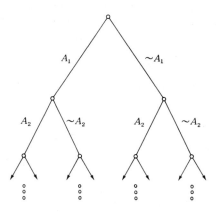

Figure 5.5

Since N_1 and N_2 are failure nodes but N is not a failure node, there must exist two ground instances C_1' and C_2' of clauses C_1 and C_2 such that C_1' and C_2' are false in $I(N_1)$ and $I(N_2)$, respectively, but both C_1' and C_2' are not falsified by $I(N)$. Therefore, C_1' must contain $\sim m_{n+1}$ and C_2' must contain m_{n+1}. Let $L_1' = \sim m_{n+1}$ and $L_2' = m_{n+1}$. By resolving upon the literals L_1' and L_2', we can obtain a resolvent C' of C_1' and C_2', namely,

$$C' = (C_1' - L_1') \cup (C_2' - L_2').$$

C' must be false in $I(N)$ since both $(C_1' - L_1')$ and $(C_2' - L_2')$ are false in $I(N)$. By Lemma 5.1, there is a resolvent C of C_1 and C_2 such that C' is a ground instance of C. Let T'' be the closed semantic tree for $(S \cup \{C\})$, obtained from T' by deleting any node or link that is below the first node where the resolvent C' is falsified. Clearly, the number of nodes in T'' is fewer than that in T'. Applying the above process on T'' again, we can obtain another resolvent of clauses in $(S \cup \{C\})$. Putting this resolvent into $(S \cup \{C\})$, we can get another smaller closed semantic tree. This process is repeated again and again until the closed semantic tree that consists of only the root node is generated. This is possible only when $\square$ is derived. Therefore, there is a deduction of $\square$ from S.

($\Leftarrow$) Conversely, suppose there is a deduction of $\square$ from S. Let R_1, $R_2, ..., R_k$ be the resolvents in the deduction. Assume S is satisfiable. Then there is a model M of S. However, if a model satisfies clauses C_1 and C_2, it must also satisfy any resolvent of C_1 and C_2 (Theorem 5.1). Therefore, M satisfies $R_1, R_2, ..., R_k$. However, this is impossible because one of these resolvents is $\square$. Hence, S must be unsatisfiable.   Q.E.D.

5.7 EXAMPLES USING THE RESOLUTION PRINCIPLE

In the previous chapters, we have used many methods to prove the inconsistency of formulas. In this section we shall use examples to demonstrate how resolution can be efficiently used to prove theorems.

Example 5.18

Let us reconsider Example 2.10. We have four statements:

(1') $P \rightarrow S$

(2') $S \rightarrow U$

(3') P

(4') U.

To show that (4') follows from (1'), (2'), and (3'), we first transform all these statements into standard forms. Thus we have

(1) $\sim P \vee S$

(2) $\sim S \vee U$

(3) P

(4) U.

We prove that U is a logical consequence of (1), (2), and (3) by refutation. We negate (4) and obtain the following proof:

(1) $\sim P \vee S$

(2) $\sim S \vee U$

(3) P

(4) $\sim U$ negation of conclusion

(5) S a resolvent of (3) and (1)

(6) U a resolvent of (5) and (2)

(7) $\square$ a resolvent of (6) and (4).

Example 5.19

Reconsider Example 2.12 in Chapter 2. We have three formulas F_1, F_2, and F_3. We want to prove that $\sim Q$ is a logical consequence of $F_1 \wedge F_2 \wedge F_3$. We negate $\sim Q$ and transform $F_1 \wedge F_2 \wedge F_3 \wedge Q$ into a standard form. This gives rise the following set of clauses:

(1) $\sim P \vee \sim Q \vee R$ $\left.\right\}$ from F_1

(2) $\sim P \vee \sim Q \vee S$

(3) P from F_2

(4) $\sim S$ from F_3

(5) Q negation of conclusion.

Using resolution, we obtain the following proof:

(6) $\sim Q \vee S$ a resolvent of (3) and (2)

(7) S a resolvent of (6) and (5)

(8) $\square$ a resolvent of (7) and (4).

Example 5.20

Consider the following set of formulas:

F_1: $(\forall x)(C(x) \to (W(x) \wedge R(x)))$

F_2: $(\exists x)(C(x) \wedge O(x))$

G: $(\exists x)(O(x) \wedge R(x))$.

Our problem is to show that G is a logical consequence of F_1 and F_2. We transform F_1, F_2, and $\sim G$ into standard forms and obtain the following five clauses:

(1) $\sim C(x) \lor W(x)$ $\biggr\}$ from F_1

(2) $\sim C(x) \lor R(x)$

(3) $C(a)$ $\biggr\}$ from F_2

(4) $O(a)$

(5) $\sim O(x) \lor \sim R(x)$ from $\sim G$.

The above set of the clauses is unsatisfiable. This can be proved by resolution as follows:

(6) $R(a)$ a resolvent of (3) and (2)

(7) $\sim R(a)$ a resolvent of (5) and (4)

(8) $\square$ a resolvent of (7) and (6).

Therefore, G is a logical consequence of F_1 and F_2.

Example 5.21

Reconsider Example 3.15. We have

F_1: $(\exists x)(P(x) \land (\forall y)(D(y) \rightarrow L(x,y)))$

F_2: $(\forall x)(P(x) \rightarrow (\forall y)(Q(y) \rightarrow \sim L(x,y)))$

G: $(\forall x)(D(x) \rightarrow \sim Q(x))$.

As usual, F_1, F_2, and $\sim G$ are transformed into the following clauses:

(1) $P(a)$ $\biggr\}$ from F_1

(2) $\sim D(y) \lor L(a,y)$

(3) $\sim P(x) \lor \sim Q(y) \lor \sim L(x,y)$ from F_2

(4) $D(b)$ $\biggr\}$ from $\sim G$.

(5) $Q(b)$

Using resolution, we obtain the following proof:

(6) $L(a,b)$ a resolvent of (4) and (2)

(7) $\sim Q(y) \lor \sim L(a,y)$ a resolvent of (3) and (1)

(8) $\sim L(a,b)$ a resolvent of (5) and (7)

(9) $\square$ a resolvent of (6) and (8).

The resolution principle is quite natural. Let us try to translate the above proof into English:

(a)   From F_1, we can assume that there is a patient a who likes every doctor (clauses (1) and (2)).

(b)   Let us assume that the conclusion is wrong. That is, assume that b is both a doctor and a quack (clauses (4) and (5)).

(c)   Since the patient a likes every doctor, a likes b (clause (6)).

(d)   Since a is a patient, a does not like any quack (clause (7)).

(e)   However, b is a quack. Therefore, a does not like b ((clause (8)).

(f)   This is impossible because of (c). Thus, we complete the proof.

Example 5.22

Premises: The custom officials searched everyone who entered this country who was not a VIP. Some of the drug pushers entered this country and they were only searched by drug pushers. No drug pusher was a VIP.

Conclusion: Some of the officials were drug pushers.

Let $E(x)$ mean "x entered this country," $V(x)$ mean "x was a VIP," $S(x, y)$ mean "y searched x," $C(x)$ mean "x was a custom official," and $P(x)$ mean "x was a drug pusher."

The premises are represented by the following formulas:

$$(\forall x)(E(x) \wedge \sim V(x) \to (\exists y)(S(x, y) \wedge C(y)))$$

$$(\exists x)(P(x) \wedge E(x) \wedge (\forall y)(S(x, y) \to P(y)))$$

$$(\forall x)(P(x) \to \sim V(x))$$

and the conclusion is

$$(\exists x)(P(x) \wedge C(x)).$$

Transforming premises into clauses, we obtain

(1) $\sim E(x) \vee V(x) \vee S(x, f(x))$

(2) $\sim E(x) \vee V(x) \vee C(f(x))$

(3) $P(a)$

(4) $E(a)$

(5) $\sim S(a, y) \vee P(y)$

(6) $\sim P(x) \vee \sim V(x).$

The negation of the conclusion is

(7) $\sim P(x) \vee \sim C(x).$

The resolution proof is as follows:

(8) $\sim V(a)$ from (3) and (6)

(9) $V(a) \lor C(f(a))$ from (2) and (4)

(10) $C(f(a))$ from (8) and (9)

(11) $V(a) \lor S(a, f(a))$ from (1) and (4)

(12) $S(a, f(a))$ from (8) and (11)

(13) $P(f(a))$ from (12) and (5)

(14) $\sim C(f(a))$ from (13) and (7)

(15) $\square$ from (10) and (14).

Thus, we have established the conclusion.

Example 5.23

The premise is that everyone who saves money earns interest. The conclusion is that if there is no interest, then nobody saves money. Let $S(x,y)$, $M(x)$, $I(x)$, and $E(x,y)$ denote "x saves y," "x is money," "x is interest," and "x earns y," respectively. Then the premise is symbolized as

$$(\forall x)((\exists y)(S(x,y) \land M(y)) \rightarrow (\exists y)(I(y) \land E(x,y)))$$

and the conclusion is

$$\sim(\exists x)\,I(x) \rightarrow (\forall x)(\forall y)(S(x,y) \rightarrow \sim M(y)).$$

Translating the premise into clauses, we have

(1) $\sim S(x,y) \lor \sim M(y) \lor I(f(x))$

(2) $\sim S(x,y) \lor \sim M(y) \lor E(x, f(x)).$

The negation of the conclusion is

(3) $\sim I(z)$

(4) $S(a,b)$

(5) $M(b).$

The proof given in the following is indeed very simple:

(6) $\sim S(x,y) \lor \sim M(y)$ from (3) and (1)

(7) $\sim M(b)$ from (6) and (4)

(8) $\square$ from (7) and (5).

Thus the conclusion is established.

Example 5.24

Premise: Students are citizens.

Conclusion: Students' votes are citizens' votes.

Let $S(x)$, $C(x)$, and $V(x,y)$ denote "x is a student," "x is a citizen," and "x is a vote of y," respectively. Then the premise and the conclusion are symbolized as follows:

$(\forall y)(S(y) \to C(y))$ (premise)

$(\forall x)((\exists y)(S(y) \wedge V(x,y)) \to (\exists z)(C(z) \wedge V(x,z)))$ (conclusion).

A standard form of the premise is

(1) $\sim S(y) \vee C(y)$.

Since

$$\sim \big((\forall x)((\exists y)(S(y) \wedge V(x,y)) \to (\exists z)(C(z) \wedge V(x,z)))\big)$$
$$= \sim \big((\forall x)((\forall y)(\sim S(y) \vee \sim V(x,y)) \vee (\exists z)(C(z) \wedge V(x,z)))\big)$$
$$= \sim \big((\forall x)(\forall y)(\exists z)(\sim S(y) \vee \sim V(x,y) \vee (C(z) \wedge V(x,z)))\big)$$
$$= (\exists x)(\exists y)(\forall z)(S(y) \wedge V(x,y) \wedge (\sim C(z) \vee \sim V(x,z)))$$

we have three clauses for the negation of the conclusion,

(2) $S(b)$

(3) $V(a,b)$

(4) $\sim C(z) \vee \sim V(a,z)$.

The proof proceeds as follows:

(5) $C(b)$ from (1) and (2)

(6) $\sim V(a,b)$ from (5) and (4)

(7) $\square$ from (6) and (3).

Again, although the above proof is quite mechanical, it is actually very natural. Indeed, we can translate the above proof into English as follows: Let us assume that b is a student, a is b's vote, and a is not a vote of any citizen. Since b is a student, b must be a citizen. Furthermore, a must not be a vote of b because b is a citizen. This is impossible. Thus we complete the proof.

5.8   DELETION STRATEGY

In the previous sections, we have shown that resolution is complete. Resolution is also more efficient than the earlier methods such as Herbrand's procedure used by Gilmore. However, unlimited applications of resolution may cause many irrelevant and redundant clauses to be generated. To see this, let us consider a simple example. Suppose we want to show, by resolution, that the set $S = \{P \vee Q, \sim P \vee Q, P \vee \sim Q, \sim P \vee \sim Q\}$ is unsatisfiable. A straightforward way of carrying out resolution on the set S is to compute all resolvents of pairs of clauses in S, add these resolvents to the set S, compute all further resolvents, and repeat this process until the empty clause $\square$ is found. That is, we generate the sequences, S^0, S^1, S^2, ..., where

$$S^0 = S,$$

$$S^n = \{\text{resolvents of } C_1 \text{ and } C_2 \mid C_1 \in (S^0 \cup \cdots \cup S^{n-1}) \text{ and } C_2 \in S^{n-1}\},$$
$$n = 1, 2, 3, \ldots.$$

This procedure is called the *level-saturation (resolution) method*. To program this method on computers, we may first list clauses of $(S^0 \cup \cdots \cup S^{n-1})$ in order, and then compute resolvents by comparing every clause $C_1 \in (S^0 \cup \cdots \cup S^{n-1})$ with a clause $C_2 \in S^{n-1}$ that is listed after C_1. When a resolvent is computed, it is appended to the end of the list so far generated. This is what we do in generating the following list of clauses:

S^0: (1) $P \vee Q$

    (2) $\sim P \vee Q$

    (3) $P \vee \sim Q$ $\left. \right\} \; S$

    (4) $\sim P \vee \sim Q$

S^1: (5) Q from (1) and (2)

    (6) P from (1) and (3)

    (7) $Q \vee \sim Q$ from (1) and (4)

    (8) $P \vee \sim P$ from (1) and (4)

    (9) $Q \vee \sim Q$ from (2) and (3)

    (10) $P \vee \sim P$ from (2) and (3)

    (11) $\sim P$ from (2) and (4)

    (12) $\sim Q$ from (3) and (4)

S^2: (13) $P \vee Q$ from (1) and (7)

(14) $P \vee Q$ from (1) and (8)

(15) $P \vee Q$ from (1) and (9)

(16) $P \vee Q$ from (1) and (10)

(17) Q from (1) and (11)

(18) P from (1) and (12)

(19) Q from (2) and (6)

(20) $\sim P \vee Q$ from (2) and (7)

(21) $\sim P \vee Q$ from (2) and (8)

(22) $\sim P \vee Q$ from (2) and (9)

(23) $\sim P \vee Q$ from (2) and (10)

(24) $\sim P$ from (2) and (12)

(25) P from (3) and (5)

(26) $P \vee \sim Q$ from (3) and (7)

(27) $P \vee \sim Q$ from (3) and (8)

(28) $P \vee \sim Q$ from (3) and (9)

(29) $P \vee \sim Q$ from (3) and (10)

(30) $\sim Q$ from (3) and (11)

(31) $\sim P$ from (4) and (5)

(32) $\sim Q$ from (4) and (6)

(33) $\sim P \vee \sim Q$ from (4) and (7)

(34) $\sim P \vee \sim Q$ from (4) and (8)

(35) $\sim P \vee \sim Q$ from (4) and (9)

(36) $\sim P \vee \sim Q$ from (4) and (10)

(37) Q from (5) and (7)

(38) Q from (5) and (9)

(39) $\square$ from (5) and (12).

Many irrelevant and redundant clauses have been generated. For example, (7), (8), (9), and (10) are tautologies. Since a tautology is true in any interpretation, if we delete a tautology from an unsatisfiable set of clauses, the remaining set must still be unsatisfiable. Therefore, a tautology is an irrelevant clause and should not be generated. If one is generated, except in a very few cases, it should be deleted. Otherwise, it may interact with other clauses and produce other unwanted redundant clauses. For example, (13)–(16), (20)–(23), (26)–(29), (33)–(36), and (37)–(38) are all such clauses. Furthermore, the clauses P, Q, $\sim P$, and $\sim Q$ are repeatedly generated. Actually, to get a proof for S, we need only generate clauses (5), (12), and (39). To solve this redundancy problem, we shall consider in this section a deletion strategy. Before we describe it, we consider the following definition.

Definition   A clause C *subsumes* a clause D if and only if there is a substitution σ such that $C\sigma \subseteq D$. D is called a *subsumed* clause.

Example 5.25

Let $C = P(x)$ and $D = P(a) \vee Q(a)$. If $\sigma = \{a/x\}$, then $C\sigma = P(a)$. Since $C\sigma \subseteq D$, C subsumes D.

It is noted that if D is identical to C, or D is an instance of C, then D is subsumed by C. *The deletion strategy is the deletion of any tautology and any subsumed clause whenever possible.* The deletion strategy is a generalization of Davis and Putnam's tautology rule given in Section 4.6. The completeness of the deletion strategy depends upon how tautologies and subsumed clauses are deleted. (See [Kowalski, 1970b].) The deletion strategy will be complete if it is used with the level-saturation method in the following way: First, list clauses of $(S^0 \cup \cdots \cup S^{n-1})$ in order; then compute resolvents by comparing every clause $C_1 \in (S^0 \cup \cdots \cup S^{n-1})$ with a clause $C_2 \in S^{n-1}$ that is listed after C_1. When a resolvent is computed, it is appended to the end of the list so far generated if it is neither a tautology nor subsumed by any clause in the list. Otherwise, it is deleted. The proof of the completeness of this deletion strategy is left as an exercise to the reader. We apply this deletion strategy to the above example and obtain the list as follows:

S^0:   (1) $P \vee Q$

     (2) $\sim P \vee Q$

                             S

     (3) $P \vee \sim Q$

     (4) $\sim P \vee \sim Q$

$S^1:$	(5)	Q	from (1) and (2)
	(6)	P	from (1) and (3)
	(7)	$\sim P$	from (2) and (4)
	(8)	$\sim Q$	from (3) and (4)
$S^2:$	(9)	$\square$	from (5) and (8).

We note that this list is much shorter than the list generated before. Therefore, the deletion strategy can improve the efficiency of the resolution principle.

In order to use the deletion strategy, we must be able to decide whether a clause is a tautology and whether a clause subsumes another clause. It is very easy to check whether a clause is a tautology by testing whether a complementary pair is present in the clause. However, the subsumption test is not so simple. We now describe an algorithm for this test.

Let C and D be clauses. Let $\theta = \{a_1/x_1, ..., a_n/x_n\}$, where $x_1, ..., x_n$ are variables occurring in D, and $a_1, ..., a_n$ are new distinct constants not occurring in C or D. Suppose $D = L_1 \vee L_2 \vee \cdots \vee L_m$. Then $D\theta = L_1\theta \vee L_2\theta \vee \cdots \vee L_m\theta$. Note that $D\theta$ is a ground clause. $\sim D\theta = \sim L_1\theta \wedge \cdots \wedge \sim L_m\theta$. The following algorithm tests whether or not C subsumes D.

Subsumption Algorithm

Step 1 Let $W = \{\sim L_1\theta, ..., \sim L_m\theta\}$.

Step 2 Set $k = 0$ and $U^0 = \{C\}$.

Step 3 If U^k contains $\square$, terminate; C subsumes D. Otherwise, let $U^{k+1} = \{\text{resolvents of } C_1 \text{ and } C_2 \mid C_1 \in U^k \text{ and } C_2 \in W\}$.

Step 4 If U^{k+1} is empty, terminate; C does not subsume D. Otherwise, set $k = k+1$ and go to Step 3.

In the above algorithm, we note that each clause in U^{k+1} is smaller by one literal than the clause in U^k from which it was obtained. Therefore, the sequence $U^0, U^1, ...,$ must eventually contain a set that contains $\square$ or is empty. The subsumption algorithm is correct. That is, C subsumes D if and only if the subsumption algorithm terminates in Step 3. This is proved as follows: If C subsumes D, there is a substitution σ such that $C\sigma \subseteq D$. Hence $C(\sigma \circ \theta) \subseteq D\theta$. Thus literals in $C(\sigma \circ \theta)$ can be resolved by using unit ground clauses in W. However $C(\sigma \circ \theta)$ is an instance of C. Therefore, literals in C can be resolved away by using unit clauses in W. This means that we will eventually find a U^k that contains $\square$. Therefore, the algorithm will terminate at Step 3. Conversely, if the algorithm terminates at Step 3, then we obtain

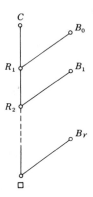

Figure 5.6

a refutation shown in Fig. 5.6, where $B_0, \ldots, B_r$ are unit clauses from W. R_1 is a resolvent of C and B_0, and R_i is a resolvent of R_{i-1} and B_{i-1}, for $i = 2, \ldots, r$.

Let σ_0 be the most general unifier obtained when resolving C and B_0, and let σ_i be the most general unifier obtained when resolving R_i and B_i, $i = 1, \ldots, r$. Then $C(\sigma_0 \circ \sigma_1 \circ \cdots \circ \sigma_r) = \{\sim B_0, \sim B_1, \ldots, \sim B_r\} \subseteq D\theta$. Let $\lambda = \sigma_0 \circ \sigma_1 \circ \cdots \circ \sigma_r$. Then $C\lambda \subseteq D\theta$. Let σ be the substitution obtained from λ by the replacement, in each component of λ, of a_i by x_i, for $i = 1, \ldots, n$. Then $C\sigma \subseteq D$. Therefore, C subsumes D.   Q.E.D.

Example 5.26

Let $C = \sim P(x) \vee Q(f(x),a)$ and $D = \sim P(h(y)) \vee Q(f(h(y)),a) \vee \sim P(z)$. Decide whether C subsumes D.

1. y and z are variables in D. Let $\theta = \{b/y, c/z\}$. Note that b and c are not in C or D. Then $D\theta = \sim P(h(b)) \vee Q(f(h(b)),a) \vee \sim P(c)$. Therefore, $\sim D\theta = P(h(b)) \wedge \sim Q(f(h(b)),a) \wedge P(c)$. Therefore

$$W = \{P(h(b)), \sim Q(f(h(b)),a), P(c)\}$$
$$U^0 = \{\sim P(x) \vee Q(f(x),a)\}.$$

2. Since U^0 does not contain $\square$, we obtain

$$U^1 = \{Q(f(h(b)), a), \sim P(h(b)), Q(f(c), a)\}.$$

3. Since U^1 is not empty and does not contain $\square$, we obtain $U^2 = \{\square\}$.

4. Since U^2 contains $\square$, we terminate the algorithm and conclude that C subsumes D.

Example 5.27

Let $C = P(x,x)$ and $D = P(f(x),y) \vee P(y, f(x))$. Determine whether or not C subsumes D.

1. x and y are variables in D. Choose new constants a and b different from any constant in C or D. Let $\theta = \{a/x, b/y\}$. Then,

$$D\theta = P(f(a), b) \vee P(b, f(a)).$$

$$\sim D\theta = \sim P(f(a), b) \wedge \sim P(b, f(a)).$$

Thus,

$$W = \{\sim P(f(a), b), \sim P(b, f(a))\}$$
$$U^0 = \{P(x, x)\}.$$

2. Since U^0 does not contain $\square$, we obtain

$$U^1 = \varnothing.$$

3. Since U^1 is empty, we terminate the algorithm and conclude that C does not subsume D.

REFERENCES

Robinson, J. A. (1965): A machine-oriented logic based on the resolution principle, *J. Assoc. Comput. Mach.* **12** 23–41.

Robinson, J. A. (1968): Generalized resolution principle, *in* "Machine Intelligence," vol. 3 (D. Michie, ed.), American Elsevier, New York, pp. 77–94.

Kowalswki, R. (1970b): "Studies in the Completeness and Efficiency of Theorem-Proving by Resolution," Ph.D. Thesis, Univ. of Edinburgh, Scotland.

Kowalski, R. and P. J. Hayes (1969): Semantic trees in automatic theorem-proving, *in* "Machine Intelligence," vol. 4 (B. Meltzer and D. Michie, eds.), American Elsevier, New York, pp. 87–101.

EXERCISES

Section 5.2

1. Prove the following set of clauses is unsatisfiable by resolution:

$$P \vee Q \vee R, \quad \sim P \vee R, \quad \sim Q, \quad \sim R.$$

2. For the set $S = \{P \vee Q, \sim Q \vee R, \sim P \vee Q, \sim R\}$ derive an empty clause from S by resolution.

Section 5.3

3. Let $\theta = \{a/x, b/y, g(x, y)/z\}$ be a substitution and $E = P(h(x), z)$. Find $E\theta$.

4. Let $\theta_1 = \{a/x, f(z)/y, y/z\}$ and $\theta_2 = \{b/x, z/y, g(x)/z\}$. Find the composition of θ_1 and θ_2.

5. Prove the following theorem: Let θ, λ, and μ all be substitutions. Then $(\theta \circ \lambda) \circ \mu = \theta \circ (\lambda \circ \mu)$ (associativity).

Section 5.4

6.   Determine whether each of the following sets is unifiable. If yes, obtain
a most general unifier.

(1) $W = \{Q(a), Q(b)\}$

(2) $W = \{Q(a, x), Q(a, a)\}$

(3) $W = \{Q(a, x, f(x)), Q(a, y, y)\}$

(4) $W = \{Q(x, y, z), Q(u, h(v, v), u)\}$

(5) $W = \{P(x_1, g(x_1), x_2, h(x_1, x_2), x_3, k(x_1, x_2, x_3)),$

$P(y_1, y_2, e(y_2), y_3, f(y_2, y_3), y_4)\}.$

Section 5.5

7.   Determine whether the following clauses have factors. If yes, give the
factors.

(1) $P(x) \vee Q(y) \vee P(f(x))$

(2) $P(x) \vee P(a) \vee Q(f(x)) \vee Q(f(a))$

(3) $P(x, y) \vee P(a, f(a))$

(4) $P(a) \vee P(b) \vee P(x)$

(5) $P(x) \vee P(f(y)) \vee Q(x, y).$

8.   Find all the possible resolvents (if any) of the following pairs of clauses:

(1) $C = \sim P(x) \vee Q(x, b), D = P(a) \vee Q(a, b)$

(2) $C = \sim P(x) \vee Q(x, x), D = \sim Q(a, f(a))$

(3) $C = \sim P(x, y, u) \vee \sim P(y, z, v) \vee \sim P(x, v, w) \vee P(u, z, w),$

$D = P(g(x, y), x, y)$

(4) $C = \sim P(v, z, v) \vee P(w, z, w), D = P(w, h(x, x), w).$

Section 5.7

9.   Reconsider Example 2.13 in Section 2.6. This time show that H_2CO_3
can be made by resolution.

10. In Exercise 14 of Chapter 2, you were asked to show that Q is a logical
consequence of P and $(P \rightarrow Q)$. Can you prove this again by using the
resolution principle?

11. Reconsider Exercise 12 of Chapter 3. Use the resolution principle this
time.

12. Consider Example 4.16. Prove that the set of clauses in this example is unsatisfiable by the resolution principle.

13. Use the resolution principle to prove that the set of clauses in Exercise 14 of Chapter 4 is unsatisfiable.

14. Prove that $(\sim Q \rightarrow \sim P)$ is a logical consequence of $(P \rightarrow Q)$ by resolution.

Section 5.8

15. Let $C = P(x, y) \vee Q(z)$ and $D = Q(a) \vee P(b, b) \vee R(u)$. Determine whether C subsumes D.

16. Let $C = P(x, y) \vee R(y, x)$ and $D = P(a, y) \vee R(z, b)$. Prove that C does not subsume D.

17. Let $C = \sim P(x) \vee P(f(x))$ and $D = \sim P(x) \vee P(f(f(x)))$. Show that C implies D, but C does not subsume D.

Chapter 6

Semantic Resolution and Lock Resolution

6.1 INTRODUCTION

In the last chapter, we considered the resolution principle as an inference rule that can be used to generate new clauses from old ones. We also have seen that unlimited applications of resolution may generate many irrelevant and redundant clauses besides useful ones. Although the deletion strategy can be used to delete some of these irrelevant and redundant clauses after they are generated, time has already been wasted by generating them. Furthermore, if useless clauses are generated, many resources such as computer time and memory are required to determine whether they are indeed irrelevant and redundant. Therefore, in order to have efficient theorem-proving procedures, we must prevent large numbers of useless clauses from being generated. This leads to the discussion of refinements of resolution.

There are many important refinements of resolution. Each refinement has its own merit. In this book, we shall consider three important refinements, namely, semantic resolution, lock resolution, and linear resolution. Semantic resolution was proposed by Slagle [1967]. It unifies Robinson's hyper-resolution [1965b], Meltzer's renamable resolution [1966], and the set-of-support strategy of Wos, Robinson, and Carson [1965]. Lock resolution was proposed by Boyer [1971]. It is a very efficient rule. Linear resolution was independently proposed by Loveland [1970a] and Luckham [1970]. It was later strengthened by Anderson and Bledsoe [1970], Reiter [1971], Loveland [1972], and Kowalski and Kuehner [1971]. Semantic resolution, lock resolution, and linear resolution are complete. That is, each can be used alone to derive the empty clause $\square$ from an unsatisfiable set of clauses.

We shall discuss semantic resolution and lock resolution in this chapter, and linear resolution in Chapter 7. In Chapter 7, we shall also consider a subcase of linear resolution, namely, input resolution. It was shown by Chang [1970a] that input resolution is equivalent to another kind of resolution, namely, unit resolution. In Chapter 8, special treatments for the equality relation will be given. Finally, in Chapter 9, some other types of proof procedures will be introduced, including variable-constrained resolution [Chang, 1972].

6.2 AN INFORMAL INTRODUCTION TO SEMANTIC RESOLUTION

Consider the following example.

Example 6.1

Let $S = \{ \sim P \vee \sim Q \vee R, P \vee R, Q \vee R, \sim R \}$. To prove the unsatisfiability of the set S, if the ordinary resolution principle is used, the following clauses will be generated.

(1)	$\sim P \vee \sim Q \vee R$	
(2)	$P \vee R$	$\left. \rule{0pt}{48pt} \right\} S$
(3)	$Q \vee R$	
(4)	$\sim R$	
(5)	$\sim Q \vee R$	from (1) and (2)
(6)	$\sim P \vee R$	from (1) and (3)
(7)	$\sim P \vee \sim Q$	from (1) and (4)
(8)	P	from (2) and (4)
(9)	Q	from (3) and (4)
(10)	$\sim Q \vee R$	from (1) and (8)
(11)	$\sim P \vee R$	from (1) and (9)
(12)	R	from (2) and (6)
(13)	$\sim Q$	from (4) and (5)
(14)	$\sim P$	from (4) and (6)
(15)	$\square$	from (4) and (12).

Among all these clauses generated, only (6) and (12) are actually used in the proof. All other clauses are irrelevant and redundant. Is it possible to have some mechanisms that can reduce the number of useless clauses generated? Indeed, there are such mechanisms and we shall now discuss them in detail.

Let us imagine that we can, in some way, divide a set of clauses into two group—S_1 and S_2. Let us further imagine that we do not allow clauses within the same group to be resolved with each other. This will obviously cut down the number of clauses generated.

For example, let us divide the clauses in Example 6.1 into S_1 (clause (2) and clause (3)) and S_2 (clause (1) and clause (4)). Since clause (1) and clause (4) are both in S_2, (1) and (4) will not be resolved. Thus we have succeeded in blocking the resolution of (1) and (4). As we shall see, this blocking will not affect our ability to deduce the empty clause $\square$.

The question is: How do we divide the clauses into two groups? In semantic resolution, we use an interpretation to divide clauses. This is why this kind of resolution is called semantic resolution. Note that in Example 6.1, (2) and (3) are all falsified by the interpretation $I = \{\sim P, \sim Q, \sim R\}$, while (1) and (4) are satisfied by I. Therefore, this interpretation can be used to divide S into S_1 and S_2, where S_1 contains all the clauses falsified by I and S_2 contains those satisfied by I. The reader should remember that we are dealing with an unsatisfiable set of clauses. No interpretation can satisfy or falsify all the clauses. Therefore, every interpretation divides S into two nonempty sets of clauses. We shall see in the next section that any interpretation can be used.

So far we have succeeded in blocking the resolution of (1) and (4). Can we impose some further restrictions on resolution so that more resolutions will be blocked? Note that in Example 6.1, (2) and (3) can still be resolved with (4). We shall show that these resolutions can also be blocked.

To block these resolutions, we can use an ingenious scheme, namely, the ordering of predicate symbols. Suppose we order all the predicate symbols in S as follows:

$$P > Q > R.$$

Suppose we further require that, when we resolve two clauses (one from S_1 and one from S_2), the resolved literal in the clause from S_1 contain the largest predicate symbol in that clause. With this restriction, we cannot resolve (2) with (4), nor (3) with (4), because R is *not* the largest literal in (2) and (3).

The foregoing ideas—the use of an interpretation to divide clauses into two groups and of an ordering to reduce the number of possible resolutions —are important concepts in semantic resolution. We shall elaborate on these ideas later. Meanwhile, we shall introduce another important concept, the clash.

Example 6.2

Consider the clauses in Example 6.1,

(1) $\sim P \vee \sim Q \vee R$

(2) $P \vee R$

(3) $Q \vee R$

(4) $\sim R$.

Let $I = \{\sim P, \sim Q, \sim R\}$ and let the ordering of predicate symbols be $P > Q > R$. Using the two restrictions discussed above, we can only generate two clauses,

(5) $\sim Q \vee R$ from (1) and (2)

(6) $\sim P \vee R$ from (1) and (3).

Both of these clauses are satisfied by I; therefore we put them into S_2 and resolve them with the clauses in S_1. We find that (5) can be resolved with (2), and (6) with (3). Both result in the same clause R:

(7) R from (3) and (5)

(8) R from (2) and (6).

The clause R is false in I, hence we put it into S_1 and resolve it with the clauses in S_2. We find that $\sim R$ is in S_2. Therefore, resolving R with $\sim R$, we obtain the empty clause $\square$.

Figure 6.1 shows the two ways in which R was derived. Both these deductions of R use clauses (1), (2), and (3); they differ only in that they use these clauses in different orders. Since only one of them is needed in the proof, it is quite wasteful to generate both of them. In order to avoid this redundancy, we introduce the clash. The idea of a clash is to generate R directly from (1), (2), and (3) without going through the "intermediate" clause (5) or (6). The set $\{(1),(2),(3)\}$ will be called a clash. A detailed description of a clash will be given in the next section. It will be also shown later that we can combine the concepts of interpretation and ordering of predicate symbols with that of the clash to restrict many resolutions.

Figure 6.1

6.3 FORMAL DEFINITIONS AND EXAMPLES OF SEMANTIC RESOLUTION

Definition Let I be an interpretation. Let P be an ordering of predicate symbols. A finite set of clauses $\{E_1, ..., E_q, N\}$, $q \geqslant 1$, is called a *semantic clash* with respect to P and I (or *PI-clash* for short), if and only if $E_1, ..., E_q$ (called *electrons*) and N (called the *nucleus*) satisfy the following conditions:

1. $E_1, ..., E_q$ are false in I.
2. Let $R_1 = N$. For each $i = 1, ..., q$, there is a resolvent R_{i+1} of R_i and E_i.
3. The literal in E_i, which is resolved upon, contains the largest predicate symbol in $E_i, i = 1, ..., q$.
4. R_{q+1} is false in I.

R_{q+1} is called a *PI-resolvent* (of the PI-clash $\{E_1, ..., E_q, N\}$).

Example 6.3

Let

$$E_1 = A_1 \vee A_3, \qquad E_2 = A_2 \vee A_3, \qquad N = \sim A_1 \vee \sim A_2 \vee A_3.$$

Let $I = \{\sim A_1, \sim A_2, \sim A_3\}$ and P be an ordering in which $A_1 > A_2 > A_3$. Then $\{E_1, E_2, N\}$ is a PI-clash. The PI-resolvent of this clash is A_3. Note that A_3 is false in I.

Example 6.4

In Example 6.3, neither $\{E_1, N\}$ nor $\{E_2, N\}$ is a PI-clash. For instance, we can see that the resolvent of E_1 and N is $\sim A_2 \vee A_3$, which is true in I. Thus $\{E_1, N\}$ is not a PI-clash. We can also see that, if we change P into an ordering in which $A_3 > A_2 > A_1$, then $\{E_1, E_2, N\}$ is not a PI-clash since the third condition for a PI-clash is not satisfied.

Example 6.5

Let

$$E_1 = \sim Q(z) \vee \sim Q(a),$$

$$E_2 = R(b) \vee S(c),$$

$$N = Q(x) \vee Q(a) \vee \sim R(y) \vee \sim R(b) \vee S(c).$$

Let $I = \{Q(a), Q(b), Q(c), \sim R(a), \sim R(b), \sim R(c), \sim S(a), \sim S(b), \sim S(c)\}$. Let P be an ordering of predicate symbols in which $Q > R > S$. Let $R_1 = N$. There is a resolvent of R_1 and E_1, namely, $\sim R(y) \vee \sim R(b) \vee S(c)$. Call this resolvent

R_2. Then there is a resolvent of R_2 and E_2, namely, $S(c)$. Denote $S(c)$ by R_3. R_3 is false in I. Since $\{E_1, E_2, N\}$ satisfies all the four conditions, it is a PI-clash. The PI-resolvent of this clash is $S(c)$.

In a PI-clash $\{E_1, E_2, ..., E_q, N\}$, E_1 is considered as the first electron, E_2 the second, ..., and E_q the last. In fact, the order of these electrons is immaterial. We can relabel any one of these electrons as the first, any of the remaining electrons as the second, and so on. No matter what order of the electrons we use, we can always get the same PI-resolvent of this clash. Therefore, we can avoid generating too many deductions of the same resolvent, as discussed in Example 6.2. This is one reason for using the concept of a clash.

Definition   Let I be an interpretation for a set S of clauses, and P be an ordering of predicate symbols appearing in S. A deduction from S is called a PI-*deduction* if and only if each clause in the deduction is either a clause in S, or a PI-resolvent.

Example 6.6

Let $S = \{A_1 \vee A_2, A_1 \vee \sim A_2, \sim A_1 \vee A_2, \sim A_1 \vee \sim A_2\}$. Let $I = \{A_1, \sim A_2\}$ and P be an ordering of the predicate symbols in which $A_1 < A_2$. The deduction shown in Fig. 6.2 is a PI-deduction of $\square$ from S. In Fig. 6.2 there are three PI-resolvents, namely, $\sim A_1$, A_2, and $\square$.

An interesting thing about semantic resolution is that we can use any ordering and any interpretation. For instance, if in Example 6.6 we let $I = \{\sim A_1, \sim A_2\}$ and $A_1 > A_2$, then we obtain a PI-deduction of $\square$ from S shown in Fig. 6.3. In fact, PI-resolution is complete; that is, from any unsatisfiable set of clauses, we can always derive the empty clause $\square$ by using some PI-resolution. We shall devote the next section to proving this.

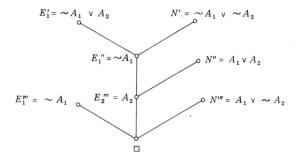

Figure 6.2

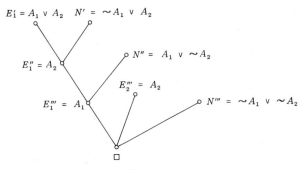

$E_1' = A_1 \lor A_2 \quad N' = {\sim}A_1 \lor A_2$

$E_1'' = A_2$

$N'' = A_1 \lor {\sim}A_2$

$E_2''' = A_2$

$E_1''' = A_1$

$N''' = {\sim}A_1 \lor {\sim}A_2$

Figure 6.3

6.4   COMPLETENESS OF SEMANTIC RESOLUTION

Lemma 6.1   If P is an ordering of predicate symbols in a finite unsatisfiable set S of ground clauses, and if I is an interpretation of S, then there is a PI-deduction of $\square$ from S.

Proof   This lemma can be proved by induction. Let A be the atom set of S. If A consists of a single element, say Q, then among the elements of S there are the clauses Q and ${\sim}Q$. Clearly, the resolvent of Q and ${\sim}Q$ is $\square$. Since one of Q and ${\sim}Q$ must be false in I, and $\square$ is false in I, $\square$ is a PI-resolvent. Therefore, Lemma 6.1 holds for this case.

Assume Lemma 6.1 holds when A consists of i elements, $1 \leqslant i \leqslant n$. To complete the induction we consider A such that A consists of exactly $n+1$ elements. There are two possible cases:

Case 1 S contains a unit clause L that is false in I. (L is a literal.) Let S' be that set obtained from S by deleting those clauses containing the literal L and by deleting ${\sim}L$ from the remaining clauses. Clearly, S' is unsatisfiable. (See the one-literal rule in Section 4.6.) Since S' contains n or fewer than n atoms, by the induction hypothesis there is a PI-deduction D' of $\square$ from S'. From the deduction D', we can obtain a PI-deduction of $\square$ from S. This is done as follows: First, for each PI-clash $\{E_1', ..., E_q', N'\}$, where $E_1', ..., E_q', N'$ are clauses attached to initial nodes of D', if N' is obtained from a clause N in S by deleting ${\sim}L$ from N, replace the clash $\{E_1', ..., E_q', N'\}$ by the PI-clash $\{E_1', ..., E_q', L, N\}$ (where $E_1', ..., E_q', L$ are electrons and N is the nucleus). Second, if E_i' is obtained from a clause E_i in S by deleting ${\sim}L$ from E_i, attach the PI-clah $\{L, E_i\}$ above the node of E_i'. After we perform the above process, it is clear that we have obtained a PI-deduction of $\square$ from S.

Case 2 S does not contain a unit clause that is false in I. In this case, choose an element B in the atom set A of S such that B contains the smallest predicate symbol. Either B or $\sim B$ must be false in I. Let L be that element in $\{B, \sim B\}$ that is false in I. Let S' be that set obtained from S by deleting those clauses containing the literal $\sim L$ and by deleting L from the remaining clauses. S' is unsatisfiable. Since S' contains n or fewer than n atoms, by the induction hypothesis there is a PI-deduction D' of $\square$ from S'. Let D_1 be the deduction obtained from D' by putting the literal L back to those clauses from which it was deleted. D_1 is still a PI-deduction since L contains the smallest predicate symbol and L is false in I. D_1 is either a PI-deduction of $\square$, or L. If it is the former, we are done. If it is the latter, consider the set $(S \cup \{L\})$. Since $(S \cup \{L\})$ contains a unit clause L that is false in I, by the proof of Case 1 given above there is a PI-deduction D_2 of $\square$ from $(S \cup \{L\})$. Combining D_1 and D_2, we can obtain a PI-deduction of $\square$ from S. This completes the proof of Lemma 6.1.   Q.E.D.

Theorem 6.1 (Completeness of PI-resolution) If P is an ordering of predicate symbols in a finite and unsatisfiable set S of clauses, and if I is an interpretation of S, then there is a PI-deduction of $\square$ from S.

Proof Since S is unsatisfiable, by Herbrand's theorem (version II) there is a finite unsatisfiable set S' of ground instances of clauses in S. By Lemma 6.1, there is a PI-deduction D' of $\square$ from S'. From the PI-deduction D', we now show that we can produce a PI-deduction of $\square$ from S. To see this, we merely attach to each node of D' a clause over or above the ground clause already there as follows: To each initial node, attach a clause in S of which the ground clause already there is an instance. Then, for each noninitial node, if clauses have been attached in this way to each of its immediate predecessor nodes and constitute a PI-clash, attach to it the PI-resolvent of which the ground clause already there is an instance. (This is possible because of the lifting lemma, Lemma 5.1.) In this fashion, a clause is attached to each node of which the ground clause already at the node is an instance. The clause attached to the terminal node must be $\square$, since the clause already there is $\square$. It is easy to see that the deduction tree, together with the attached clauses, is a PI-deduction of $\square$ from S. This completes the proof.

It is noted that, without losing the completeness property, the deletion strategy given in Section 5.8 can be used together with PI-resolution.

6.5 HYPERRESOLUTION AND THE SET-OF-SUPPORT STRATEGY: SPECIAL CASES OF SEMANTIC RESOLUTION

In this section, we shall introduce two special kinds of interpretations to be used in semantic resolution. One interpretation leads to *hyperresolution* and the other leads to the *set-of-support strategy*.

A   Hyperresolution

Let us consider an interpretation I in which every literal is the negation of an atom. If this interpretation is used, every electron and every PI-resolvent must contain only atoms. Similarly, if every literal in I is an atom, then every electron and every PI-resolvent must contain only negations of atoms. Hyperresolution is based on these considerations.

Definition   A clause is called *positive* if it does not contain any negation sign. A clause is called *negative* if every literal of it contains the negation sign. A clause is called *mixed* if it is neither positive nor negative.

Definition   A *positive hyperresolution* is a special case of PI-resolution in which every literal in the interpretation I contains the negation sign. It is called positive hyperresolution because all the electrons and PI-resolvents in this case are positive.

Definition   A *negative hyperresolution* is a special case of PI-resolution in which the interpretation I does not contain any negation sign. It is called negative hyperresolution because all the electrons and PI-resolvents in this case are negative.

From Theorem 6.1, we can see that both positive and negative hyperresolutions are complete.

Example 6.7

For the following unsatisfiable set of clauses,

$$S = \{Q(a) \vee R(x),$$
$$\sim Q(x) \vee R(x),$$
$$\sim R(a) \vee \sim S(a),$$
$$S(x)\}$$

let P be an ordering defined as $R < Q < S$. Then we obtain positive and negative hyperdeductions shown in Fig. 6.4a and Fig. 6.4b, respectively.

It is often the case that the axioms of a theorem are represented by some positive and mixed clauses, and the negation of the conclusion is represented by a negative clause. In this case, the positive hyperresolution roughly corresponds to "thinking forward," while the negative hyperresolution corresponds to "thinking backward." This can be best explained by the following example.

Example 6.8

Let the axioms be Q, R, and $\sim Q \vee \sim R \vee S$. Let the negation of the conclusion be $\sim S$. Then we have the following four clauses:

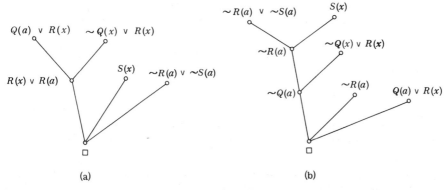

Figure 6.4

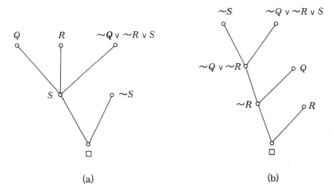

Figure 6.5

(1) Q

(2) R

(3) $\sim Q \vee \sim R \vee S$

(4) $\sim S.$

Let the ordering of predicate symbols be $Q > R > S$. Then Fig. 6.5a is a positive hyperdeduction of $\square$. The reader can see that the deduction starts with axioms and derives S, which is contradictory to the negation of the conclusion $\sim S$. In Fig. 6.5b, the deduction is a negative hyperdeduction of $\square$. This time the deduction starts from the negation of the conclusion. Assuming the conclusion is false, we derive a contradiction from this assumption.

B   The Set-of-Support Strategy

The set-of-support strategy was proposed by Wos, Robinson, and Carson [1965]. As discussed in the previous chapters, a theorem consists of a set of axioms $A_1, A_2, ..., A_n$ and a conclusion B. To prove the theorem, we are essentially proving that $A_1 \wedge A_2 \wedge \cdots \wedge A_n \wedge \sim B$ is unsatisfiable. Since $A_1 \wedge A_2 \wedge \cdots \wedge A_n$ is usually satisfiable, it might be wise to avoid resolving clauses in $A_1 \wedge A_2 \wedge \cdots \wedge A_n$. This is what the set-of-support strategy tries to accomplish.

Definition A subset T of a set S of clauses is called a *set of support* of S if $S - T$ is satisfiable. A *set-of-support resolution* is a resolution of two clauses that are not both from $S - T$. A *set-of-support deduction* is a deduction in which every resolution is a set-of-support resolution.

We shall now prove that the set-of-support resolution is complete. This can be easily done by using Theorem 6.1.

Theorem 6.2 (Completeness of the Set-of-Support Strategy) If S is a finite unsatisfiable set of clauses and T is a subset of S such that $S - T$ is satisfiable, then there is a set-of-support deduction of $\square$ from S with T as a set of support.

Proof Since $S - T$ is satisfiable, there is an interpretation I that satisfies $S - T$. Choose any ordering P of predicate symbols in S. By Theorem 6.1, there is a PI-deduction D of $\square$ from S. From D, consider each PI-clash $\{E_1, E_2, ..., E_q, N\}$. The PI-resolvent of this clash is obtained by first resolving E_1 and N, then resolving E_2 with this resolvent, and so on. Every resolvent involves an electron E_i. However electron is false in I. Therefore, for every corresponding resolution, the two clauses cannot both come from $S - T$. Thus, the deduction D can be transformed into a set-of-support deduction of $\square$ from S.   Q.E.D.

Example 6.9

Let S be a set consisting of the following clauses:

(1) $P(g(x_1, y_1), x_1, y_1)$

(2) $\sim P(x_2, h(x_2, y_2), y_2)$

(3) $\sim P(x_3, y_3, u_3) \vee P(y_3, z_3, v_3) \vee \sim P(x_3, v_3, w_3) \vee P(u_3, z_3, w_3)$

(4) $\sim P(k(x_4), x_4, k(x_4))$.

Let T be a set consisting of clause (4). Then the following deduction is a set-of-support deduction with T as the set of support. We see that no resolution is performed among (1), (2), and (3).

(5) $\sim P(x_3, y_3, k(z_3)) \vee P(y_3, z_3, v_3) \vee \sim P(x_3, v_3, k(z_3))$

                                        a resolvent of (4) and (3)

(6) $\sim P(x_3, y_3, k(h(y_3, v_3))) \lor \sim P(x_3, v_3, k(h(y_3, v_3)))$

a resolvent of (5) and (2)

(7)   □   a resolvent of (6) and (1).

6.6 SEMANTIC RESOLUTION USING ORDERED CLAUSES

In this section, we shall consider some orderings other than orderings of predicate symbols. For the propositional logic, orderings of predicate symbols used in conjunction with semantic resolution are quite effective. For any electron E, we have no difficulty in singling out a literal in E to be resolved upon. For example, consider $E_1 = A_1 \lor A_3$, $E_2 = A_2 \lor A_3$, and $N = \sim A_1 \lor \sim A_2 \lor A_3$ in Example 6.3. Using $I = \{\sim A_1, \sim A_2, \sim A_3\}$ and the ordering P in which $A_1 > A_2 > A_3$, we can easily decide that the literals A_1 and A_2 in E_1 and E_2, respectively, have to be resolved upon. However, for the first-order logic where variables are involved, in an electron E there may be more than one literal that contains the same largest predicate symbol in E. Any one of these literals is a candidate to be resolved upon. This is best illustrated by the following example.

Example 6.10

Consider

$$E = Q(a) \lor Q(b) \lor Q(c) \lor Q(d)$$

$$N = \sim Q(x).$$

Let I be an interpretation in which every literal is negative, and P be any ordering of predicate symbols. Thus, clearly, $\{E, N\}$ is a PI-clash. Since all the four literals in E contain the same predicate symbol Q, every one of them can be resolved with N. Therefore, from the PI-clash $\{E, N\}$, we can obtain four PI-resolvents, namely, $Q(b) \lor Q(c) \lor Q(d)$, $Q(a) \lor Q(c) \lor Q(d)$, $Q(a) \lor Q(b) \lor Q(d)$, and $Q(a) \lor Q(b) \lor Q(c)$.

From the above example we have seen that, using orderings of predicate symbols, we may not be able to single out *uniquely* a literal in an electron. Thus we may be forced to generate *more than one* semantic resolvent from a semantic clash. In order to remedy this situation, we consider *ordered clauses* [Reiter, 1971; and Slagle and Norton 1971].

The idea of ordered clauses is to consider a clause as a *sequence* of literals rather than a set of literals. In so doing, we set the order of all literals in a clause. We agree that a literal L_2 is greater than a literal L_1 in a clause just in case L_2 follows L_1 in the sequence specified by the

clause. Thus, the last literal in an ordered clause will always be considered the largest literal in the clause. For example, if we consider
$A_3 \vee A_1 \vee A_2$ as an ordered clause, then A_2 is the largest literal. We now
formally give the following definitions.

Definition   An *ordered clause* is a sequence of distinct literals.

As a clause, an ordered clause is also interpreted as a disjunction of all
the literals in the ordered clause. The only difference is that the order of
literals in a clause is immaterial, while the order of literals in an ordered
clause is deliberately specified. An ordered clause will also be written as a
disjunction of literals. This will not be confused since we shall always use
the word "ordered" for ordered clauses.

Definition   A literal L_2 is said to be *greater than* a literal L_1 in an
ordered clause (or L_1 is *smaller than* L_2) if and only if L_2 follows L_1 in the
sequence specified by the ordered clause.

We shall consider semantic resolution using ordered clauses. However, we
first define ordered resolution for ordered clauses.

Definition   If two or more literals (with the same sign) of an ordered
clause C have a most general unifier σ, then the ordered clause obtained
from the sequence $C\sigma$ by deleting any literal that is identical to a smaller
literal in the sequence is called an *ordered factor* of C.

Example 6.11

Consider the ordered clause $C = \underline{P(x)} \vee Q(x) \vee \underline{P(a)}$. The first and the last
literals (underlined) of C have a most general unifier $\sigma = \{a/x\}$. Hence,
$C\sigma = P(a) \vee Q(a) \vee P(a)$. In this sequence $C\sigma$, there are two identical literals.
Since the second $P(a)$ is identical to a smaller literal, namely, the first $P(a)$,
we delete the second $P(a)$ from $C\sigma$ and obtain the sequence $P(a) \vee Q(a)$,
which is an ordered factor of C. We note that $Q(a) \vee P(a)$ is *not* an
ordered factor of C since it is obtained from $C\sigma$ by deleting the first $P(a)$.

Definition   Let C_1 and C_2 be ordered clauses with no variables in
common. Let L_1 and L_2 be two literals in C_1 and C_2, respectively. If L_1
and $\sim L_2$ have a most general unifier σ, and if C is the ordered clause
obtained by concatenating the sequences $C_1\sigma$ and $C_2\sigma$, removing $L_1\sigma$ and
$L_2\sigma$, and deleting any literal that is identical to a smaller literal in the
remaining sequence, then C is called an *ordered binary resolvent* of C_1 against
C_2. The literals L_1 and L_2 are called the *literals resolved upon*.

The above definition is the same as the definition of a binary resolvent
given in Section 5.5 of Chapter 5, except that the order of literals in an

ordered binary resolvent is deliberately specified. Note that we use the word
"against" to emphasize that an ordered binary resolvent of C_1 against C_2
is not the same as one of C_2 against C_1.

Example 6.12

Consider the ordered clauses

$$C_1 = P(x) \vee Q(x) \vee R(x) \quad \text{and} \quad C_2 = \sim P(a) \vee Q(a).$$

Choose $L_1 = P(x)$ and $L_2 = \sim P(a)$. Since $\sim L_2 = P(a)$, L_1 and $\sim L_2$ have
a most general unifier $\sigma = \{a/x\}$. First, concatenating the sequences $C_1\sigma$
and $C_2\sigma$, we obtain the sequence, $P(a) \vee Q(a) \vee R(a) \vee \sim P(a) \vee Q(a)$.
Second, removing $L_1\sigma$ and $L_2\sigma$ from the above sequence, we obtain the
sequence $Q(a) \vee R(a) \vee Q(a)$. Finally, since in $Q(a) \vee R(a) \vee Q(a)$ the second
$Q(a)$ is identical to a smaller literal, namely, the first $Q(a)$, we delete the
second $Q(a)$ and obtain $Q(a) \vee R(a)$. The sequence $Q(a) \vee R(a)$ is an ordered
binary resolvent of C_1 against C_2. $P(x)$ and $\sim P(a)$ are the literals resolved
upon.

Definition An *ordered resolvent* of an ordered clause C_1 against an
ordered clause C_2 is one of the following ordered binary resolvents:

1. an ordered binary resolvent of C_1 against C_2;
2. an ordered binary resolvent of C_1 against an ordered factor of C_2;
3. an ordered binary resolvent of an ordered factor of C_1 against C_2;
4. an ordered binary resolvent of an ordered factor of C_1 against an
ordered factor of C_2.

Example 6.13

Consider the ordered clauses

$$C_1 = P(x) \vee Q(x) \vee R(x) \vee P(a),$$
$$C_2 = \sim P(a) \vee Q(a).$$

An ordered factor of C_1 is $C_1' = P(a) \vee Q(a) \vee R(a)$. An ordered binary
resolvent of C_1' against C_2 is $Q(a) \vee R(a)$. Therefore, $Q(a) \vee R(a)$ is an
ordered resolvent of C_1 against C_2.

Ordered resolution is an inference rule that generates ordered resolvents
from a set of ordered clauses. It is not difficult to see that ordered resolution
is complete. That is, ordered resolution will always generate the empty
clause $\square$ from an unsatisfiable set of ordered clauses. Proof of this is left
to the reader as an exercise.

We have defined ordered resolution. Now, we shall consider semantic resolution for ordered clauses. We shall still use the concepts of interpretation and clash. However, we shall use the concept of ordered clauses rather than that of ordering of predicate symbols. As discussed in the beginning of this section, this has an advantage because now we can always uniquely single out a literal in an electron to resolve upon.

Definition   Let I be an interpretation. A finite sequence of ordered clauses, $(E_1, E_2, ..., E_q, N), q \geqslant 1$, is called an *ordered semantic clash* with respect to I (or *OI-clash* for short) if and only if $E_1, E_2, ..., E_q$ (called *ordered electrons*) and N (called the *ordered nucleus*) satisfy the following conditions:

1. $E_1, E_2, ..., E_q$ are false in I.
2. Let $R_q = N$. For each $i = q, q-1, ..., 1$, there is an ordered resolvent R_{i-1} of E_i against R_i.
3. The literal in E_i that is resolved upon is the "last literal" in $E_i, i = 1, ..., q$; the literal in R_i that is resolved upon is the "largest literal" that has an instance true in I.
4. R_0 is false in I.

R_0 is called an *OI-resolvent* of the *OI*-clash $(E_1, E_2, ..., E_q, N)$.

Except for condition (3), the above definition is similar to that of the *PI*-resolvent given in Section 6.3. We now give an example to illustrate how an *OI*-clash can be found.

Example 6.14

Consider the ordered clauses

(1) $A_1 \vee A_2$

(2) $A_1 \vee A_3$

(3) $\sim A_3 \vee \sim A_2 \vee A_1$.

Let $I = \{\sim A_1, \sim A_2, \sim A_3\}$. Since (1) and (2) are false in I, they can be used as ordered electrons. Since (3) has a literal that is true in I, (3) can be used as an ordered nucleus. Therefore, let $N = \sim A_3 \vee \sim A_2 \vee A_1$. In N, there are two literals, namely, $\sim A_3$ and $\sim A_2$, that are true in I. Therefore, we need two ordered electrons. Let $R_2 = N$, i.e., $R_2 = \sim A_3 \vee \sim A_2 \vee A_1$. In R_2, the largest literal that has an instance (itself in this case) true in I is $\sim A_2$. Since $\sim A_2$ and the last literal of (1) can be resolved upon, we have $E_2 = A_1 \vee A_2$. The ordered resolvent R_1 of E_2 against R_2 is given as $R_1 = A_1 \vee \sim A_3$. Now, in R_1, the largest literal that has an instance (itself in this case) true in I is $\sim A_3$. Since $\sim A_3$ and the last literal of (2) can be resolved upon, we have $E_1 = A_1 \vee A_3$. The ordered resolvent R_0 of E_1 against R_1 is given as $R_0 = A_1$. Since R_0 is false in I, we conclude that (E_1, E_2, N), or $((2), (1), (3))$, is an *OI*-clash, and A_1 is the *OI*-resolvent of this clash.

From the above example, we can see that the order of the ordered electrons is determined by the order of literals in the ordered nucleus. Therefore, for the above example, ((1), (2), (3)) would not be an *OI*-clash.

Example 6.15

Consider the clauses given in Example 6.10,

(1) $Q(a) \lor Q(b) \lor Q(c) \lor Q(d)$

(2) $\sim Q(x)$.

Suppose that we treat (1) and (2) as ordered clauses. Let *I* be an interpretation in which every literal is negative. Since (1) is false in *I*, (1) can be used as an electron. Resolving the last literal of (1) against $\sim Q(x)$, we obtain an ordered resolvent $Q(a) \lor Q(b) \lor Q(c)$, which is false in *I*. Therefore, ((1),(2)) is an *OI*-clash with (1) and (2) as the ordered electron and the ordered nucleus, respectively. There is only one *OI*-resolvent of this clash, as opposed to four *PI*-resolvents using an ordering of predicate symbols.

Definition Let *I* be an interpretation for a set *S* of ordered clauses. A deduction from *S* is called an *OI*-deduction if and only if each ordered clause in the deduction is either an ordered clause in *S* or an *OI*-resolvent.

Example 6.16

Consider the following ordered clauses:

(1) $Q(a) \lor R(x)$

(2) $\sim Q(x) \lor R(x)$

(3) $\sim R(x) \lor \sim S(a)$

(4) $S(x)$.

Let *I* be an interpretation in which every literal is negative. Then the deduction shown in Fig. 6.6 is an *OI*-deduction of $\square$ from *S*.

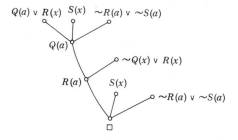

Figure 6.6

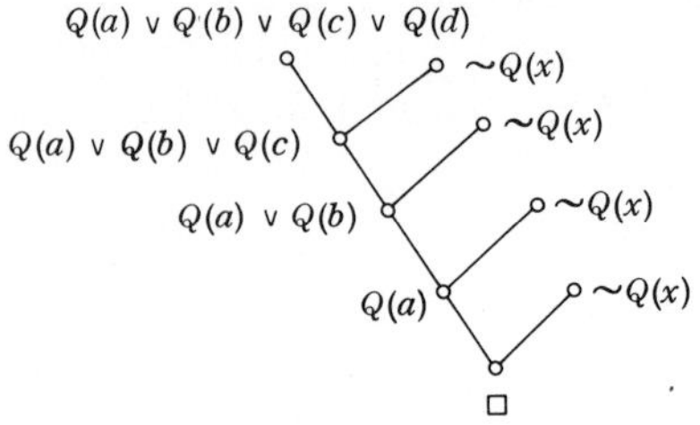

Figure 6.7

Example 6.17

Consider the set S consisting of the ordered clauses $Q(a) \vee Q(b) \vee Q(c) \vee Q(d)$ and $\sim Q(x)$, as given in Example 6.10. S is unsatisfiable. Let I be an interpretation in which every literal is negative. Then the deduction shown in Fig. 6.7 is an OI-deduction of $\square$ from S. Only four ordered OI-resolvents have been generated. However, if PI-resolution were used, 40 PI-resolvents would be generated by the level-saturation method before $\square$ could be found.

Slagle and Norton [1971] experimented with OI-resolution; their results indicated that it is quite efficient. Many theorems can be proved by OI-resolution; however, unfortunately, OI-resolution is *not* complete. The following counterexample is given by Anderson [1971].

Example 6.18

Consider the following set S of ordered clauses:

(1)　　$P \vee Q$

(2)　　$Q \vee R$

(3)　　$R \vee W$　　　$\left.\right\}$ S.

(4)　　$\sim R \vee \sim P$

(5)　　$\sim W \vee \sim Q$

(6)　　$\sim Q \vee \sim R$

Let I be an interpretation in which every literal is negative. Thus, clauses (1)–(3) can be used as ordered electrons, and clauses (4)–(6) can be used as an ordered nucleus. From clauses (1)–(6), we can obtain the following OI-resolvents:

(7)　　$R \vee P$　　　from OI-clash $((3), (1), (5))$

(8)　　$P \vee Q$　　　from OI-clash $((1), (2), (6))$.

From clauses (1)–(8), we can obtain the following OI-resolvent:

(9) $Q \vee R$ from OI-clash ((2), (7), (4)).

We note that clauses (8) and (9) are in S. Therefore, from clauses (1)–(9) we can not produce any new OI-resolvents. That is, $\square$ cannot be produced by OI-resolution. However, S is unsatisfiable. Hence, OI-resolution is not complete.

6.7   IMPLEMENTATION OF SEMANTIC RESOLUTION

In this section, we shall consider how PI-resolution is implemented. Although OI-resolution is not complete, we can use the concept of ordered clauses to implement PI-resolution. We shall say that a *positive ordered clause* is an ordered clause that does not contain any negation sign, and a *negative ordered clause* is an ordered clause of which every literal contains the negation sign. A *nonpositive (nonnegative) ordered clause* is an ordered clause that is not positive (negative). In order to relieve us from specifying an interpretation every time PI-resolution is used, we shall consider only positive hyperresolution. (Negative hyperresolution can be treated in a similar manner.) Thus we need consider only positive ordered clauses as candidates for electrons and nonpositive ones for the nucleus. *We agree that for any nonpositive ordered clauses, negative literals are put after positive literals. Let S be a set of ordered clauses. Let P be an ordering of predicate symbols in S.* The following algorithm can compute positive hyperresolvents.

Generation of Positive Hyperresolvents

Step 0   Let M and N be the sets of all positive and nonpositive ordered clauses in S, respectively.

Step 1   Set $j = 1$.

Step 2   Let $A_0 = \emptyset$ and $B_0 = N$.

Step 3   Set $i = 0$.

Step 4   If A_i contains $\square$, terminate; a contradiction is found. Otherwise, go to the next step.

Step 5   If B_i is empty, go to Step 8. Otherwise, go to the next step.

Step 6   Compute the set

$W_{i+1} = \{$ordered resolvents of C_1 against C_2, where C_1 is an ordered clause or an ordered factor of an ordered clause

in M, C_2 is an ordered clause in B_i, the resolved literal
of C_1 contains the "largest" predicate symbol in C_1, and
the resolved literal of C_2 is the "last" literal of C_2}.

Let A_{i+1} and B_{i+1} be the sets of all the positive and nonpositive ordered
clauses in W_{i+1}, respectively.

Step 7   Set $i = i+1$ and go to Step 4.

Step 8   Set $T = A_0 \cup \cdots \cup A_i$ and $M = T \cup M$.

Step 9   Set $j = j+1$.

Step 10   Compute the set

$R = \{$ ordered resolvents of C_1 against C_2, where C_1 is an ordered
clause or an ordered factor of an ordered clause in T, C_2 is
an ordered clause in N, and the resolved literal of C_1
contains the "largest" predicate symbol in C_1}.

(Note that in the above definition of R, the resolved literal of C_2 can be
any literal of C_2, not necessarily the last literal.) Let A_0 and B_0 be the
sets of all the positive and nonpositive ordered clauses in R, respectively.

Step 11   Go to Step 3.

In the above algorithm, for each cycle, that is for each j, B_i will eventually
be empty since the maximum number of negative literals in any ordered
clause of B_i decreases by one as i increases by one. All the clauses in each
A_i are positive hyperresolvents. It is not hard to see that if S is unsatisfiable,
then $\square$ can be generated by the above algorithm. The deletion strategy can
be also incorporated into the above algorithm without losing the completeness property. That is, for T and M obtained in Step 8 of the algorithm, any
clause in T or M subsumed by other clauses in T or M can be deleted.
(Note that no tautologies are in T or M since all clauses in T or M are
positive.) However, in the following example, we do not use the deletion
strategy.

Example 6.19

Let S be a set of ordered clauses given as $S = \{P_6 \vee P_4, P_5 \vee P_4, P_4 \vee P_1,$
$\sim P_1 \vee \sim P_2, P_3 \vee \sim P_6, \sim P_4, P_2 \vee \sim P_5 \vee \sim P_3\}$. For   simplicity,   we
denote $P_1, P_2, ..., P_6$ by $1, 2, ..., 6$, respectively. Therefore, S is written as
$S = \{6 \vee 4, 5 \vee 4, 4 \vee 1, \sim 1 \vee \sim 2, 3 \vee \sim 6, \sim 4, 2 \vee \sim 5 \vee \sim 3\}$. Let P be
an ordering of predicate symbols in which $1 < 2 < 3 < 4 < 5 < 6$. From S,
we obtain

$$M = \{6 \vee 4, 5 \vee 4, 4 \vee 1\}$$
$$N = \{\sim 1 \vee \sim 2, 3 \vee \sim 6, \sim 4, 2 \vee \sim 5 \vee \sim 3\}.$$

Now, we apply the above algorithm.

1. $j = 1$.

   a. $A_0 = \varnothing$

     $B_0 = N = \{\sim 1 \vee \sim 2, 3 \vee \sim 6, \sim 4, 2 \vee \sim 5 \vee \sim 3\}$.

   b. Since neither A_0 contains $\square$ nor B_0 is empty, we apply Step 6 and obtain

     $W_1 = \{4 \vee 3, 1\}$

     $A_1 = \{4 \vee 3, 1\}$

     $B_1 = \varnothing$.

   c. Since B_1 is empty, we go to Step 8 to set

     $T = A_0 \cup A_1 = \{4 \vee 3, 1\}$

     $M = T \cup M = \{4 \vee 3, 1, 6 \vee 4, 5 \vee 4, 4 \vee 1\}$.

2. $j = 2$.

   a. Computing R, we obtain $R = \{3, \sim 2\}$. From R, we obtain $A_0 = \{3\}$, $B_0 = \{\sim 2\}$.

   b. Since neither A_0 contains $\square$ nor B_0 is empty, we apply Step 6 and obtain $W_1 = \varnothing$, $A_1 = \varnothing$, $B_1 = \varnothing$.

   c. Since B_1 is empty, we let

     $T = A_0 \cup A_1 = \{3\}$

     $M = T \cup M = \{3, 4 \vee 3, 1, 6 \vee 4, 5 \vee 4, 4 \vee 1\}$.

3. $j = 3$.

   a. Applying Step 10, we obtain $R = \{2 \vee \sim 5\}$, $A_0 = \varnothing$, $B_0 = \{2 \vee \sim 5\}$.

   b. Since neither A_0 contains $\square$ nor B_0 is empty, we apply Step 6 and obtain $W_1 = \{4 \vee 2\}$, $A_1 = \{4 \vee 2\}$, $B_1 = \varnothing$.

   c. Since B_1 is empty, we let

     $T = A_0 \cup A_1 = \{4 \vee 2\}$

     $M = T \cup M = \{4 \vee 2, 3, 4 \vee 3, 1, 6 \vee 4, 5 \vee 4, 4 \vee 1\}$.

4. $j = 4$.

   a. Applying Step 10, we obtain $R = \{2\}$, $A_0 = \{2\}$, $B_0 = \varnothing$.

   b. Since B_0 is empty, we let

     $T = A_0 = \{2\}$

     $M = T \cup M = \{2, 4 \vee 2, 3, 4 \vee 3, 1, 6 \vee 4, 5 \vee 4, 4 \vee 1\}$.

5. $j = 5$.

  a. Applying Step 10, we obtain $R = \{\sim 1\}$, $A_0 = \varnothing$, $B_0 = \{\sim 1\}$.

  b. Since neither A_0 contains $\square$ nor B_0 is empty, we apply Step 6 and obtain $W_1 = \{\square\}$, $A_1 = \{\square\}$, $B_1 = \varnothing$.

  c. Since A_1 contains $\square$, we terminate the algorithm; a contradiction is found.

We have generated six positive hyperresolvents, namely, $\square$, 2, $4 \vee 2$, 3, $4 \vee 3$ and 1. To get the hyperdeduction of $\square$, we first trace the ancestry of $\square$ as shown in Fig. 6.8, which is then easily transformed into the hyperdeduction of $\square$ shown in Fig. 6.9. We note that all hyperresolvents generated by the algorithm are used in the proof. In this example, neither irrelevant nor redundant resolvents have been generated by the algorithm.

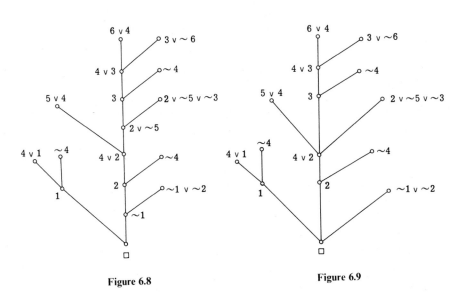

Figure 6.8   **Figure 6.9**

6.8 LOCK RESOLUTION

    Lock resolution was introduced by Boyer [1971]. It is a refinement of resolution which uses a concept similar to that of ordered clauses. Given a set S of clauses, essentially, the idea of lock resolution is to use indices to order literals of clauses in S. That is, it involves arbitrarily indexing each occurrence of a literal in S with an integer; different occurrences of the same literal in S may be indexed differently. Resolution is then permitted only on literals of lowest index in each clause. The literals in resolvents inherit their indices from their parent clauses. If a literal in a resolvent has more than one possible inherited index, the lowest index is assigned to this literal. Before we give a formal definition of lock resolution, we first consider the following example.

Example 6.20

Consider the following two clauses:

(1) $_1P \vee {}_2Q$

(2) $_3 \sim P \vee {}_4Q.$

The integer beneath a literal is the index associated with that literal. The literals P and Q of clause (1) have indices 1 and 2, respectively. We shall use $_1P$ and $_2Q$ to refer to the literals P and Q of clause (1), respectively. Similarly, we shall use $_3 \sim P$ and $_4Q$ to refer to the literals $\sim P$ and Q of clause (2), respectively. Since the index of $_1P$ is lower than that of $_2Q$, $_1P$ is permitted to be resolved upon. Similarly, since the index of $_3 \sim P$ is lower than that of $_4Q$, $_3 \sim P$ is permitted to be resolved upon. Thus, resolving clauses (1) and (2) upon $_1P$ and $_3 \sim P$, we obtain the following clause:

(3) $_2Q \vee {}_4Q.$

Now $_2Q$ and $_4Q$ mean the same literal Q. Since 2 is lower than 4, Q is indexed with 2. Therefore, we obtain

(4) $_2Q.$

Clause (4) is called a lock resolvent of clauses (1) and (2). Note that if the literals of clause (2) are indexed differently as

(2*) $_4 \sim P \vee {}_3Q,$

then the literal in clause (2*) that is permitted to be resolved upon is $_3Q$. However, $_1P$ and $_3Q$ cannot be resolved. Therefore, there is no lock resolvent of clauses (1) and (2*).

Definition   Let C be a clause that has every one of its literals indexed with integers. If two or more literals (with the same sign) of C have a most general unifier σ, then the clause obtained from $C\sigma$ by deleting any literal that is identical to a literal of lower index is called a *lock factor* of C.

Example 6.21

Let $C = {}_2P(x) \vee {}_8Q(a) \vee {}_{11}P(a) \vee {}_5Q(x)$. $_2P(x)$ and $_{11}P(a)$ have a most general unifier $\sigma = \{a/x\}$. Thus $C\sigma = {}_2P(a) \vee {}_8Q(a) \vee {}_{11}P(a) \vee {}_5Q(a)$. In $C\sigma$, except for its index, $_2P(a)$ is identical to $_{11}P(a)$. Since $_{11}P(a)$ has higher index than does $_2P(a)$, $_{11}P(a)$ will be deleted. Similarly, since $_8Q(a)$ has higher index than does $_5Q(a)$, $_8Q(a)$ will be deleted. Therefore, deleting $_{11}P(a)$ and $_8Q(a)$, we obtain $_2P(a) \vee {_5Q(a)}$, which is a lock factor of C.

In a clause, if there is more than one of the same literal, we always keep only the literal with the lowest index and delete the other identical literals. This operation is called *merging low* for identical literals. For instance, in the above example, we have merged low for $_2P(a)$ and $_{11}P(a)$, and also for $_8Q(a)$ and $_5Q(a)$.

Definition Let C_1 and C_2 be two clauses with no variable in common and with every literal in them indexed. Let L_1 and L_2 be two literals of "lowest" index in C_1 and C_2, respectively. If L_1 and $\sim L_2$ have a most general unifier σ, and if C is the clause obtained from $(C_1\sigma \vee C_2\sigma)$ by removing $L_1\sigma$ and $L_2\sigma$ and by merging low for any identical literals in the remaining clause, then C is called a *binary lock resolvent* of C_1 and C_2. The literals L_1 and L_2 are called the *literals resolved upon*.

Example 6.22

Consider the clauses: $C_1 = {_1P(x)} \vee {_2Q(x)} \vee {_3R(x)}$ and $C_2 = {_4\sim P(a)} \vee {_5Q(a)}$. Since $_1P(x)$ and $_4\sim P(a)$ have the lowest indices in C_1 and C_2, respectively, we choose $L_1 = P(x)$ and $L_2 = \sim P(a)$. L_1 and $\sim L_2$ have a most general unifier $\sigma = \{a/x\}$. Therefore, $(C_1\sigma \vee C_2\sigma) = {_1P(a)} \vee {_2Q(a)} \vee {_3R(a)} \vee {_4\sim P(a)} \vee {_5Q(a)}$. Removing $L_1\sigma$ and $L_2\sigma$, that is, $_1P(a)$ and $_4\sim P(a)$, from $(C_1\sigma \vee C_2\sigma)$, we obtain $_2Q(a) \vee {_3R(a)} \vee {_5Q(a)}$. Now, $_2Q(a)$ and $_5Q(a)$ are identical literals with different indices. Merging low for $_2Q(a)$ and $_5Q(a)$ in $_2Q(a) \vee {_3R(a)} \vee {_5Q(a)}$, we obtain $_2Q(a) \vee {_3R(a)}$, which is a binary lock resolvent of C_1 and C_2. $_1P(x)$ and $_4\sim P(a)$ are the literals resolved upon.

Definition Let C_1 and C_2 be two clauses with every literal in them indexed. A *lock resolvent* of C_1 and C_2 is one of the following binary lock resolvents:

1. a binary lock resolvent of C_1 and C_2;
2. a binary lock resolvent of C_1 and a lock factor of C_2;
3. a binary lock resolvent of a lock factor of C_1 and C_2;
4. a binary lock resolvent of a lock factor of C_1 and a lock factor of C_2.

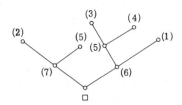

Figure 6.10

Definition   Let S be a set of clauses, where every literal in S is indexed with an integer. A deduction from S is called a *lock deduction* if and only if every clause in the deduction is either a clause in S or a lock resolvent.

Example 6.23

Consider the following set S of clauses:

(1) $_1P \lor {}_2Q$ ⎫

(2) $_3P \lor {}_4{\sim}Q$ ⎪ $S.$

(3) $_6{\sim}P \lor {}_5Q$ ⎬

(4) $_8{\sim}P \lor {}_7{\sim}Q$ ⎭

From clauses (1)–(4), there is only one lock resolvent:

(5) $_6{\sim}P$ from (3) and (4).

From clauses (1)–(5), there are only two lock resolvents:

(6) $_2Q$ from (1) and (5)

(7) $_4{\sim}Q$ from (2) and (5).

Finally, resolving clauses (6) and (7), we obtain

(8) $\square$.

Thus we obtain a lock deduction of $\square$ shown in Fig. 6.10. We note that altogether only three lock resolvents were generated. However, if ordinary (unrefined) resolution were used, 37 resolvents would be generated by the level-saturation method before $\square$ could be generated. (See Section 5.8.)

Example 6.24

Consider the following set of clauses:

(1) $_5P(y,a) \lor {}_1P(f(y),y)$ ⎫

(2) $_6P(y,a) \lor {}_2P(y,f(y))$ ⎪

(3) $_8{\sim}P(x,y) \lor {}_3P(f(y),y)$ ⎬ $S.$

(4) $_9{\sim}P(x,y) \lor {}_4P(y,f(y))$ ⎪

(5) $_{10}{\sim}P(x,y) \lor {}_7{\sim}P(y,a)$ ⎭

From S, we can obtain the following lock deduction of $\square$:

(6) $_5P(a,a) \lor {}_{10}{\sim}P(x,f(a))$ a lock resolvent of (1) and (5)

(7) $_8{\sim}P(x,a) \lor {}_{10}{\sim}P(y,f(a))$ a lock resolvent of (3) and (5)

(8) $_{10}\sim P(x, f(a)) \vee {}_{10}\sim P(y, f(a))$ a lock resolvent of (6) and (7)

(9) $_6 P(a, a)$ a lock resolvent of (8) and (2)

(10) $_9 \sim P(x, a)$ a lock resolvent of (8) and (4)

(11)   □   a lock resolvent of (9) and (10).

6.9 COMPLETENESS OF LOCK RESOLUTION

In this section, we shall prove the completeness of lock resolution. As usual, this is done by first proving the completeness of lock resolution for the ground case, which is then "lifted" to the general case. For a clause C that has every one of its literals indexed, any instance of C is obtained by first making substitutions for variables and then merging low for identical literals. That is, if σ is a substitution, first we obtain $C\sigma$ and then merge low for identical literals in $C\sigma$ to obtain an instance of C. For example, let $C = {}_1 P(x) \vee {}_2 P(y)$. If $\sigma = \{a/x, a/y\}$, then an instance of C is $_1 P(a)$. By this convention, the following lifting lemma is obvious. We leave its proof to the reader. (Use Lemma 5.1.)

Lemma 6.2   Let C_1 and C_2 be two clauses with every literal of them indexed. If C_1' and C_2' are instances of C_1 and C_2, respectively, and if C' is a lock resolvent of C_1' and C_2', then there is a lock resolvent C of C_1 and C_2 such that C' is an instance of C.

We now prove the completeness of lock resolution for the ground case.

Lemma 6.3   Let S be a set of "ground" clauses, where every literal in S is indexed with an integer. If S is unsatisfiable, then there is a lock deduction of the empty clause □ from S.

Proof   Let $k(S)$ be defined to be the total number of appearances of literals in S minus the number of clauses in S. $k(S)$ is called the *excess literal parameter* by Anderson and Bledsoe [1970]. We prove Lemma 6.3 by induction on $k(S)$. If □ is in S, Lemma 6.3 is obvious. Assume □ is not in S. If $k(S) = 0$, then S consists of only unit clauses. Since S is unsatisfiable, there is a literal L such that $_i L$ and $_j \sim L$ are in S, where i and j are the indices of $_i L$ and $_j \sim L$, respectively. Clearly, □ is a lock resolvent of $_i L$ and $_j \sim L$. Thus, Lemma 6.3 holds for $k(S) = 0$. Assume Lemma 6.3 holds when $k(S) < n$. To complete the induction, we consider the case where $k(S) = n$ and $n > 0$. Since $k(S) > 0$, there is at least one nonunit clause in S. Let r be the highest index among all the nonunit clauses of S. Let $C = C' \vee {}_r L$ be a nonunit clause that has a literal $_r L$ indexed with r, where C' is a nonempty clause. Let $S_1 = (S - \{C\}) \cup \{C'\}$ and $S_2 = (S - \{C\}) \cup \{{}_r L\}$. Clearly, S_1 and

S_2 are unsatisfiable. However, $k(S_1) < n$ and $k(S_2) < n$. Hence, by the induction hypothesis, there are lock deductions D_1' and D_2' of $\square$ from S_1 and S_2, respectively. Let D_1 be the deduction obtained from D_1' by putting $_rL$ back with C'. Clearly, D_1 is also a lock deduction since r is the highest index in a clause attached to any noninitial node of D_1. D_1 is a lock deduction of $\square$ or $_rL$ from S. If it is the former, we are done, If it is the latter, combining D_1 and D_2', we obtain a lock deduction of $\square$ from S. This completes the proof of Lemma 6.3.

Theorem 6.3 **(Completeness of Lock Resolution)** Let S be a set of clauses, where every literal in S is indexed with an integer. If S is unsatisfiable, then there is a lock deduction of the empty clause $\square$ from S.

Proof Since S is unsatisfiable, by Herbrand's theorem (vertion II) there is a finite unsatisfiable set S' of ground instances of clauses in S. By Lemma 6.3, there is a lock deduction D' of $\square$ from S'. Using Lemma 6.2 and a process similar to that given in the proof of Theorem 6.1, we can transform D' to produce a lock deduction of $\square$ from S. This completes the proof.

Although lock resolution itself is complete, it is not compatible with most other resolution strategies. (See [Boyer, 1971].) For example, the combination of lock resolution and the deletion strategy is not complete. Lock resolution is not compatible with the set-of-support strategy either. (See Exercises 20 and 21 of this chapter.) Lock resolution is a very restrictive refinement of resolution. Boyer has given many examples to show that lock resolution is indeed a very efficient inference rule.

REFERENCES

Anderson, R. (1971): *SIGART Newsletter*, October.

Anderson, R., and W. W. Bledsoe (1970): A linear format for resolution with merging and a new technique for establishing completeness, *J. Assoc. Comput. Mach.* **17** 525–534.

Boyer, R. S. (1971): *Locking: a restriction of resolution*, Ph.D. Thesis, University of Texas at Austin, Texas.

Chang, C. L. (1970a): The unit proof and the input proof in theorem proving, *J. Assoc. Comput. Mach.* **17** 698–707.

Chang, C. L. (1972): Theorem proving with variable-constrained resolution, *Information Sci.* **4** 217–231.

Kowalski, R., and D. Kuehner (1971): Linear resolution with selection function, *Artificial Intelligence* **2** 227–260.

Loveland, D. W. (1970a): A linear format for resolution, *Proc. IRIA Symp. Automatic Demonstration, Versailles, France, 1968*, Springer-Varlag, New York, pp. 147–162.

Loveland, D. W. (1970b): Some linear Herbrand proof procedures: an analysis, Dept. of Computer Science, Carnegie-Mellon University.

Loveland, D. W. (1972): A unifying view of some linear Herbrand procedures, *J. Assoc. Comput. Mach.* **19** 366–384.

Luckham, D. (1970): Refinements in resolution theory, *Proc. IRIA Symp. Automatic Demonstration, Versailles, France, 1968*, Springer-Verlag, New York, pp. 163–190.

Meltzer, B. (1966): Theorem-proving for computers: some results on resolution and renaming, *Computer J.* **8** 341–343.

Reiter, R. (1971): Two results on ordering for resolution with merging and linear format, *J. Assoc. Comput. Mach.* **18** 630–646.

Robinson, J. A. (1965b): Automatic deduction with hyper-resolution, *Internat. J. Comput. Math.* **1** 227–234.

Slagle, J. R. (1967): Automatic theorem proving with renamable and semantic resolution, *J. Assoc. Comput. Mach.* **14** 687–697.

Slagle, J. R., and L. Norton (1971): Experiments with an automated theorem prover having partial ordering rules, Division of Computer Research and Technology, National Inst. Health, Bethesda, Maryland.

Wos, L., G. A. Robinson, and D. F. Carson (1965): Efficiency and completeness of the set of support strategy in theorem proving, *J. Assoc. Comput. Mach.* **12** 536–541.

EXERCISES

Section 6.3

1. Determine whether the following sets of clauses, interpretations, and orderings of atoms constitute *PI*-clashes. If yes, find the *PI*-resolvents. If no, give reasons.

   (a) $E_1 = P, E_2 = Q, E_3 = R, N = \sim P \vee \sim Q \vee \sim R \vee \sim W$

        $I = \{\sim P, \sim Q, \sim R, W\}$

        $P < Q < R < W.$

   (b) $E_1 = A_1 \vee A_3, E_2 = A_2 \vee A_3, N = \sim A_1 \vee \sim A_2 \vee A_3$

        $I = \{\sim A_1, \sim A_2, \sim A_3\}$

        $A_1 > A_2 > A_3.$

   (c) $E_1 = A_1 \vee A_3, E_2 = A_2 \vee A_3, N = \sim A_1 \vee \sim A_2 \vee A_3$

        $I = \{\sim A_1, \sim A_2, \sim A_3\}$

        $A_3 > A_1 > A_2.$

   (d) $E_1 = P(a), E_2 = Q(a,b) \vee S(b,c), N = \sim P(x) \vee \sim Q(x,y) \vee R(x)$

        $I = \{\sim P(a), P(b), \sim Q(a,b), \sim S(b,c), \sim R(a)\}$

        $P > Q > R > S.$

2. Prove the following theorem: For the ground case, the nucleus N of a *PI*-clash is true in I, and for the general (nonground) case, there exists at least one instance N' of N such that N' is true in I.

3. Let $S = \{P, Q \vee \sim P, R \vee \sim P, \sim P \vee \sim Q \vee \sim R\}$. Prove that S is unsatisfiable by semantic resolution. The following cases are to be considered:

   (a) $I = \{\sim P, \sim Q, \sim R\}; P > Q > R$

   (b) $I = \{P, Q, R\}; Q > P > R$

   (c) $I = \{\sim P, \sim Q, R\}; R > Q > P$.

Section 6.4

4. Prove Theorem 6.1 by using the semantic tree technique, that is, the technique given in the proof of Theorem 5.3. (Consult [Kowalski and Hayes, 1969].)

5. Prove that the deletion strategy is compatable with PI-resolution, that is, that the combination of the deletion strategy and PI-resolution is complete.

Section 6.5

6. Let $S = \{P \vee Q, \sim P \vee Q, P \vee \sim Q, \sim P \vee \sim Q\}$. Let $P > Q$. Prove that S is unsatisfiable by (a) positive hyperresolution, and (b) negative hyperresolution.

7. Consider Example 5.16. In that example, we have the following set S of clauses:

$$S = \{\sim T(x, y, u, v) \vee P(x, y, u, v), \sim P(x, y, u, v) \vee E(x, y, v, u, v, y),$$

$$T(a, b, c, d), \sim E(a, b, d, c, d, b)\}.$$

Prove that S is unsatisfiable by (a) positive hyperresolution, and (b) negative hyperresolution. Translate every step of each proof into English. Which deduction corresponds to thinking forward (thinking backward)?

8. Let

   (1) $M(a, s(c), s(b))$

   (2) $P(a)$

   (3) $M(x, x, s(x))$

   (4) $\sim M(x, y, z) \vee M(y, x, z)$

   (5) $\sim M(x, y, z) \vee D(x, z)$

   (6) $\sim P(x) \vee \sim M(y, z, u) \vee \sim D(x, u) \vee D(x, y) \vee D(x, z)$

   (7) $\sim D(a, b)$

$\left. \right\} S.$

Let the set of support consist of clause (7). Prove the unsatisfiability
of S by the set-of-support strategy.

9. Reconsider Example 5.21. We have the following set of clauses:

(1) $P(a)$

(2) $\sim D(y) \vee L(a,y)$

(3) $\sim P(x) \vee \sim Q(y) \vee \sim L(x,y)$

(4) $D(b)$

(5) $Q(b)$.

Let the set of support consist of clause (4) and clause (5). Prove the
unsatisfiability of the above set of clauses. Translate the entire proof
into English.

Section 6.6

10. Consider the following set of ordered clauses:

$$S = \{P \vee Q, \sim P \vee Q, P \vee \sim Q, \sim P \vee \sim Q\}.$$

Let $I = \{P,Q\}$. Prove that S is unsatisfiable by OI-resolution.

11. In Example 6.18, we gave a counterexample to show that OI-resolution
is not complete. Can you give another counterexample?

Section 6.7

12. Let $S = \{\sim P \vee \sim Q \vee \sim R, P, Q, R\}$.

a. Use the algorithm mentioned in Section 6.7 to prove that S is
unsatisfiable.

b. Use the same algorithm again without the concept of ordered
clauses, that is, in Step 6, the resolved literal of C_2 is not restricted
to the last literal of C_2.

c. Compare the number of clauses generated in each case.

13. Use the algorithm of Section 6.7 to prove the theorems given in (a)
Example 5.21, and (b) Example 5.22.

14. Show that if a set S of ordered clauses is unsatisfiable, then $\square$ can
be generated by the algorithm of Section 6.7.

Section 6.8

15. Let $S = \{P, Q, R, W, \sim P \vee \sim Q \vee \sim R \vee \sim W\}$.

a. If ordinary resolution is used, how many resolvents will be generated
from S by the level-saturation method before $\square$ can be generated?

b. Index the literals of S. If lock resolution is used, how many lock resolvents will be generated from S by the level-saturation method before $\square$ can be generated?

c. Compare the number of clauses generated in each case.

16. Consider the set S of clauses given in Example 6.18. Let us index literals of S as follows:

(1) $\quad {}_7P \vee {}_1Q$ (2) $\quad {}_8Q \vee {}_2R$

(3) $\quad {}_9R \vee {}_3W$ (4) $\quad {}_{10}\sim R \vee {}_4\sim P$

(5) $\quad {}_{11}\sim W \vee {}_5\sim Q$ (6) $\quad {}_{12}\sim Q \vee {}_6\sim R.$

Deduce the empty clause $\square$ from S by lock resolution.

17. Consider the set of clauses (this set is taken from Example 5.22):

(1) $\quad {}_1\sim E(x) \vee {}_2V(x) \vee {}_3S(x, f(x))$

(2) $\quad {}_4\sim E(x) \vee {}_5V(x) \vee {}_6C(f(x))$

(3) $\quad {}_7P(a)$

(4) $\quad {}_8E(a)$

(5) $\quad {}_9\sim S(a, y) \vee {}_{10}P(y)$

(6) $\quad {}_{11}\sim P(x) \vee {}_{12}\sim V(x)$

(7) $\quad {}_{13}\sim P(x) \vee {}_{14}\sim C(x).$

Derive the empty clause $\square$ from the above set by lock resolution.

Section 6.9

18. Prove Lemma 6.2.
19. For any arbitrary set S of clauses,

(a) give a method to index every literal in S so that the set of all lock resolvents from S is equal to the set of all ordinary (unrefined) resolvents from S, that is, simulate resolution by lock resolution;

(b) simulate positive hyperresolution by lock resolution.

20. Give a counterexample to show that the combination of the deletion strategy and lock resolution is not complete.

21. Give a counterexample to show that the combination of the set-of-support strategy and lock resolution is not complete.

Chapter 7

Linear Resolution

7.1 INTRODUCTION

When proving an identity, we often start with the left-hand side of the identity, apply an inference rule to it to obtain some other expression, then apply some inference rule again to the expression we have just obtained until the left-side expression is identical to the right-side expression of the identity. The idea of linear resolution is similar to this kind of chain reasoning. It starts with a clause, resolves it against a clause to obtain a resolvent, and resolves this resolvent against some clause until the empty clause □ is obtained.

The special appeal of linear deduction is its simple structure. Furthermore, linear resolution is complete and is compatible with the set-of-support strategy. In addition, as will be shown, heuristic methods can be conveniently applied with it.

Linear resolution was independently proposed by Loveland [1970a] and Luckham [1970]. It was later strengthened by Anderson and Bledsoe [1970], Yates *et al.* [1970], Reiter [1971], Loveland [1972], and Kowalski and Kuehner [1971]. In this chapter, we shall consider the version given by Kowalski and Kuehner [1971] and Loveland [1972] because it can be easily implemented on a digital computer.

7.2   LINEAR RESOLUTION

Definition   Given a set S of clauses and a clause C_0 in S, a *linear deduction* of C_n from S with *top clause* C_0 is a deduction of the form shown in Fig. 7.1 where

1. for $i = 0, 1, ..., n-1, C_{i+1}$ is a resolvent of C_i (called a *center clause*) and B_i (called a *side clause*), and
2. each B_i is either in S, or is a C_j for some $j, j < i$.

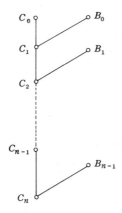

Figure 7.1

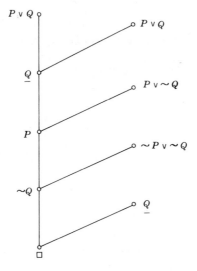

Figure 7.2

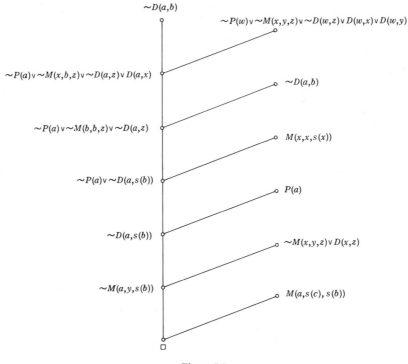

Figure 7.3

Example 7.1

Let $S = \{P \vee Q, \sim P \vee Q, P \vee \sim Q, \sim P \vee \sim Q\}$. Then Fig. 7.2 is a linear deduction of $\square$ from S with top clause $P \vee Q$. There are four side clauses. Three of them are from S, while one of them, namely, Q (underlined), is a resolvent derived before $\square$ is deduced.

Example 7.2

Consider the set of clauses $S = \{M(a, s(c), s(b)), \quad P(a), \quad M(x, x, s(x)), \sim M(x, y, z) \vee M(y, x, z), \quad \sim M(x, y, z) \vee D(x, z), \quad \sim P(x) \vee \sim M(y, z, u) \vee \sim D(x, u) \vee D(x, y) \vee D(x, z), \sim D(a, b)\}$. Figure 7.3 is a linear deduction of $\square$ from S.

The linear deduction introduced in this section is a primitive one and is later modified to the *OL-deduction* discussed in Section 7.4. The proof of the completeness of linear resolution is given in Section 7.5.

7.3 INPUT RESOLUTION AND UNIT RESOLUTION

When we consider a refinement of resolution, we would like it to be complete, that is, to guarantee that the empty clause can always be derived from an unsatisfiable set of clauses. However, efficiency is also important in mechanical theorem proving. Sometimes, we may like to trade completeness for efficiency. That is, there may be some refinements of resolution that are efficient but incomplete. If a refinement of resolution is efficient and powerful enough to prove a large class of theorems, even though it is not complete, it may still be useful. In this section, we shall consider some incomplete but efficient refinements of resolution, namely, input resolution and unit resolution. Input resolution is a subcase of linear resolution. It is much easier to implement and, although not complete, is more efficient than linear resolution. We shall show that input resolution is equivalent to unit resolution. That is, theorems that can be proved by input resolution can also be proved by unit resolution, and vice versa.

Given a set S of clauses, since S is the original input set, we shall call each member of S an *input clause*.

Definition   An *input resolution* is a resolution in which one of the two parent clauses is an input clause. An *input deduction* (to emphasize the input set S, we sometimes shall say *S-input deduction*) is a deduction in which every resolution is an input resolution. An *input refutation* is an input deduction of $\Box$ from S.

An input deduction is actually a linear deduction in which every side clause is an input clause, and hence is a subcase of linear deduction.

Example 7.3

Consider the following set of clauses:

$\sim P(x, y, u) \lor \sim P(y, z, v) \lor \sim P(x, v, w) \lor P(u, z, w)$

$P(g(x, y), x, y)$

$P(x, h(x, y), y)$

$\sim P(k(x), x, k(x))$.

Figure 7.4 shows an input refutation from this set.

Definition   A *unit resolution* is a resolution in which a resolvent is obtained by using at least one unit parent clause, or a unit factor of a parent clause. A *unit deduction* is a deduction in which every resolution is a unit resolution. A *unit refutation* is a unit deduction of $\Box$.

Unit resolution is essentially an extension of the one-literal rule of Davis and Putnam (see Section 4.6) to the first-order logic. This rule is especially important, for in order to deduce $\Box$ from a given set of clauses, one must obtain successively shorter clauses, and unit resolution provides a

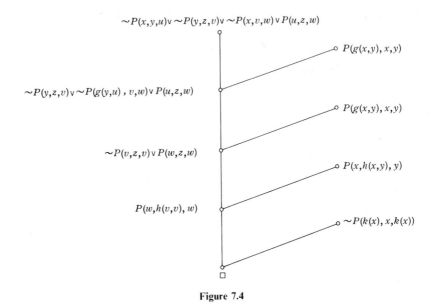

Figure 7.4

means for progressing rapidly toward shorter clauses. Unit resolution was used intensively by Wos, Carson, and Robinson [1964]. We now prove that unit resolution and input resolution are equivalent [Chang, 1970].

Lemma 7.1   There is a unit refutation from a set S of ground clauses if and only if there is an input refutation from S.

Proof   Lemma 7.1 is proved by induction. Let A be the atom set of S. If A consists of a single element, say Q, then among the elements of S, there exist unit clauses Q and $\sim Q$. Clearly, the resolvent of Q and $\sim Q$ is the empty clause $\square$. This deduction is both a unit and an input refutation. Hence, Lemma 7.1 holds for this case. Assume Lemma 7.1 holds when A consists of i elements, $1 \leqslant i \leqslant n$. To complete the induction, we consider A such that A consists of exactly $n+1$ elements.

($\Rightarrow$)   If there is a unit refutation from S, then S must contain at least one unit clause, say L, where L is a literal. Let S' be that set obtained from S by deleting those clauses containing the literal L and by deleting $\sim L$ from the remaining clauses. Clearly, since there is a unit refutation from S, there must be a unit refutation from S'. (Proof of this is left as an exercise for the reader.) But S' contains n or fewer than n atoms. By the induction hypothesis, there is an S'-input refutation D' from S'. Let D be the deduction obtained from D' by putting the literal $\sim L$ back into the clauses from which it was deleted. Let T be the clause attached to the root node of D. Clearly, D is an

S-input deduction of T from S. T must be either $\square$ or $\sim L$. If T is $\square$, we are done. If T is $\sim L$, we can obtain $\square$ by resolving T with L, which is an input clause. Thus, the refutation obtained from D and the resolution of T and L is an input refutation from S. Therefore we have completed the proof of the first part of Lemma 7.1.

($\Leftarrow$)   Conversely, if there is an input refutation from S, then S must contain at least one unit clause, say L, where L is a literal.

Let

$$S' = \{S \cup [\mathrm{Res}(C, L) \mid C \in S]\} - \{\text{all clauses that contain } L \text{ or } \sim L\}$$

where $\mathrm{Res}(C, L)$ denotes a resolvent of a clause C and the clause L. Since there is an S-input refutation from S, there must be an S'-input refutation from S'. But S' contains n or fewer than n atoms. By the induction hypothesis, there is a unit refutation from S'. However, every clause in S' is either a member of S or a resolvent obtained by applying a unit resolution on the clause L and a clause in S. Therefore, there is a unit refutation from S. This completes the proof of the second half of Lemma 7.1.

Theorem 7.1   (Equivalence of Unit and Input Resolutions)   There is a unit refutation from a set S of clauses if and only if there is an input refutation from S.

Proof   ($\Rightarrow$)   If there is a unit refutation D_1 from S, then from D_1 we can obtain a ground unit refutation D_1' by replacing the clause C attached to each node of D_1 by an appropriate ground instance of C. Let S' be the set of ground clauses attached to the initial nodes of D_1'. By Lemma 7.1, there is an input refutation D_2' from S'. From D_2', using the lifting lemma (Lemma 5.1), we can obtain an input refutation D_2 from S.

($\Leftarrow$)   The proof of the second half of Theorem 7.1 is the same as the above proof except that the words "unit" and "input" are interchanged.   Q.E.D.

Given a set S of clauses, if S has an input refutation, then by Theorem 7.1 we can also derive $\square$ from S by applying unit resolution. Since unit resolution is easier to program than input resolution, we have implemented only unit resolution. A LISP program implementing unit resolution is described in Appendix A. This program has been shown to be quite efficient.

7.4   LINEAR RESOLUTION USING ORDERED CLAUSES AND THE INFORMATION OF RESOLVED LITERALS

We can strengthen linear resolution by introducing two concepts into it. One of them is the concept of ordered clauses. As discussed in Chapter 6, incorporation of this concept into semantic resolution greatly increases its efficiency. We shall discover later that this is also true for linear resolution. Furthermore, unlike semantic resolution, incorporation of the concept of ordered clauses into linear resolution will *not* destroy its completeness. Another concept is one that uses the information of literals resolved upon. In resolution, when a resolvent is obtained, literals resolved upon are deleted. Actually, the information provided by these literals is very useful. As discovered by Loveland [1968, 1969a, b, 1972] and Kowalski and Kuehner [1971], this information can be used to improve linear resolution.

Before introducing the mechanism of using the information of resolved literals, let us first examine a rather simple case. Consider a set S of clauses defined as $S = \{P \vee Q, P \vee \sim Q, \sim P \vee Q, \sim P \vee \sim Q\}$. Fig. 7.5 shows a linear refutation from S.

Note that one of the side clauses (clause P) is not an input clause. In fact it is easy to see that there is no unit proof for this set of clauses. Thus, according to Theorem 7.1, there is no input proof either. That is, the usage of a center clause as a side clause is inevitable.

It would be nice if we could find a necessary and sufficient condition under which a side clause must be a center clause generated previously. That is, a side clause is a previously generated center clause when and only when the condition is satisfied.

We shall show that if the information of resolved literals is appropriately

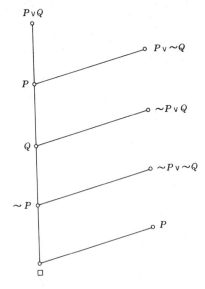

Figure 7.5

recorded and the concept of ordered clauses is used, we can clearly define this condition. It will be obvious later that using this condition will cut down the number of possible resolutions. Actually, it can also allow us to do more. If we have determined that a side clause must be a center clause previously generated, we do not even have to determine which center clause to use. We simply perform an operation on the clause and obtain a new clause. Therefore, we do not have to store previous center clauses in the computer memory. This greatly simplifies the computer implementation of linear resolution.

We now discuss the mechanism of recording the information of resolved literals. Suppose C_1 and C_2 are two ordered clauses given as

$$C_1 = P \vee Q$$
$$C_2 = \sim Q \vee R.$$

There is an ordered resolvent (namely, $P \vee R$) of C_1 against C_2, with Q and $\sim Q$ being the literals resolved upon. Since Q and $\sim Q$ are complementary to each other, we need only record one of them. Suppose we agree to record Q, the last literal of C_1. Then the ordered resolvent can be represented by $P \vee \boxed{Q} \vee R$, where the framed literal is the literal resolved upon. This is how we store the information of resolved literals. That is, literals resolved upon are indicated by framed literals in an ordered clause. The framed literals are merely for recording those literals that have been resolved upon; they do not participate in resolution.

In the above scheme, if a framed literal is not followed by any unframed literal, we shall delete this framed literal. That we can do this will be explained later.

The algorithm that employs both the concept of ordered clauses and the information of resolved literals is called OL-deduction (ordered linear deduction). Before presenting the precise algorithm, we first describe it by an example.

Example 7.4

Consider the set of ordered clauses defined as $S = \{P \vee Q, P \vee \sim Q, \sim P \vee Q, \sim P \vee \sim Q\}$. A linear refutation using the information of resolved literals and the ordered clause concept is shown in Fig. 7.6, where the center clauses are obtained as follows:

1. We start with the ordered clause $P \vee Q$ as the top clause. The last literal of this ordered clause is Q, which can be resolved with $P \vee \sim Q$ upon $\sim Q$. Therefore, resolving $P \vee Q$ against $P \vee \sim Q$ and recording the literal Q, we obtain an ordered resolvent that is represented by $P \vee \boxed{Q}$. Since $\boxed{Q}$ is not followed by any unframed literal, it is deleted. Therefore, we obtain P, as shown in Fig. 7.6.

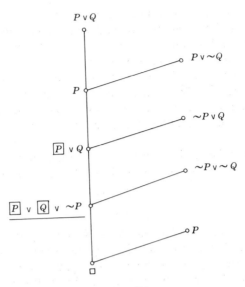

Figure 7.6

2. The last literal of the ordered clause obtained in (1) is P, which can be resolved against $\sim P \vee Q$ upon $\sim P$. Therefore, resolving P against $\sim P \vee Q$ and framing the literal P, we obtain $\boxed{P} \vee Q$.

3. The last literal of $\boxed{P} \vee Q$ is Q, which can be resolved with $\sim P \vee \sim Q$. Thus, resolving $\boxed{P} \vee Q$ against $\sim P \vee \sim Q$ and recording Q, we obtain $\boxed{P} \vee \boxed{Q} \vee \sim P$.

4. Now we have obtained the clause $\boxed{P} \vee \boxed{Q} \vee \sim P$. This clause has an interesting feature. Note that the last literal of $\boxed{P} \vee \boxed{Q} \vee \sim P$ is $\sim P$, which is complementary to one of its framed literals, namely $\boxed{P}$. This is the condition that we were trying to find before. In other words, the side clause at this step must be a center clause. In this case, the clause is P. Resolving $\boxed{P} \vee \boxed{Q} \vee \sim P$ with P upon $\sim P$, we obtain $\boxed{P} \vee \boxed{Q} \vee \boxed{\sim P}$. However, none of these framed literals is followed by unframed literals. Therefore, deleting them, we obtain $\square$.

As emphasized before, the clause $\boxed{P} \vee \boxed{Q} \vee \sim P$ is a special one because the last literal of this clause is complementary to one of its framed literals. Every time this kind of clause is generated, we shall use a center clause as the side clause. We shall call this kind of clause a reducible clause.

Definition   An ordered clause C is a *reducible ordered clause* if and only if the last literal of C is unifiable with the negation of a framed literal of C.

Actually, whenever a reducible ordered clause is generated, we do not have to search the memory for a deduced center clause to resolve it with. Instead, we may simply delete the last literal from this ordered clause. For example, when $\boxed{P} \vee \boxed{Q} \vee \sim P$ is generated, we simply delete $\sim P$ from $\boxed{P} \vee \boxed{Q} \vee \sim P$ and obtain $\boxed{P} \vee \boxed{Q}$. Since $\boxed{P}$ and $\boxed{Q}$ are not followed by any unframed literal, they are deleted and thus we obtain $\square$. We shall call this kind of operation the *reduction* of a reducible ordered clause.

Definition Let C be a reducible ordered clause. Let the last literal L be unifiable with the negation of some framed literal with a most general unifier σ. The *reduced ordered clause of C* is the ordered clause obtained from $C\sigma$ by deleting $L\sigma$ and every subsequent framed literal not followed by any unframed literal.

Once the reduction of reducible clauses is incorporated into OL-deduction, we do not have to store intermediate clauses any more. This important aspect of OL-deduction makes it very suitable for computer implementation. Note that the detection of a reducible ordered clause already narrows down the choice of resolutions. The introduction of the reduction mechanism effectively reduces it to only one resolution.

OL-deduction is essentially the same as what Loveland [1968, 1969a, b, 1972] has called the model elimination method. OL-deduction is also essentially a special *SL*-deduction of Kowalski and Kuehner [1971]. We now formally define OL-deduction. In the following, an ordered clause may contain both framed and unframed literals. To take this into consideration, we repeat the definitions of an ordered factor, an ordered binary resolvent, and an ordered resolvent given in Section 6.6.

In an ordered clause, if there is more than one occurrence of the same unframed literal, we always keep only the leftmost one and delete the other identical unframed literals. This operation is called *merging left* for identical unframed literals. For instance, merging left for identical unframed literals in $P \vee Q \vee \boxed{R} \vee S \vee Q \vee P$, we obtain $P \vee Q \vee \boxed{R} \vee S$.

Definition If two or more unframed literals (with the same sign) of an ordered clause C have a most general unifier σ, the ordered clause obtained from the sequence $C\sigma$ by merging left for any identical literals in $C\sigma$ and by deleting every framed literal not followed by an unframed literal in the remaining clause is called an *ordered factor* of C.

Definition Let C_1 and C_2 be two ordered clauses with no variables in common and L_1 and L_2 be two unframed literals in C_1 and C_2, respectively. Let L_1 and $\sim L_2$ have a most general unifier σ. Let C^* be the ordered clause obtained by concatenating the sequence $C_1\sigma$ and $C_2\sigma$, framing $L_1\sigma$, removing $L_2\sigma$, and merging left for any identical unframed literals in the remaining sequence. Let C be obtained from C^* by removing every framed literal not followed by any unframed literal in C^*. C is called an *ordered binary resolvent* of C_1 against C_2. The literals L_1 and L_2 are called the *literals resolved upon*.

Definition   An *ordered resolvent* of an ordered clause C_1 against an ordered clause C_2 is any of the following ordered binary resolvents:

1. an ordered binary resolvent of C_1 against C_2;
2. an ordered binary resolvent of C_1 against an ordered factor of C_2;
3. an ordered binary resolvent of an ordered factor of C_1 against C_2;
4. an ordered binary resolvent of an ordered factor of C_1 against an ordered factor of C_2.

Definition   Given a set S of ordered clauses and an ordered clause C_0 in S, an OL-deduction of C_n from S with top ordered clause C_0 is a deduction of the form shown in Fig. 7.7, which satisfies the following conditions:

1. For $i = 0, 1, 2, \ldots, n-1, C_{i+1}$ is an ordered resolvent of C_i (called a *center ordered clause*) against B_i (called a *side ordered clause*); the literal resolved upon in C_i (or an ordered factor of C_i) is the *last* literal.

2. Each B_i is either an ordered clause in S or an instance of some $C_j, j < i$. B_i is an instance of some $C_j, j < i$, if and only if C_i is a reducible ordered clause. In this case, C_{i+1} is the reduced ordered clause of C_i.

3. No tautology is in the deduction.

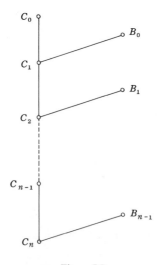

Figure 7.7

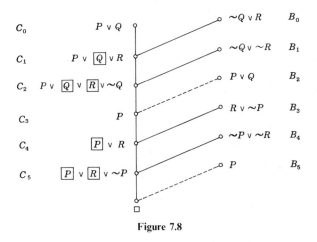

Figure 7.8

The definition of an ordered clause can be used to prove the following lemma. We leave the proof to the reader.

Lemma 7.2 In an OL-deduction, if C_i is a reducible ordered clause, then there exists a center ordered clause $C_j, j < i$, such that the reduced ordered clause C_{i+1} of C_i is an ordered resolvent of C_i against an instance of C_j.

Thus the conditions in the OL-deduction can also be stated as follows:

1. Every B_i is either in S or an instance of some $C_j, j < i$.
2. If C_i is a reducible ordered clause, then C_{i+1} is the reduced ordered clause of C_i. Otherwise, C_{i+1} is an ordered resolvent of C_i with B_i in S where the literal resolved upon in C_i is the *last* literal.
3. No tautology is in the deduction.

Definition An OL-refutation is an OL-deduction of $\Box$.

Example 7.5

Consider a set of ordered clauses $S = \{P \vee Q, \sim Q \vee R, R \vee \sim P, \sim Q \vee \sim R, \sim P \vee \sim R\}$. Then Fig. 7.8 is an OL-deduction of $\Box$ from S with top ordered clause $P \vee Q$. Note that C_2 and C_5 are reducible ordered clauses. Although no resolution is needed at those steps, the reader can find out what the side clauses would be by those indicated by dashed lines. For example, B_2 is $P \vee Q$, which is the top clause C_0, and B_5 is P, which is C_3 deduced in this refutation.

Example 7.6

In Example 7.2, we presented an example of a linear deduction without the concept of ordered clauses and the information of resolved literals. In Fig. 7.9, we prove that example again by using an OL-deduction.

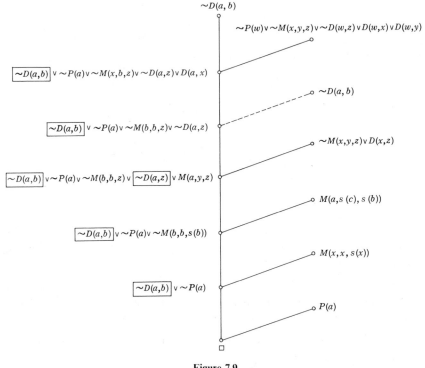

Figure 7.9

7.5   COMPLETENESS OF LINEAR RESOLUTION

In this section, we shall prove the completeness of OL-deduction. We first have to prove the following lemma.

Lemma 7.3   If C is a ground ordered clause in an unsatisfiable set S of ground ordered clauses, and if $S - \{C\}$ is satisfiable, then there exists an OL-refutation from S with top ordered clause C.

Proof   We prove this lemma by induction on the number of elements in the atom set of S. Let A be the atom set of S. If A consists of a single element, say Q, then among the elements of S there are ordered clauses Q and $\sim Q$. Clearly the resolvent of Q and $\sim Q$ is $\square$. Since $S - \{C\}$ is satisfiable, one of Q and $\sim Q$ must be C. Thus Lemma 7.3 holds for this case.

Assume Lemma 7.3 holds when A consists of i elements, $1 \leqslant i \leqslant n$. To complete the induction, we consider A such that A consists of exactly $n+1$ elements.

Case 1 C is a unit clause. Let $C \triangleq L$, where L is a literal. Let S' be the set obtained from S by deleting those ordered clauses containing L and by deleting $\sim L$ from the remaining ordered clauses. S' must be unsatisfiable. Let T' be an unsatisfiable subset of S' such that every proper subset of T' is satisfiable. (This subset can be obtained by exhaustively considering all possible subsets of S'.) T' must contain some ordered clause E' that was obtained from a clause in S by deleting the literal $\sim L$, for otherwise T' would be a subset of $S-\{C\}$. Thus $S-\{C\}$ would be unsatisfiable, violating the fact that it is satisfiable. We now have an ordered clause E' in the unsatisfiable set T' of clauses, and $T'-\{E'\}$ is satisfiable. Since T' contains n or fewer than n atoms, by the induction hypothesis there is an OL-refutation D' from T' with top ordered clause E' as shown in Fig. 7.10a. By adding the literal $\sim L$, in its proper place, back to all ordered clauses from which it was deleted except for the top most ordered clause E', and obtaining E' by resolving L against E, we obtain a deduction D of either $\square$ or $(\sim L)$ as shown in Fig. 7.10b.

In Fig. 7.10b, the symbol $(\vee \sim L)$ means the addition of the literal $\sim L$, in its proper place, back to the ordered clauses from which it was deleted. We now have to show that D is also an OL-deduction.

Note that no clause in D is a tautology since D is obtained from D', which contains neither a tautology nor the literal L. Note also that if $\sim L$ does not appear as the last literal of a center clause, then the resolution in D is exactly the same as that in D'. If $\sim L$ does appear as the last literal of a center clause, then $\sim L$ should be resolved upon. In this case, we replace that part of D shown in Fig. 7.11a by the deduction shown in Fig. 7.11b. It is obvious that $\boxed{L} \vee C_i \vee (\sim L)$ is reducible, and $\boxed{L} \vee C_i$ is obtained by the reduction operation.

After these replacements have been made whenever required in Fig. 7.10b,

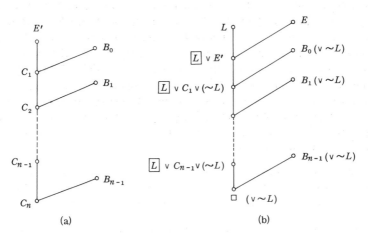

Figure 7.10

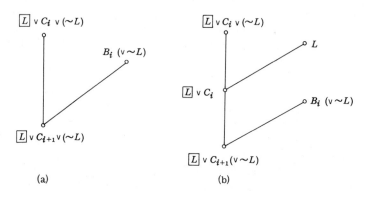

Figure 7.11

an OL-deduction D^* of $\square$ or $\boxed{L} \vee \sim L$ from S with top ordered clause C is obtained. If D^* is a deduction of $\square$, we are done. Otherwise, D^*, together with the resolution of $\boxed{L} \vee \sim L$ against L, is a desired OL-refutation from S with top ordered clause C.

Case 2 C is not a unit clause. In this case let L be the first (leftmost) literal in C, that is, $C \triangleq L \vee C'$, where C' is a nonempty ordered clause. Let S' be that set obtained from S by deleting those ordered clauses containing $\sim L$ and by deleting L from the remaining ordered clauses. S' is unsatisfiable. We shall show that $S' - \{C'\}$ is satisfiable. Let I be an interpretation that satisfies $S - \{C\}$. (Such an I exists because $S - \{C\}$ is satisfiable.) Since S is unsatisfiable, C must be false in I. Therefore, L is false in I. Consequently $S' - \{C'\}$ is true in I. Thus $S' - \{C'\}$ is satisfiable. Since S' contains n or fewer than n atoms, by the induction hypothesis there is an OL-deduction D' of $\square$ from S' with top ordered clause C'. Adding the literal L, in its proper place, to all ordered clauses from which it was deleted, we obtain an OL-deduction D_1 of L from S with top ordered clause C. Now $\{L\} \cup (S - \{C\})$ is unsatisfiable and $S - \{C\}$ is satisfiable. As proved in Case 1, there is an OL-deduction D_2 of $\square$ from $\{L\} \cup (S - \{C\})$ with top ordered clause $\{L\}$. Putting D_1 on top of D_2, we obtain an OL-refutation from S with top ordered clause C. This completes the proof.

Theorem 7.2   (Completeness of OL-deduction).   If C is an ordered clause in an unsatisfiable set S of ordered clauses and if $S - \{C\}$ is satisfiable, then there is an OL-refutation from S with top ordered clause C.

Proof Since S is unsatisfiable and $S - \{C\}$ is satisfiable, by Herbrand's theorem (version II) there is a finite set S' of ground instances of ordered clauses in S and a ground instance C' of C such that S' is unsatisfiable, C' is in S', and $S' - \{C'\}$ is satisfiable. By Lemma 7.3, there is an OL-refutation D' from S' with top ordered clause C'. Using the lifting lemma (similar to Lemma 5.1), from D' we can obtain an OL-refutation from S with top ordered clause C. This completes the proof.

It can be proved easily that set-of-support resolution is implied by OL-resolution. That is, Theorem 7.2 implies Theorem 6.2. We leave this proof as an exercise for the reader.

7.6   LINEAR DEDUCTION AND TREE SEARCHING

In the remaining sections, we shall discuss a very important problem of linear deduction, namely, how to implement linear deduction efficiently.

Let us assume that we have chosen a top clause C_0. We then can single out all the possible side clauses. After resolving C_0 with these possible side clauses, we obtain resolvents $R_1, \ldots, R_m$. Every $R_i, 1 \leq i \leq m$, is a possible center clause that may lead to a proof. If some R_i is the empty clause, we are done. Otherwise, for every i, we find all possible side clauses that can be resolved with R_i and continue this process until an empty clause is generated. In order to describe the above process conveniently, a linear deduction tree such as the one shown in Fig. 7.7 will be represented by a path as shown in Fig. 7.12. That is, a side clause B_i is attached to an arc of the path, and C_i is a resolvent of C_{i-1} and B_{i-1}. In the sequel, we shall implement the OL-deduction.

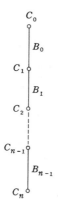

Figure 7.12

Let us first consider a simple example. (In this example, tautologies are included for illustrative purposes.) Consider the following set of clauses:

(1) $P \vee Q$

(2) $\sim P \vee Q$

(3) $P \vee \sim Q$

(4) $\sim P \vee \sim Q$.

Let us choose clause (1) to be the top clause C_0. The ordered clauses that can be the side clauses of C_0 are (3) and (4). Resolving C_0 with clauses (3) and (4), we obtain the tree shown in Fig. 7.13. We have generated two

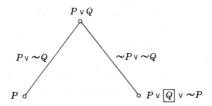

Figure 7.13

center clauses. For each such clause, we can single out all possible side clauses to be resolved and further develop the tree in Fig. 7.13 to that in Fig. 7.14.

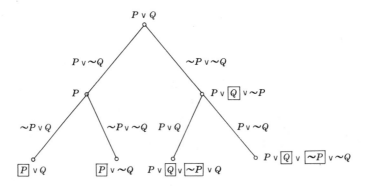

Figure 7.14

When an ordered resolvent is generated, we always check whether it is reducible. If it is, we always reduce it to the reduced ordered clause. In Fig. 7.14, $P \vee \boxed{Q} \vee \boxed{\sim P} \vee \sim Q$ is reducible. Its reduced ordered clause is P.

Therefore, the tree in Fig. 7.14 can be simplified to the tree shown in Fig. 7.15.

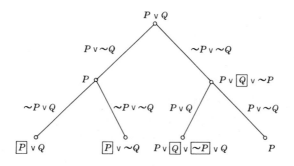

Figure 7.15

Now there are four possible center clauses. Again, for every center clause, we can single out all possible clauses to resolve. Thus, we obtain the tree shown in Fig. 7.16. The ordered resolvents $\boxed{P} \vee \boxed{Q} \vee \sim P$ and

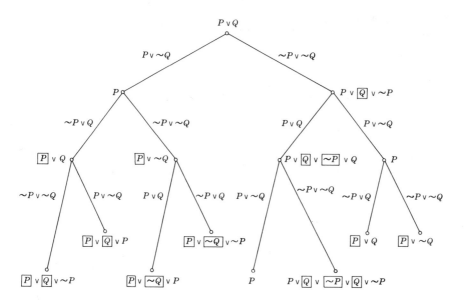

Figure 7.16

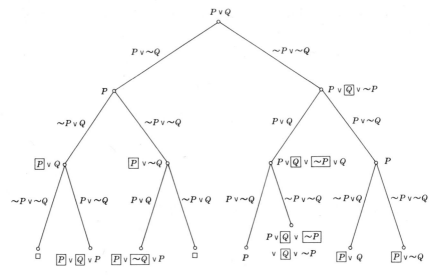

Figure 7.17

$\boxed{P} \vee \boxed{\sim Q} \vee \sim P$ are reducible. Reducing them, we obtain the tree shown in Fig. 7.17. Since the empty clause has been generated, the process is terminated. The leftmost branch of the tree in Fig. 7.17 corresponds to the OL-refutation shown in Fig. 7.18.

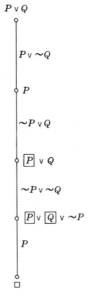

Figure 7.18

The reader can now see that the problem of finding an OL-refutation can be viewed as a tree-searching problem. Many techniques developed for tree searching can be applied to OL-deduction. Thanks to the persistent effort of researchers in tree searching, we have seen great progress made in this field. For a comprehensive study of general tree searching techniques, see [Slagle, 1971] and [Nilsson, 1971].

In the above example, the method of developing the tree in Fig. 7.17 is called the *breadth-first* method. In tree-searching terminology, C_0 is the *top node* of the tree. The side clauses may be considered as *operators*, which can be used to develop the successor nodes of a node. Let C be an ordered clause that denotes a node. Single out all the possible side clauses and resolve them with C to obtain ordered resolvents $R_1, ..., R_m$. $R_1, ..., R_m$ will be called the *immediate successor nodes* of C. When all the possible immediate successor nodes of a node C are generated, we say that node C has been *expanded*. That is, expanding a node means generating all its immediate successor nodes.

Suppose that S is a set of ordered clauses to be proved. Suppose that C_0 is an ordered clause in S that is selected to be the top clause. Then, the breadth-first method is described as follows.

The Breadth-First Method

Step 1   Let CLIST $= (C_0)$.

Step 2   If CLIST is empty, terminate without a proof. Otherwise, continue.

Step 3   Let C be the first ordered clause in CLIST. Delete C from CLIST.

Step 4   Find all the ordered clauses in S that can be side clauses of C. If no such side clauses exist, go to Step 2. Otherwise, resolve C with all these side clauses. Let $R_1, ..., R_m$ denote these ordered resolvents. Let R_i^* be the reduced ordered clause of R_i if R_i is reducible. If R_i is not reducible, let $R_i^* = R_i$.

Step 5   If some R_q^* is an empty clause, $1 \leqslant q \leqslant m$, terminate with a proof; otherwise, continue.

Step 6   Put $R_1^*, ..., R_m^*$ (in an arbitrary order) at the end of CLIST and go to Step 2.

A minimal proof from S with top clause C_0 is an OL-refutation (from S with top clause C_0) that involves the smallest number of resolutions. The breadth-first method will always find such a minimal proof if one exists. However, obviously, many irrelevant and redundant clauses may be generated by this breadth-first method before the empty clause is found. This is evident in the above example. Another way to search a tree is the *depth-first* method. The idea is that, unlike the breadth-first method which expands nodes from top to bottom, the depth-first method expands nodes from left to right. The depth-first method is almost the same as the breadth-first method except that in Step 5 we put $R_1*, \ldots, R_m*$ at the beginning of CLIST, instead of at the end of CLIST. In addition, we must provide a *depth bound* for the depth-first method in order to prevent us from expanding nodes along a wrong path (branch), that is, a path that cannot lead to the empty clause. The depth of a clause in an OL-deduction is defined as follows.

Definition   In an OL-deduction with top ordered clause C_0, the *depth of* C_0 is 0. If the depth of a certain ordered clause C is k and R is an ordered resolvent of C and some side clause, then the *depth of* R is $(k+1)$. The *length* of a proof (refutation) with top clause C_0 is the depth of the empty clause.

Let $d*$ be a threshold number specified beforehand. The depth-first method is now described as follows:

The Depth-First Method

Step 1   Let CLIST $= (C_0)$.

Step 2   If CLIST is empty, terminate without a proof. Otherwise, continue.

Step 3   Let C be the first ordered clause in CLIST. Delete C from CLIST. If the depth of C is greater than $d*$, go to Step 2. Otherwise, continue.

Step 4   Find all the ordered clauses in S that can be side clauses of C. If no such side clauses exist, go to Step 2. Otherwise, resolve C with all of its side clauses. Let $R_1, \ldots, R_m$ denote these ordered resolvents. Let R_i* be the reduced ordered clause of R_i if R_i is reducible. Otherwise, let $R_i* = R_i$.

Step 5   If some R_q* is an empty clause, $1 \leq q \leq m$, terminate with a proof. Otherwise, continue.

Step 6   Put $R_1*, \ldots, R_m*$ (in an arbitrary order) at the beginning of CLIST and go to Step 2.

It is obvious that applying the depth-first method with $d* = 2$ on the tree in Fig. 7.17 will generate the tree shown in Fig. 7.19. Clearly, the tree in Fig. 7.19 is smaller than the one in Fig. 7.17.

Although, in general, the depth-first method searches a smaller tree than the breadth-first method does, we would like to improve it still futher. We notice that in the depth-first method, when a node is chosen for expansion, it is completely expanded. That is, all the possible successor nodes of the node are generated. However, we can modify this and generate only one successor node at a time. Let us consider Fig. 7.19. The top node of the

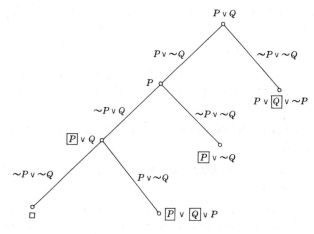

Figure 7.19

tree is $P \vee Q$. There are two successor nodes, namely, P and $P \vee \boxed{Q} \vee \sim P$, of this top node. Suppose only P is generated. Now, P is a center clause. There are two successor nodes of node P. Suppose only $\boxed{P} \vee Q$ is generated. Again, $\boxed{P} \vee Q$ has two successor nodes, namely $\square$ and $\boxed{P} \vee \boxed{Q} \vee P$. If only $\square$ is generated, we obtain an empty clause. The entire process is depicted by the tree shown in Fig. 7.20, which is even smaller than the tree in Fig. 7.19. Because of the above consideration, we give a modified depth-first method as follows, where d^* is a prespecified threshold number.

A Modified Depth-First Method

Step 1 Find all ordered clauses in S that can be side clauses of C_0. If no such side clauses exist, terminate without a proof. Otherwise, let $B_0{}^1, \ldots, B_0{}^r$

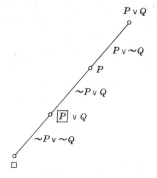

Figure 7.20

be all these side clauses. Construct pairs $(C_0, B_0{}^1), ..., (C_0, B_0{}^r)$. Let CLIST be a list of all these pairs arranged in an arbitrary order.

Step 2   If CLIST is empty, terminate without a proof. Otherwise, continue.

Step 3   Let (C, B) be the first pair in CLIST. Delete (C, B) from CLIST. If the depth of C is greater than d^*, go to Step 2. Otherwise, continue.

Step 4   Resolve C with B. Let $R_1, ..., R_m$ denote all the ordered resolvents of C against B. Let $R_i{}^*$ be the reduced ordered clause of R_i if R_i is reducible. Otherwise, let $R_i{}^* = R_i$.

Step 5   If some $R_q{}^*$ is the empty clause, $1 \leqslant q \leqslant m$, terminate with a proof. Otherwise, continue.

Step 6   For each $i = 1, 2, ..., m$, find ordered clauses in S that can be side clauses of $R_i{}^*$. If no such side clauses exist, delete $R_i{}^*$. Otherwise, let $B_{i1}, B_{i2}, ..., B_{is_i}$ be all these side clauses. Construct pairs $(R_i{}^*, B_{i1}), ...,$ $(R_i{}^*, B_{is_i})$. Put all these pairs (in an arbitrary order) at the beginning of CLIST and go to Step 2.

In most situations, both the ordinary and the modified depth-first methods are better than the breadth-first method. If the set S of ordered clauses has a proof with a length less than d^*, both versions of the depth-first method guarantee finding a proof. Although the modified depth-first method is more difficult to implement than the ordinary depth-first method, it is clearly superior. In the next section, we assume that the modified depth-first method is used.

7.7   HEURISTICS IN TREE SEARCHING

In this section, we shall discuss "heuristics" that may be used with the modified depth-first method to speed up the searching. The use of a heuristic may not guarantee finding a proof even though there is one. Nevertheless, it is often a necessary tool for proving a large or difficult theorem, that is, a theorem with a long proof. There are many heuristics in the field of theorem proving and we shall now discuss some of them. We note that, although the following heuristics are introduced to be used with the modified depth-first method, they can also be applied to other kinds of resolution, not necessarily only to linear resolution.

A   Deletion Strategy

An ordered clause C_1 is said to *subsume* another ordered clause C_2 if and only if the clause consisting of the unframed literals of C_1 subsumes the clause consisting of the unframed literals of C_2. An ordered clause is said to be a *tautology* if and only if it contains a complementary pair of unframed literals. Since tautologies and subsumed clauses are irrelevant and redundant, we should delete them whenever possible. In the modified depth-first method, at the end of Step 4 we should check whether R_i^* is a tautology, $i = 1, \dots, m$. If R_i^* is a tautology, it should be deleted. In addition, if there is a pair (C, B) in CLIST such that R_i^* is subsumed by C, then R_i^* should also be deleted.

B   The Fewest-Literal Preference Strategy

In the modified depth-first method, pairs of CLIST are arranged in an arbitrary order. Actually, we should arrange the pairs of CLIST in such a way that the potentially "good" pairs appear before the "bad" ones. Thus, in Step 3 of the method, when we take the first pair from CLIST, we can be sure that we have a good one that can lead to an empty clause promptly.

Let the length of an ordered clause E, denoted as length(E), be defined as the number of unframed literals in E. Suppose (C, B) is a pair in CLIST. Let R denote an ordered resolvent of C against B. Since length(R) indicates the minimum number of resolutions necessary to lead R to an empty clause, we may use length(R) as an indicator of the goodness of the pair (C, B). That is, the smaller the value of length(R), the better the pair (C, B). Actually, in practice, we do not have to calculate R in order to calculate length(R). A mere estimation of length(R) will be sufficient. Clearly,

$$\text{length}(R) \leqslant \text{length}(C) + \text{length}(B) - 2. \tag{7.1}$$

Moreover, since we only need a relative number, we may simply use (length(C)+length(B)) as a measurement of goodness. Let

$$f(C, B) = \text{length}(C) + \text{length}(B). \tag{7.2}$$

Then $f(C, B)$ will be used to evaluate the goodness of the pair (C, B). In the modified depth-first method, just before Step 3 we may arrange pairs of CLIST by putting pairs with small values of f before those with large values of f. This ordering scheme was proposed by Slagle [1965], and is called the *fewest-literal preference strategy*.

C   Use of Heuristic Evaluation Functions

In the above discussion, the only measurement of the goodness of a clause is the number of literals it contains. This is, of course, far too simple. For a pair (C, B) in CLIST of the modified depth-first method, let us define $h^*(C, B)$ to be the number of resolutions in a minimal proof (refutation) with top ordered clause C and with first side clause B. If $h^*(C, B)$ is known, then we could use $h^*(C, B)$ to order pairs of CLIST. However, $h^*(C, B)$ is generally not known. Therefore, $h^*(C, B)$ has to be estimated. Let $h(C, B)$ be an estimate of $h^*(C, B)$. Usually, $h(C, B)$ is obtained from a set of known data. In the sequel, we shall consider how to obtain an $h(C, B)$. Suppose $h(C, B)$ can be expressed as

$$h(C, B) = w_0 + w_1 f_1(C, B) + \cdots + w_n f_n(C, B) \tag{7.3}$$

where each $f_i, i = 1, \ldots, n$, is a real-valued function of C and B, and w_i is a weight associated with f_i. Each f_i is called a *feature* of the pair (C, B). Usually, features are selected ad hoc by the experimenter who believes that they are relevant to $h^*(C, B)$. For example, we may use some of the features suggested in Table 7.1. Actually, $h(C, B)$ can be a function, linear or non-

TABLE 7.1

1. number of unframed literals in C
2. number of framed literals in C
3. number of side clauses of C
4. number of constants in the last literal of C
5. number of function symbols in the last literal of C
6. number of framed literals of C, each of which subsumes the last literal of C
7. number of distinct variables in C and B
8. length(C) + length$(B) - 2$
9. number of constants in $C/(1 + $ number of variables of $C)$
10. number of constants in $C/(1 + $ number of distinct variables of $C)$
11. depth of C
12. number of unframed literals that are in both C and B
13. number of literals in B, each of which has a framed complement in C
14. number of distinct predicate symbols in C and B

linear, of $f_1(C, B), \ldots, f_n(C, B)$. For illustrative purposes, we shall only consider linear functions in the sequel. The function $h(C, B)$ may be used to order pairs of CLIST, that is, to put pairs with small values of h before those with large values of h. We shall call this $h(C, B)$ a *heuristic evaluation function*, or *evaluation function* for short. In the next section, we shall give a method for obtaining an $h(C, B)$.

7.8 ESTIMATIONS OF EVALUATION FUNCTIONS

As discussed in Section 7.7C, we assume that

$$h(C, B) = w_0 + w_1 f_1(C, B) + \cdots + w_n f_n(C, B).$$

In this section, we shall discuss the following two important questions:

1. Given a set of features, how can we determine the appropriate values of $w_0, \ldots, w_n$?

2. How can we know that a set of features is good (relevant to $h^*(C, B)$) or not?

We shall now answer the first question. Let us assume that we have collected a set of pairs, $(C_1, B_1), \dots, (C_q, B_q)$. Also assume that we know the values of $h^*(C_1, B_1), \dots, h^*(C_q, B_q)$. Then the problem is to determine $w_0, w_1, \dots, w_n$ such that

$$\sum_{i=1}^{q} [h^*(C_i, B_i) - (w_0 + w_1 f_1(C_i, B_i) + \cdots + w_n f_n(C_i, B_i))]^2 \quad (7.4)$$

is minimum. This is the so called *least-square estimation* [Draper and Smith, 1966]. Let the matrices F, H, and W be defined as

$$F = \begin{bmatrix} 1 & f_1(C_1, B_1) & \cdots & f_n(C_1, B_1) \\ 1 & f_1(C_2, B_2) & \cdots & f_n(C_2, B_2) \\ \vdots & & & \\ 1 & f_1(C_q, B_q) & \cdots & f_n(C_q, B_q) \end{bmatrix}$$

$$H = \begin{bmatrix} h^*(C_1, B_1) \\ h^*(C_2, B_2) \\ \vdots \\ h^*(C_q, B_q) \end{bmatrix}, \qquad W = \begin{bmatrix} w_0 \\ w_1 \\ \vdots \\ w_n \end{bmatrix}.$$

Then the W that minimizes the expression (7.4) is given by

$$W = (F' F)^{-1} F' H \qquad (7.5)$$

where F' is the transpose of F and $(F' F)^{-1}$ is the inverse of $(F' F)$. Substituting W into (7.3), an evaluation function can be obtained.

Example 7.7

Consider the tree shown in Fig. 7.17. Let us redraw the tree in Fig. 7.21, where each ordered clause is numbered. For this example, suppose we use the features defined as:

$f_1(C, B) = \text{length}(C) + \text{length}(B) - 2$

$f_2(C, B) = \text{number of framed literals in } C$

$f_3(C, B) = \text{number of unframed literals that are in both } C \text{ and } B$

$f_4(C, B) = \text{number of literals in } B, \text{ each of which has a framed complement in } C$

$f_5(C, B) = \text{number of framed literals in } C, \text{ each of which subsumes the last literal of } C.$

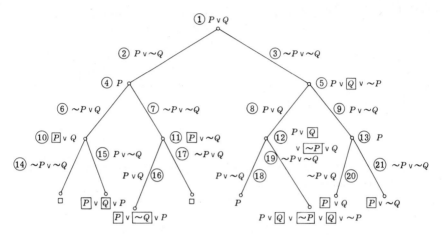

Figure 7.21

Calculating the values of the features of each pair in the tree of Fig. 7.21, we obtain Table 7.2.

TABLE 7.2

i	pair(C_i, B_i)	$f_1(C_i, B_i)$	$f_2(C_i, B_i)$	$f_3(C_i, B_i)$	$f_4(C_i, B_i)$	$f_5(C_i, B_i)$	$h^*(C_i, B_i)$
1	(1, 2)	2	0	1	0	0	3
2	(1, 3)	2	0	0	0	0	4
3	(4, 6)	1	0	0	0	0	2
4	(4, 7)	1	0	0	0	0	2
5	(5, 8)	2	1	1	0	0	4
6	(5, 9)	2	1	1	1	0	3
7	(10, 14)	1	1	0	1	0	1
8	(10, 15)	1	1	0	0	0	3
9	(11, 16)	1	1	0	0	0	3
10	(11, 17)	1	1	0	1	0	1
11	(12, 18)	2	2	1	2	1	3
12	(12, 19)	2	2	0	1	1	4
13	(13, 20)	1	0	0	0	0	2
14	(13, 21)	1	0	0	0	0	2

Computing formula (7.5) on a computer, we obtain

$$w_0 = 0.30 \qquad w_1 = 1.68$$
$$w_2 = 0.76 \qquad w_3 = -0.24 \qquad (7.6)$$
$$w_4 = -1.44 \qquad w_5 = 0.60.$$

Thus we have

$$h(C, B) = 0.30 + 1.68 f_1(C, B) + 0.76 f_2(C, B) - 0.24 f_3(C, B)$$
$$- 1.44 f_4(C, B) + 0.60 f_5(C, B). \tag{7.7}$$

In the modified depth-first method, just before Step 3, if we use the above function $h(C, B)$ to order pairs of CLIST, then the tree that is searched by the modified depth-first method will be the one shown in either Fig. 7.22a or b.

In the above example, we showed how to obtain an evaluation function. The evaluation function obtained in Example 7.7 is only for sets of *ground* clauses. It was based upon one example. Obviously, the more examples we use to estimate an evaluation function, the better it will be. Nevertheless, we shall now apply the evaluation function obtained in Example 7.7 to a new example to see how good the estimation is.

Example 7.8

Let us consider the following set of ordered clauses:

(1) $\sim P_1 \vee \sim P_2 \vee P_3$

(2) P_1

(3) P_2

(4) $\sim P_1 \vee \sim P_3$

(5) $\sim P_3 \vee P_4$

(6) $\sim P_3 \vee P_5$

(7) $\sim P_4 \vee P_5$.

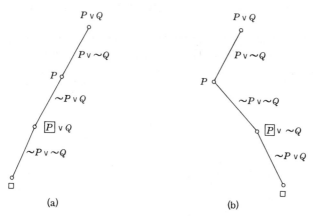

Figure 7.22

Let us use clause (1) as the top clause. There are three ordered clauses that can be side clauses of clause (1). They are clauses (4), (5), and (6). We now use $h(C, B)$ defined by Eq. (7.7) to calculate the following:

$$h(1, 4) = 0.30 + 1.68 \times (3 + 2 - 2) + 0.76 \times 0 - 0.24 \times 1$$
$$- 1.44 \times 0 - 0.60 \times 0 = 3.42$$
$$h(1, 5) = 0.30 + 1.68 \times (3 + 2 - 2) + 0.76 \times 0 - 0.24 \times 0$$
$$- 1.44 \times 0 - 0.60 \times 0 = 3.66$$
$$h(1, 6) = 0.30 + 1.68 \times (3 + 2 - 2) + 0.76 \times 0 - 0.24 \times 0$$
$$- 1.44 \times 0 - 0.60 \times 0 = 3.66.$$

Since $h(1, 4)$ is the smallest, we choose clause (4) as the side clause of clause (1).

(7) $\sim P_1 \vee \sim P_2$ from (1) and (4).

The rest of the proof is trivial. The reader can easily see that the following steps will complete the proof:

(8) $\sim P_1$ from (7) and (3)

(9) $\square$ from (8) and (2).

For this example, no irrelevant clauses were generated. We note that should the evaluation function make us choose clause (5) or (6) as a side clause of clause (1), irrelevant clauses would be generated. Thus, we have shown that the evaluation function is useful for this example.

We can now answer the second question: How can we know that a set of features is good? Evidently, if a set of features is good, then after using a large number of examples to estimate an evaluation function, the function should generalize to new examples. That is, the evaluation function should give good performance on new examples. If it does not generalize to new problems, then the experimenter should analyze the data and consider new features. For an excellent discussion of estimations of evaluations in theorem proving, the reader is encouraged to read [Slagle and Farrell, 1971].

REFERENCES

Anderson, R., and W. W. Bledsoe (1970): A linear format for resolution with merging and a new technique for establishing completeness, *J. Assoc. Comput. Mach.* **17** 525–534.

Chang, C. L. (1970): The unit proof and the input proof in theorem proving, *J. Assoc. Comput. Mach.* **17** 698–707.

Draper, N. R., and H. Smith (1966): "Applied Regression Analysis," Wiley, New York.

Kowalski, R., and D. Kuhner (1971): Linear resolution with selection function, *Artificial Intelligence* **2** 227–260.

Loveland, D. W. (1968): Mechanical theorem proving by model elimination, *J. Assoc. Comput. Mach.* **15** 236–251.

Loveland, D. W. (1969a): A simplified format for the model elimination theorem-proving procedure, *J. Assoc. Comput. Mach.* **16** 349–363.

Loveland, D. W. (1969b): Theorem provers combining model elimination and resolution, *in:* "Machine Intelligence," vol. 4 (B. Meltzer and D. Michie, eds.), American Elsevier, New York, pp. 73–86.

Loveland, D. W. (1970a): A linear format for resolution, *Proc. IRIA Symp. Automatic Demonstration*, Versailles, France, 1968, Springer-Verlag, New York, pp. 147–162.

Loveland, D. W. (1970b): Some linear Herbrand proof procedures: an analysis, Dept. of Computer Science, Carnegie-Mellon University, Dec. 1970.

Loveland, D. W. (1972): A unifying view of some linear Herbrand procedures, *J. Assoc. Comput. Mach.* **19** 366–384.

Luckham, D. (1970): Refinement theorems in resolution theory, *Proc. IRIA Symp. Automatic Demonstration*, Versailles, France, 1968, Springer-Verlag, New York, pp. 193–190.

Nilsson, N. J. (1971): "Problem-Solving Methods in Artificial Intelligence," McGraw-Hill, New York.

Reiter, R. (1971): Two results on orderings for resolution with merging and linear format, *J. Assoc. Comput. Mach.* **18** 630–646.

Slagle, J. R. (1965): A proposed preference strategy using sufficiency resolution for answering question, UCRL-14361, Lawrence Radiation Lab., Livermore, Calif.

Slagle, J. R. (1971): "Artificial Intelligence, the Heuristic Programming Approach," McGraw-Hill, New York.

Slagle, J. R., and C. D. Farrell (1971): Experiments in automatic learning for a multipurpose heuristic program, *Comm. ACM*, **14** 91–99.

Wos, L., D. E. Carson, and G. A. Robinson (1964): The unit preference strategy in theorem proving, *Proc. AFIPS 1964 Fall Joint Computer Conf.* **26** 615–621.

Yates, R., B. Raphael, and J. Hart (1970): Resolution graphs, *Artificial Intelligence* **1** 257–289.

EXERCISES

Section 7.2

1. Use linear resolution to prove that the following sets of clauses are unsatisfiable:

   a. $S = \{\sim P \vee Q, P \vee R, \sim Q, \sim R\}$

   b. $S = \{P \vee Q \vee R, \sim P, \sim Q, \sim R \vee W, \sim W\}$

   c. $S = \{P \vee Q, Q \vee R, R \vee W, \sim R \vee \sim P, \sim W \vee \sim Q, \sim Q \vee \sim R\}$.

2. Use linear resolution to prove the unsatisfiability of the following set of clauses:

   a. $\sim E(x) \vee V(x) \vee S(x, f(x))$ b. $\sim E(x) \vee V(x) \vee C(f(x))$

   c. $P(a)$ d. $E(a)$

   e. $\sim S(a, y) \vee P(y)$ f. $\sim P(x) \vee \sim V(x)$

   g. $\sim P(x) \vee \sim C(x)$.

   Use clause (g) as the top clause.

3. Prove the unsatisfiability of the following clauses by linear resolution. Use the last clause as the top clause.

   a. $P(x, I(x), e)$

   b. $\sim S(x) \vee \sim S(y) \vee \sim P(x, I(y), z) \vee S(z)$

   c. $S(a)$

   d. $\sim S(e)$.

Section 7.3

4. Prove the unsatisfiability of the following sets of clauses by input resolution.

   a.
   $$S = \begin{cases} P(I(x), x, e), \\ P(e, x, x), \\ \sim P(x, y, u) \vee \sim P(y, z, v) \vee \sim P(u, z, w) \vee P(x, v, w), \\ \sim P(x, y, u) \vee \sim P(y, z, v) \vee \sim P(x, v, w) \vee P(u, z, w), \\ \sim P(a, x, e). \end{cases}$$

   b.
   $$S = \begin{cases} D(x, x), \\ \sim D(x, y) \vee \sim D(y, z) \vee D(x, z), \\ P(x) \vee D(g(x), x), \\ P(x) \vee L(l, g(x)), \\ P(x) \vee L(g(x), x), \\ L(l, a), \\ \sim P(x) \vee \sim D(x, a), \\ \sim L(l, x) \vee \sim L(x, a) \vee P(f(x)), \\ \sim L(l, x) \vee \sim L(x, a) \vee D(f(x), x). \end{cases}$$

5. Consider Exercise 4 again. This time use unit resolution.
6. Let S be an unsatisfiable set of ground clauses. Let L be a unit clause in S. Let S' be that set obtained from S by deleting those clauses containing L and by deleting $\sim L$ from the remaining clauses. Prove that if there is a unit refutation from S, there must be a unit refutation from S'.

Section 7.4

7. Prove the unsatisfiability of the following set of ordered clauses by OL-deduction:

$P(e, x, x)$

$P(x, e, x)$

$P(x, I(x), e)$

$P(I(x), x, e)$

$S(b)$

$\sim S(x) \vee \sim S(y) \vee \sim P(x, I(y), z) \vee S(z)$

$\sim P(x, y, u) \vee \sim P(y, z, v) \vee \sim P(x, v, w) \vee P(u, z, w)$

$\sim P(x, y, u) \vee \sim P(y, z, v) \vee \sim P(u, z, w) \vee P(x, v, w)$

$\sim S(I(b))$.

8. Prove the unsatisfiability of the following set of clauses by OL-deduction:

$P(I(x), x, e)$

$P(e, x, x)$

$\sim P(x, y, u) \vee \sim P(y, z, u) \vee \sim P(u, z, w) \vee P(x, v, w)$

$\sim P(x, y, u) \vee \sim P(y, z, v) \vee \sim P(x, v, w) \vee P(u, z, w)$

$\sim P(a, e, a)$.

9. Prove Lemma 7.2.
10. Prove that Theorem 7.2 implies Theorem 6.2.

Section 7.6

11. Consider the following set of ordered clauses:

$$S = \{P \vee Q, P \vee \sim Q \vee R, P \vee \sim Q \vee \sim R, \sim P \vee R, \sim P \vee \sim R\}.$$

Let $P \vee Q$ be the top ordered clause. Let $d^* = 6$. Prove that S is unsatisfiable by (a) the breadth-first method, (b) the depth-first method, and (c) the modified depth-first method.

Section 7.8

12. Let S be the set of following ordered clauses:

$P \vee Q \vee R$

$P \vee Q \vee \sim R$

$P \vee \sim Q \vee R$

$P \vee \sim Q \vee \sim R$

$\sim P \vee Q \vee R$

$\sim P \vee Q \vee \sim R$

$\sim P \vee \sim Q \vee R$

$\sim P \vee \sim Q \vee \sim R$.

Let $P \vee Q \vee R$ be the top ordered clause. Using the evaluation function defined by Eq. (7.7) to order pairs of CLIST, find the tree that will be searched by the modified depth-first method.

Chapter 8

The Equality Relation

8.1 INTRODUCTION

Equality is a very important relation, and many theorems can be easily symbolized through its use. For example, one can employ only one predicate, that of equality, one function, the successor function, and one constant 0, to formalize number theory [Meltzer, 1968a]. The equality relation has many special properties: it is reflexive, symmetric, and transitive. In addition, we can also substitute equals for equals. When the equality relation is used to symbolize a theorem, besides axioms for the specific theorem itself, we usually need a collection of extra axioms describing these properties of equality. This is explained in the following example.

Example 8.1

Consider the theorem: If $x \circ x = e$ for all x in a group G, where $\circ$ is a binary operator and e is the identity in G, then G is commutative. (See Example 4.3.) In the following, A_1, A_2, A_3, and A_4 are four axioms of the group G, and B is the conclusion of the theorem:

A_1: $x, y \in G$ implies that $x \circ y \in G$ (closure property).

A_2: $x, y, z \in G$ implies that $x \circ (y \circ z) = (x \circ y) \circ z$ (associativity property).

A_3: $x \circ e = e \circ x = x$ for all $x \in G$ (identity property).

A_4:   For every $x \in G$, there exists an element $x^{-1} \in G$ such that $x \circ x^{-1} = x^{-1} \circ x = e$ (inverse property).

B: If $x \circ x = e$ for all $x \in G$, then G is commutative, that is, $u \circ v = v \circ u$ for all $u, v \in G$.

Suppose we use the predicate of equality to symbolize this theorem. Suppose we also adopt the customary convention and use $x = y, x \neq y$, and $(x \circ y)$ to denote the literals $= (x, y)$ and $\sim \; = (x, y)$, and the function $\circ (x, y)$, respectively. Then the above axioms and conclusion can be represented by

A_1': $(\forall x)(\forall y)(\exists z)(x \circ y = z)$

A_2': $(\forall x)(\forall y)(\forall z)(x \circ (y \circ z) = (x \circ y) \circ z)$

A_3': $(\forall x)((x \circ e = x) \wedge (e \circ x = x))$

A_4': $(\forall x)((x \circ i(x) = e) \wedge (i(x) \circ x = e))$

B': $(\forall x)(x \circ x = e) \rightarrow ((\forall u)(\forall v)(u \circ v = v \circ u))$.

Now, as we did in Chapter 4, negating B' and transforming $A_1' \wedge A_2' \wedge A_3' \wedge A_4' \wedge \sim B'$ into a standard form, we obtain the set S consisting of the following clauses, where f, a, and b are Skolem functions:

(1) $x \circ y = f(x, y)$

(2) $x \circ (y \circ z) = (x \circ y) \circ z$

(3) $x \circ e = x$

(4) $e \circ x = x$

(5) $x \circ i(x) = e$

(6) $i(x) \circ x = e$

(7) $x \circ x = e$

(8) $a \circ b \neq b \circ a$.

The above set S is somewhat simpler than the one obtained in Example 4.3. However, using *resolution alone*, we *cannot* formally prove that S is unsatisfiable. The trouble lies in that the properties of equality are not explicitly specified in S. To completely axiomatize the theorem, we need extra axioms to describe the properties of equality. These extra axioms for the theorem are given as follows:

(9) $x = x$ reflexivity of equality

(10) $x \neq y \vee y = x$ symmetry of equality

(11) $x \neq y \vee y \neq z \vee x = z$ transitivity of equality

(12) $x \neq y \vee x \neq u \vee y = u$

(13) $x \neq y \vee u \neq x \vee u = y$

(14) $x \neq y \vee f(x,z) = f(y,z)$

(15) $x \neq y \vee f(z,x) = f(z,y)$ substitutivity of equality.

(16) $x \neq y \vee x \circ z = y \circ z$

(17) $x \neq y \vee z \circ x = z \circ y$

(18) $x \neq y \vee i(x) = i(y)$

Now, let K be the set consisting of clauses (9)–(18). Then $(S \cup K)$ is an unsatisfiable set that can be proved by resolution.

From the above example, if only resolution is used, we need extra axioms for the properties of the equality relation. That is, every time equality is used, we have to provide axioms that specify the reflexive, symmetric, transitive, and substitutive properties of equality. Such an approach is very clumsy. It not only requires more clauses for the representation of a theorem, but also tends to cause the generation of numerous useless resolvents. In order to curtail this problem and to dispense with the troublesome axioms of equality, such as clauses (9)–(18), many remedies have been proposed. For instance, Darlington [1968b] used a second-order equality substitution axiom, Robinson [1968a] introduced a generalized resolution principle that provides a built-in equality, and Robinson and Wos [1969] proposed paramodulation to handle equality. Also, there are the system given by Sibert [1969] and E-resolution proposed by Morris [1969] for the treatment of equality. In this chapter, we shall only consider paramodulation, because it is simple and natural. We shall see later that paramodulation can replace those axioms concerning the symmetric, transitive, and substitutive properties of equality. We shall also see that in conjunction with resolution, paramodulation can be used to prove theorems in a very natural and efficient way.

8.2 UNSATISFIABILITY UNDER SPECIAL CLASSES OF MODELS

We know that a set S of clauses is unsatisfiable if and only if S is false under *all* its interpretations (H-interpretations). However, there are many interesting sets of clauses that are true in some interpretations, but false in others. For example, the set S in Example 8.1 is such a set. Even though S is not unsatisfiable, S is false in every interpretation that satisfies K. In this section, we would like to characterize such a set S. This is done by restricting ourselves to special classes of models. As we did before, we shall organize the following material around the concepts of unsatisfiability and refutation; the formula to be refuted is taken to be in standard form. In general, we give the following definition.

Definition   Given a set S of clauses, let W be the set of all interpretations of S. Let Q be a nonempty subset of W. Then S is said to be *Q-unsatisfiable* if and only if S is false in every element of Q.

Of course, in order to make the above definition meaningful, we must have a way to specify Q. If Q is finite, we may just list all the elements of Q. Otherwise, Q is usually defined by the conjunction of axioms of a theory. For example, consider Example 8.1 again. Let Q be the class of all interpretations which satisfy K, that is, all models of K. Then S is false in every element of Q since $(S \cup K)$ is unsatisfiable. Therefore, S is Q-unsatisfiable. In fact, for this case, since K is the set of axioms of the equality theory and since Q is the set of all models of K, S will be called *E-unsatisfiable*, where E stands for equality.

Many authors have considered special classes of models of different theories. For example, models of equality theory have been considered in [Robinson and Wos, 1969], those of partial ordering and set theories in [Slagle, 1971a], and those of theories specified by two-literal clauses in [Dixon 1971a, b]. In this book, since treatments of equality theory and other theories follow a similar approach, we shall consider only models of equality theory.

We now formally define E-unsatisfiable sets as follows.

Definition   An *E-interpretation* I of a set S of clauses is an interpretation of S satisfying the following four conditions. Let α, β, and γ be any terms in the Herbrand universe of S, and let L be a literal in I. Then

  1. $(\alpha = \alpha) \in I$;
  2. if $(\alpha = \beta) \in I$, then $(\beta = \alpha) \in I$;
  3. if $(\alpha = \beta) \in I$ and $(\beta = \gamma) \in I$, then $(\alpha = \gamma) \in I$;
  4. if $(\alpha = \beta) \in I$ and L' is the result of replacing some one occurrence of α in L by β, then $L' \in I$.

Actually, the reader can easily see that an E-interpretation is just an interpretation that satisfies the reflexive, symmetric, transitive, and substitutive axioms of equality, that is, a model of equality theory.

Definition   A set S of clauses is called *E-satisfiable* if and only if there is an E-interpretation that satisfies all clauses in S. Otherwise, S is called *E-unsatisfiable*.

Example 8.2

Consider $S \triangleq \{P(a), \sim P(b), a = b\}$. There are 64 interpretations of S. Among them, only the following six are E-interpretations:

$$\{P(a), \quad P(b), \quad a = a, b = b, a = b, b = a\}$$
$$\{\sim P(a), \sim P(b), a = a, b = b, a = b, b = a\}$$
$$\{P(a), \quad P(b), \quad a = a, b = b, a \neq b, b \neq a\}$$
$$\{P(a), \quad \sim P(b), a = a, b = b, a \neq b, b \neq a\}$$
$$\{\sim P(a), P(b), \quad a = a, b = b, a \neq b, b \neq a\}$$
$$\{\sim P(a), \sim P(b), a = a, b = b, a \neq b, b \neq a\}.$$

Clearly, S is false in every one of the above six E-interpretations. Therefore, S is E-unsatisfiable. It is noted that S is not unsatisfiable, since an interpretation containing $P(a)$, $\sim P(b)$, and $a = b$ will satisfy S.

In the following, we shall extend Herbrand's theorem to E-unsatisfiable sets.

Definition Let S be a set of clauses. The *set F of the functionally reflexive axioms* for S is the set defined as

$$F \triangleq \{f(x_1, \ldots, x_n) = f(x_1, \ldots, x_n)\}$$

for all n-place function symbols f occurring in S.

Definition Let S be a set of clauses. Then the *set of equality axioms* for S is the set consisting of the following clauses:

(1) $\quad x = x.$

(2) $\quad x \neq y \vee y = x.$

(3) $\quad x \neq y \vee y \neq z \vee x = z.$

(4) $\quad x_j \neq x_0 \vee \sim P(x_1, \ldots, x_j, \ldots, x_n) \vee P(x_1, \ldots, x_0, \ldots, x_n)$ for $j = 1, \ldots, n$, for every n-place predicate symbol P occurring in S.

(5) $\quad x_j \neq x_0 \vee f(x_1, \ldots, x_j, \ldots, x_n) = f(x_1, \ldots, x_0, \ldots, x_n)$ for $j = 1, \ldots, n$, for every n-place function symbol f occurring in S.

Theorem 8.1 Let S be a set of clauses and K be the set of equality axioms for S. Then S is E-unsatisfiable if and only if $(S \cup K)$ is unsatisfiable.

Proof ($\Rightarrow$) Suppose S is E-unsatisfiable, but $(S \cup K)$ is satisfiable. Then there exists an interpretation I that satisfies $(S \cup K)$. Since I satisfies K,

I satisfies all the four conditions for an *E*-interpretation. Therefore *I* is an
E-interpretation. But *I* satisfies *S*. This contradicts the assumption that *S*
is *E*-unsatisfiable. Hence $(S \cup K)$ must be unsatisfiable.

($\Leftarrow$) Suppose $(S \cup K)$ is unsatisfiable, but *S* is *E*-satisfiable. Then there
exists an *E*-interpretation I_E such that I_E satisfies *S*. Clearly, I_E satisfies *K*
also. Hence I_E satisfies $(S \cup K)$. This contradicts the assumption that $(S \cup K)$
is unsatisfiable. Hence *S* must be *E*-unsatisfiable.   Q.E.D.

Theorem 8.2 A finite set *S* of clauses is *E*-unsatisfiable if and only if
there is a finite set *S'* of ground instances of clauses in *S* such that *S'* is
E-unsatisfiable.

Proof ($\Rightarrow$) Let *K* be the set of equality axioms for *S*. Then, by Theorem
8.1, $(S \cup K)$ is unsatisfiable. By Herbrand's theorem, there is a finite set *S'* of
ground instances of clauses in *S* such that $(S' \cup K)$ is unsatisfiable. Hence,
by Theorem 8.1, *S'* is *E*-unsatisfiable.

($\Leftarrow$) Since *S'* is *E*-unsatisfiable, every *E*-interpretation falsifies *S'*, that is,
every *E*-interpretation falsifies *S*. Therefore *S* is *E*-unsatisfiable.   Q.E.D.

8.3 PARAMODULATION—AN INFERENCE RULE FOR EQUALITY

As discussed in the previous chapters, $\square$ can be deduced from an *unsatisfiable* set of clauses by resolution. In this section, we shall introduce
paramodulation. It will be shown later that using both resolution and paramodulation, we can always deduce $\square$ from an *E-unsatisfiable* set of clauses.

Consider the following clauses

C_1: $P(a)$

C_2: $a = b$.

Since C_2 is an equality $a = b$, we substitute *b* for *a* in C_1 to obtain a
clause

C_3: $P(b)$.

In general, the equality substitution rule can be stated as follows: If a
clause C_1 contains a term *t* and if a unit clause C_2 is $t = s$, then infer a
clause by substituting *s* for one single occurrence of *t* in C_1.

Paramodulation is essentially an extension of the above equality substitution rule. It can be applied to any pair of clauses (not necessarily unit
clauses). For convenience, if an expression *E* (for example, a clause, a literal,
or a term) contains a term *t*, we shall sometimes denote *E* by *E*[*t*]. Also,
if one single occurrence of *t* is replaced by a term *s*, the result will be

denoted by $E[s]$. *By this convention, paramodulation for ground clauses can now be stated as follows: If C_1 is $L[t] \lor C_1'$, where $L[t]$ is a literal containing a term t, and C_1' is a clause, and if C_2 is $t = s \lor C_2'$, where C_2' is a clause, then infer the following clause, called a paramodulant:*

$$L[s] \cup C_1' \cup C_2'.$$

Example 8.3

Consider the following clauses

C_1: $P(a) \lor Q(b)$

C_2: $a = b \lor R(b)$.

Let L be $P(a)$, C_1' be $Q(b)$, and C_2' be $R(b)$. Since L contains a, L is denoted by $L[a]$. Clause C_2 contains an equality literal $a = b$. Therefore, $L[b]$ is $P[b]$. Hence a paramodulant of C_1 and C_2 is

$$P(b) \lor Q(b) \lor R(b).$$

Of course, paramodulation can be also defined on general clauses. In this case, we usually need to make instantiation before paramodulation is applied. Consider the clauses

C_1: $P(x) \lor Q(b)$

C_2: $a = b \lor R(b)$.

Clause C_1 does not contain the term a. However, if x is replaced by a, we obtain an instance of C_1,

C_1': $P(a) \lor Q(b)$.

Now, from C_1' and C_2, we can obtain a paramodulant

C_3': $P(b) \lor Q(b) \lor R(b)$.

C_3' will also be called a paramodulant of C_1 and C_2.

We now give a formal definition of paramodulation.

Definition   Let C_1 and C_2 be two clauses (called *parent* clauses) with no variables in common. If C_1 is $L[t] \lor C_1'$, and C_2 is $r = s \lor C_2'$, where $L[t]$ is a literal containing the term t and C_1' and C_2' are clauses, and if t and r have a most general unifier σ, then infer

$$L\sigma[s\sigma] \cup C_1'\sigma \cup C_2'\sigma,$$

where $L\sigma[s\sigma]$ denotes the result obtained by replacing one single occurrence of $t\sigma$ in $L\sigma$ by $s\sigma$. The above inferred clause is called a *binary paramodulant* of C_1 and C_2; the literals L and $r = s$ are called the *literals paramodulated upon*. Sometimes, we also say that we apply paramodulation from C_2 into C_1.

Example 8.4

Consider the clauses

C_1: $P(g(f(x))) \vee Q(x)$

C_2: $f(g(b)) = a \vee R(g(c))$.

For this example, L is $P(g(f(x)))$, C_1' is $Q(x)$, r is $f(g(b))$, s is a, and C_2' is $R(g(c))$. L contains the term $f(x)$ that can be unified with r. Let t be $f(x)$. A most general unifier of t and r is $\sigma = \{g(b)/x\}$. Therefore, $L\sigma[t\sigma]$ is $P(g(f(g(b))))$. Hence $L\sigma[s\sigma]$ is $P(g(a))$. Since $C_1'\sigma$ is $Q(g(b))$ and $C_2'\sigma$ is $R(g(c))$, we obtain a binary paramodulant,

$$P(g(a)) \vee Q(g(b)) \vee R(g(c)).$$

The literals paramodulated upon are $P(g(f(x)))$ and $f(g(b)) = a$.

Definition A *paramodulant* of clauses C_1 and C_2 (called *parent* clauses) is one of the following binary paramodulants:

1. a binary paramodulant of C_1 and C_2;
2. a binary paramodulant of C_1 and a factor of C_2;
3. a binary paramodulant of a factor of C_1 and C_2;
4. a binary paramodulant of a factor of C_1 and a factor of C_2.

Paramodulation is an inference rule for the equality relation. This rule was introduced by Robinson and Wos [1969]. It can be shown that the combination of resolution and paramodulation is complete for E-unsatisfiable sets of clauses. That is, using resolution and paramodulation, we can always generate the empty clause $\square$ from an E-unsatisfiable set of clauses. Furthermore, many refinements of paramodulation are still complete when used with resolution. This will be discussed in the following sections.

8.4 HYPERPARAMODULATION

As in resolution, the purpose of refining paramodulation is to increase efficiency. Many refinements of paramodulation have been considered by Chang [1970b], Chang and Slagle [1971], Kowalski [1970a], Wos and Robinson [1970], and Anderson [1971]. The refinements of paramodulation are somewhat weaker than those of resolution. There are no complete counterparts to *PI*-resolution (Section 6.3) and OL-resolution (Section 7.4) in paramodulation. But, fortunately, hyperresolution and linear resolution can be extended to paramodulation. In this section, we shall first consider hyperparamodulation. Then, in the following sections, we shall discuss unit, input, and linear paramodulations.

Definition Let P be an ordering of the predicate symbols that includes the predicate symbols in clauses C_1 and C_2. Then a paramodulant of C_1 and C_2 is called a *P-hyperparamodulant* if and only if the following two conditions are satisfied:

1. C_1 and C_2 are positive clauses.
2. The literals paramodulated upon in C_1 and C_2 contain the largest predicate symbols in C_1 and C_2, respectively.

On the other hand, a P-hyperresolvent is a PI-resolvent where every literal in the interpretation I contains the negation sign. It is noted that in this case electrons and P-hyperresolvents are positive.

Definition Let P be an ordering of the predicate symbols in a set S of clauses. Then a *P-hyperdeduction with resolution and paramodulation* is a deduction in which every clause is a clause in S, or a P-hyperresolvent, or a P-hyperparamodulant. A *P-hyperrefutation with resolution and paramodulation* is a P-hyperdeduction of $\square$ with resolution and paramodulation.

Example 8.5

Consider $S \triangleq \{ \sim Q(a) \lor \sim R(a) \lor a = b,\ Q(a) \lor a = b,\ R(a),\ f(a) \neq f(b),\ f(x) = f(x) \}$. Let P be the ordering of the predicate symbols defined by $Q > R > \ =$. For this example, we obtain a P-hyperrefutation with resolution and paramodulation shown in Fig. 8.1, where $\curvearrowright$ indicates that a P-hyperparamodulation is applied from left to right and $\frown$ indicates that a P-hyperresolution is applied.

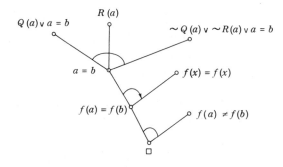

Figure 8.1

Theorem 8.3 (Completeness of P-hyperparamodulation) If P is an ordering of the predicate symbols in a finite E-unsatisfiable set S of clauses, then there is a P-hyperrefutation with resolution and paramodulation from $(S \cup \{x = x\} \cup F)$, where F is the set of the functionally reflexive axioms for S.

Proof Let K be the set of equality axioms for S. Since S is E-unsatisfiable, by Theorem 8.1 $(S \cup K)$ is unsatisfiable. Let I be the interpretation of S in which every literal is negative. By Theorem 6.1, there is a PI-refuatation D (with resolution only) from $(S \cup K)$. We now show that every clause in D can also be deduced from $(S \cup \{x = x\} \cup F)$. For each PI-clash $(E_1, E_2, ..., E_q, N)$ in D, since $E_1, E_2, ..., E_q$ are positive, $E_1, E_2, ..., E_q$ must be in S or $\{x = x\}$, for no clause in K other than $x = x$ is positive. If the nucleus N is in S, then the PI-resolvent of the clash is the desired one. If N is in K (N is not $x = x$), then we have the following four possible cases:

Case 1 N is $x \neq y \vee y = x$. In this case $q = 1$ and E_1 is $t_1 = t_2 \vee E_1{}'$, where t_1 and t_2 are terms (not necessarily ground terms) and $E_1{}'$ is a clause. The resolvent of the clash is $t_2 = t_1 \vee E_1{}'$, which is also a P-hyperparamodulant of E_1 and $x = x$.

Case 2 N is $x \neq y \vee y \neq z \vee x = z$. In this case $q = 2$, E_1 is $t_1 = t_2 \vee E_1{}'$, and E_2 is $s_1 = s_2 \vee E_2{}'$, where $E_1{}'$ and $E_2{}'$ are clauses and t_1, t_2, s_1, and s_2 are terms such that there is a most general unifier σ for which $t_2\sigma = s_1\sigma$. The resolvent of the clash is $t_1\sigma = s_2\sigma \vee E_1{}'\sigma \vee E_2{}'\sigma$, which is also a P-hyperparamodulant of E_1 and E_2.

Case 3 N is of the form $x_j \neq x_0 \vee \sim P(x_1, ..., x_j, ..., x_n) \vee P(x_1, ..., x_0, ..., x_n)$. In this case, $q = 2$, E_1 is $t_j = t_0 \vee E_1{}'$, and E_2 is $P(s_1, ..., s_j, ..., s_n) \vee E_2{}'$, where $E_1{}'$ and $E_2{}'$ are clauses and $t_j, t_0, s_1, ..., s_j, ..., s_n$ are terms such that there is a most general unifier σ for which $t_j\sigma = s_j\sigma$. The resolvent of the clash is $E_1{}'\sigma \vee E_2{}'\sigma \vee P(s_1\sigma, ..., s_{j-1}\sigma, t_0\sigma, s_{j+1}\sigma, ..., s_n\sigma)$, which is also a P-hyperparamodulant of E_1 and E_2.

Case 4 N is of the form $x_j \neq x_0 \vee f(x_1, ..., x_j, ..., x_n) = f(x_1, ..., x_0, ..., x_n)$. In this case $q = 1$ and E_1 is $t_j = t_0 \vee E_1{}'$, where $E_1{}'$ is a clause and t_j and t_0 are terms. The resolvent of the clash is $E_1{}' \vee f(x_1, ..., t_j, ..., x_n) = f(x_1, ..., t_0, ..., x_n)$, which is also a P-hyperparamodulant of E_1 and $f(x_1, ..., x_n) = f(x_1, ..., x_n)$.

Thus we have shown that every clause in the refutation D can be derived from $(S \cup \{x = x\} \cup F)$ by using P-hyperresolution and P-hyperparamodulation. This completes the proof.

Theorem 8.3 says that for resolution we need consider only the clashes whose electrons are positive, and for paramodulation we need consider only positive clauses. This deduction procedure would eliminate as much as possible unnecessary computation, especially when the input set of clauses contains only a few positive clauses. In addition, the ordering P of the predicate symbols may play an important role in the deduction procedure. If we wish, we may use more resolution than paramodulation (or vice versa) by letting the equality symbol be the smallest (or the largest) predicate symbol in the ordering P. If the equality symbol $=$ is the smallest predicate symbol in the ordering P, we obtain a very interesting corollary of Theorem 8.3.

Corollary 8.1   If P is an ordering of the predicate symbols in a finite E-unsatisfiable set S of clauses such that the equality symbol $=$ is the smallest in P, and if F is the set of the functionally reflexive axioms for S, then $\square$ can be deduced from $(S \cup \{x = x\} \cup F)$ by resolution and para-modulation in which

1. all resolvents are P-hyperresolvents, and
2. paramodulation is applied only from clauses consisting entirely of positive equality literals into positive clauses.

It is noted that, if we use P-hyperparamodulation and P-hyperresolution, the set F of the functionally reflexive axioms is required to obtain a proof. For example, consider $S \triangleq \{a = b, f(a) \neq f(b)\}$. S is E-unsatisfiable. But there is no P-hyperrefutation with resolution and paramodulation from $(S \cup \{x = x\})$. However, there is a P-hyperrefutation with resolution and para-modulation from $(S \cup \{x = x\} \cup F)$, where $F \triangleq \{f(x) = f(x)\}$.

8.5   INPUT AND UNIT PARAMODULATIONS

The material presented in this section is an extension of that given in Section 7.3. Let S be the original input set of clauses. As defined before, every member of S is called an *input* clause. In the following, let R and P stand for resolution and paramodulation, respectively.

An *input paramodulation* is a paramodulation in which one of the two parent clauses is an input clause.

A *unit paramodulation* is a paramodulation in which a paramodulant is obtained by using at least a unit parent clause or a unit factor of a parent clause.

An *input (unit) deduction by P* is a deduction that is obtained by employ-ing input (unit) paramodulation only. An *input (unit) deduction by R and P* is a deduction that is obtained by employing both input (unit) resolution and input (unit) paramodulation.

An *input (unit) refutation* is an input (unit) deduction of the empty clause.

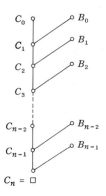

Figure 8.2

Lemma 8.1   If a set S of ground clauses has an input refutation by R and P, then it has a unit refutation by R and P.

Proof   As shown in Fig. 8.2, let D be an input refutation from S, where $C_0, B_0, ..., B_{n-1}$ are input clauses. Let D_i be that portion of D such that D_i consists of $C_i, C_{i+1}, ..., C_n$ and $B_i, B_{i+1}, ..., B_{n-1}$ $(i = 0, 1, 2, ..., n-1)$. If all the B_i $(i = 0, 1, ..., n-1)$ are unit ground clauses, we are done; otherwise, the input refutation D can be transformed into a unit refutation by the following procedure:

1. Let r be the largest integer such that B_r is a nonunit ground clause and B_i is a unit ground clause for $i = r+1, r+2, ..., n-1$. Let $U_{r+1} = \{B_{r+1}, B_{r+2}, ..., B_{n-1}\}$.
2. Let L_r be the literal in B_r involved in resolution or paramodulation.
3. Reduce B_r to the unit clause L_r by using the set U_{r+1}. This can be done since $B_r - \{L_r\}$ is a subclause of C_{r+1}, and C_{r+1} can be reduced to the empty clause by using U_{r+1}.
4. Let C_{r+1}' be the resolvent or paramodulant of C_r and L_r, where C_{r+1}' corresponds to C_{r+1}. Clearly, C_{r+1}' is a subclause of C_{r+1}. C_{r+1}' can be reduced to the empty clause $\square$ by U_{r+1} since C_{r+1} can be reduced to $\square$ by U_{r+1}. Let D_{r+1}' be such a refutation.
5. Replace D_{r+1} by D_{r+1}' and B_r by L_r.
6. Let $U_r = U_{r+1} \cup \{L_r\}$ and $r = r-1$ and go back to Step 2.

By repeating the above process, finally we will get a clause C_1' that is a resolvent or a paramodulant of C_0 and L_0. But C_1' can be reduced to $\square$ by using U_1. Since U_1 is a set of unit ground clauses and each L_i is obtained from B_i by using unit resolution and unit paramodulation, it is clear that we can get a unit refutation from S. This completes our proof.

Theorem 8.4 If a set S of clauses has an input refutation by R and P, then S, together with its functionally reflexive axioms, has a unit refutation by R and P.

Proof Since S has an input refutation D_1 by R and P, there is a ground substitution θ (substituting ground terms for variables) such that $D_1\theta$ is an input refutation by R and P. Let S' be the set of all the ground clauses at the initial nodes of $D_1\theta$. By Lemma 8.1, S' has a ground unit refutation D_2' by R and P. From the unit refutation D_2', we can obtain a unit refutation from S and its functionally reflexive axioms. This is done as follows: For the ground clause C', at each initial node of D_2' replace C' by a clause C in S such that C' is an instance of C. (If C' is a ground unit clause, C must be a unit clause or have a unit factor.) For the ground clause B' at each non-initial node N of D_2', if N_1 and N_2 are the two predecessor nodes of N, and if B_1 and B_2 are the clauses that have been obtained in this way for N_1 and N_2, respectively, then we replace the ground clause B' and the node N as follows: If B' is a unit resolvent, replace B' by a unit resolvent B of B_1 and B_2 such that B' is an instance of B. (If B' is a ground unit clause, B must be a unit clause or have a unit factor.) If B' is a unit paramodulant of B_1' and B_2', where B_1' and B_2' are ground clauses attached to N_1 and N_2, respectively, then assume B' is derived by paramodulating B_1' into B_2'. In this case, paramodulating zero or more appropriate functionally reflexive axioms into B_2, we can obtain an instance B_2* of B_2 such that, when B_1 is paramodulated into B_2*, we can obtain a clause B that has B' as an instance. (If B' is a ground unit clause, B must be a unit clause or have a unit factor.) Replace B' and N by the deduction of B. (Note that the deduction of B is a unit deduction.) In this fashion, we can obtain a unit refutation by R and P from S and its functionally reflexive axioms. This completes the proof.

We suspect that the converse of Theorem 8.4 is also true. At present, we do not have a proof for it. However, if S consists of only unit clauses, it will be shown in the following that S has an input refutation. This is very important in many systems. For example, in a system of rewriting rules [Quinlan and Hunt, 1968], every rewriting rule can be considered as a unit clause (axiom). In this case, we can use an input strategy.

In the figures that follow, $\curvearrowleft$ indicates that paramodulation is applied from right to left, while $\curvearrowright$ indicates it is applied from left to right. $\cap$ indicates that resolution is applied. We now introduce a lemma whose proof is given in Appendix B.

Lemma 8.2 In the deduction by R and P shown in Fig. 8.3a, let the ground unit clause M be derived by paramodulating the ground clause $s_1 = s_2$ into the ground unit clause M', and let the ground clause L be derived (using resolution or paramodulation) from the ground unit clauses L' and M. Then

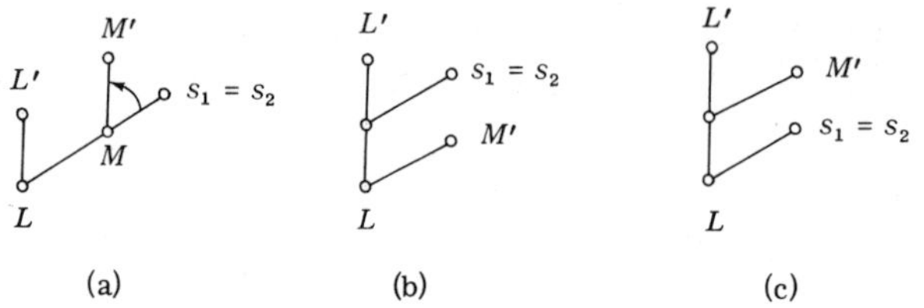

Figure 8.3

the deduction shown in Fig. 8.3a can be transformed into either the deduction shown in Fig. 8.3b or that shown in Fig. 8.3c.

A *branch* of a deduction tree is a path that links an initial node to the root of the tree. For example, the bold line in Fig. 8.4a is a branch. A node N of a deduction tree is called a *side node* to a branch if the node N is not on the branch and is joined to the branch by a single link. For example, nodes ① and ② in Fig. 8.4a are side nodes to the branch shown by the bold line. A clause is called a *side clause to a branch B* if it is attached to a side node to branch B.

Lemma 8.3 If C is a clause in an E-unsatisfiable set S of ground unit clauses that contains all the ground instances of $x = x$ and if $S - \{C\}$ is E-satisfiable, then S has an input refutation by R and P with top clause C.

Proof By Theorem 8.3, S has a refutation D by R and P. (We note that since S consists of only ground unit clauses, in the refutation D resolution appears only at the last step where the empty clause $\square$ is obtained.) Since $S - \{C\}$ is E-satisfiable, C must appear at least once in D. The following procedure transforms D into an input refutation with top clause C:

1. Let B be a branch starting at a node attached to any occurrence of C. Without loss of generality, we can assume that in the refutation D, any paramodulations off the branch B are applied from right to left.
2. Check whether there is a noninitial side node to B. If there is none, then the tree we have is an input refutation with top clause C. Otherwise, pick one such N.
3. Let N_1 and N_2 be the immediate predecessor nodes of N.
4. Applying Lemma 8.2 on N, N_1, and N_2, we can eliminate the node N.
5. Go back to Step 2.

By applying the above procedure on D, we can obtain an input refutation by R and P with top clause C. This completes the proof.

Example 8.6

In Fig. 8.4, using Lemma 8.2, we illustrate a sequence of transformations that convert a noninput refutation into an input refutation.

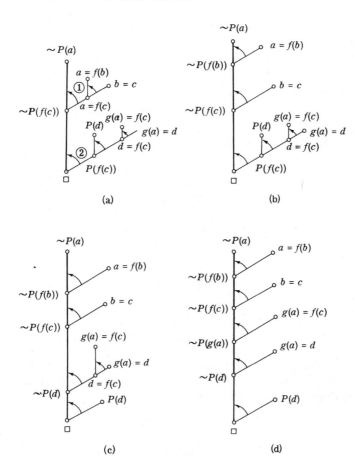

Figure 8.4

Theorem 8.5   If C is a clause in an E-unsatisfiable set S of unit clauses including $x = x$ and the functionally reflexive axioms, and if $S - \{C\}$ is E-satisfiable, then S has an input refutation by R and P with top clause C.

Proof   Since S is E-unsatisfiable and $S - \{C\}$ is E-satisfiable, by Theorem 8.2 there is a finite set S' of ground instances of clauses in S and a ground instance C' of C such that C' is in S', S' is E-unsatisfiable, and $S' - \{C'\}$ is E-satisfiable. Add all the ground instances of $x = x$ into S'. Hence, by Lemma 8.3, S' has an input refutation D' by R and P with top clause C'. Using a (lifting) proof similar to that of Theorem 8.4, from the input refutation D' we can obtain an input refutation D from S by R and P with top clause C.

Q.E.D.

8.6   LINEAR PARAMODULATION

Definition   Given a set S of clauses and a clause C_0 in S, a *linear deduction of C_n from S with top clause C_0 by R and P* is a deduction of the form shown in Fig. 8.5, where

1. For $i = 0, \ldots, n-1$, C_{i+1} is a resolvent or a paramodulant of C_i (called a *center* clause) and B_i (called a *side* clause).
2. Each B_i is either in S, or is a C_j for some $j < i$.

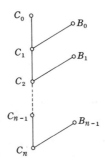

Figure 8.5

We note that a linear deduction by R and P is an input deduction by R and P if B_i is in S for $i = 0, \ldots, n-1$. A *linear refutation by R and P* is a linear deduction of $\square$ by R and P.

Example 8.7

Let $S \triangleq \{ \sim Q(c) \vee c = d, \ \sim Q(c) \vee f(c) \neq f(d), \ Q(c) \vee a = b, \ Q(c) \vee f(a) \neq f(b), \ f(x) = f(x) \}$. Then a linear refutation from S with top clause $\sim Q(c) \vee c = d$ is shown in Fig. 8.6. There are six side clauses. Five of them are from S, while one of them, namely $\sim Q(c)$, is a clause derived before $\square$.

Lemma 8.4   If C is a clause in an E-unsatisfiable set S of ground clauses that contains all the ground instances of $x = x$, and if $S - \{C\}$ is E-satisfiable, then S has a linear refutation by R and P with top clause C.

Proof   Let $k(S)$ be defined to be the total number of appearances of literals in S minus the number of clauses in S. We prove Lemma 8.4 by induction on $k(S)$. If $k(S) = 0$, then S consists of only ground unit clauses. Thus, by Lemma 8.3, S has an input refutation by R and P with top clause C. Since an input refutation is a linear refutation, Lemma 8.4 holds for $k(S) = 0$. Assume Lemma 8.4 holds when $k(S) < n$. To complete the induction, we consider the case where $k(S) = n$ and $n > 0$. Since $k(S) > 0$, there is a nonunit clause in S. If C is a nonunit clause, let $C = A \vee L$, where A is a

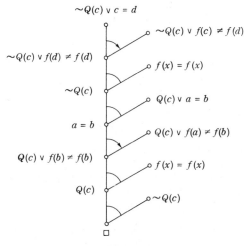

Figure 8.6

nonempty ground clause and L is a ground literal. Let $S_1 = (S - \{C\}) \cup \{A\}$ and $S_2 = (S - \{C\}) \cup \{L\}$. Clearly, S_1 and S_2 are E-unsatisfiable, and $S_1 - \{A\}$ and $S_2 - \{L\}$ are E-satisfiable. However, $k(S_1) < n$ and $k(S_2) < n$. Hence, by the induction hypothesis, S_1 and S_2 have linear refutations D_1 and D_2 by R and P with top clause A and L, respectively. Let D_1' be the linear deduction obtained from D_1 by putting L back with A. If D_1' is a linear refutation by R and P, we are done. Otherwise, D_1' is a linear deduction of L from S by R and P. By combining D_1' and D_2, we obtain a linear refutation by R and P with top clause C from S. Therefore, Lemma 8.4 holds for this case. If C is a unit clause, C cannot be a ground instance of $x = x$ since $(S - \{C\})$ is E-satisfiable and S is E-unsatisfiable. In this case, we choose a nonunit clause B in $(S - \{C\})$. Without loss of generality, we can assume that $(S - \{B\})$ is E-satisfiable, for otherwise we can delete B from S. Since $(S - \{C\})$ is E-satisfiable and B is in $S - \{C\}$, we can choose a literal L in B such that $B = A \vee L$ and $((S - \{B\}) \cup \{A\}) - \{C\}$ is E-satisfiable. Let $S_1 = (S - \{B\}) \cup \{A\}$ and $S_2 = (S - \{B\}) \cup \{L\}$. Clearly, S_1 and S_2 are E-unsatisfiable, and $(S_1 - \{C\})$ and $(S_2 - \{L\})$ are E-satisfiable. However, $k(S_1) < n$ and $k(S_2) < n$. Again, using the induction hypothesis, we can show that there is a linear refutation from S by R and P with top clause C. This completes the proof.

Now, using reasoning similar to that in the proof of Theorem 8.5, we can lift Lemma 8.4 to the general (nonground) case.

Theorem 8.6   (Completeness of Linear Paramodulation)   If C is a clause in an E-unsatisfiable set S of clauses including $x = x$ and the functionally reflexive axioms, and if $S - \{C\}$ is E-satisfiable, then S has a linear refutation by R and P with top clause C.

REFERENCES

Anderson, R. (1970): Completeness results for E-resolution, *Proc. AFIPS 1970 Spring Joint Computer Conf.*, pp. 653–656.

Anderson, R. (1971): Completeness of the locking restriction for paramodulation, Dept. of Computer Science, Univ. of Houston, Texas.

Chang, C. L. (1970b): Renamable paramodulation for automatic theorem proving with equality, *Artificial Intelligence* **1** 247–256.

Chang, C. L., and J. R. Slagle (1971): Completeness of linear refutation for theories with equality, *J. Assoc. Comput. Mach.* **18** 126–136.

Darlinton, J. L. (1968b): Automatic theorem proving with equality substitutions and mathematical induction, *in:* "Machine Intelligence," vol. 3 (B. Meltzer and D. Michie, eds.), American Elsevier, New York, pp. 113–127.

Dixon, J. (1971b): Z-resolution: theorem-proving with compiled axioms, to appear in *J. Assoc. Comput. Mach.*

Dixon, J. (1971c): Experiments with a Z-resolution program, Division of Computer Research and Technology, National Institutes of Health, Bethesda, Maryland.

Kowalski, R. (1970a): The case for using equality axioms in automatic demonstration, *Proc. Symp. Automatic Demonstration*, Springer-Verlag, New York, pp. 112–127.

Meltzer, B. (1968a): A new look at mathematics and its mechanization, *in:* "Machine Intelligence," vol. 3 (D. Michie, ed.), American Elsevier, New York, pp. 63–70.

Morris, J. B. (1969): E-resolution: Extension of resolution to include the equality, *Proc. Inter. Joint Conf. Artificial Intelligence*, Washington, D.C., pp. 287–294.

Norton, L. M. (1971): Experiments with a heuristic theorem-proving for the predicate calculus with equality, *Artificial Intelligence*, **2** 261–284.

Quinlan, J. R., and E. B. Hunt (1968): A formal deductive problem-solving system, *J. Assoc. Comput. Mach.* **15** 625–646.

Robinson, J. A. (1968a): The generalized resolution principle, *in:* "Machine Intelligence," vol. 3 (D. Michie, ed.), American Elsevier, New York, pp. 77–94.

Robinson, G. A., and L. Wos (1969): Paramodulation and theorem proving in first order theories with equality, *in:* "Machine Intelligence," vol. 4 (B. Meltzer and D. Michie, eds.), American Elsevier, New York, pp. 135–150.

Sibert, E. E. (1969): A machine oriented logic incorporating the equality relation, *in:* "Machine Intelligence," vol. 4 (B. Meltzer and D. Michie, eds.), American Elsevier, New York, pp. 103–134.

Slagle, J. R. (1971a): Automatic theorem proving with built-in theories including equality, partial ordering and sets, to appear in *J. Assoc. Comput. Mach.*

Slagle, J. R., and L. Norton (1971): Experiments with an automated theorem proving having partial ordering rules, Division of Computer Research and Tech., National Inst. of Health, Bethesda, Maryland.

Wos, L. and G. A. Robinson (1970): Paramodulation and set of support, *Proc. Symp. Automatic Demonstration*, Springer-Verlag, New York, pp. 276–310.

Wos, L., G. A. Robinson, D. F. Carson, and L. Shalla (1967): The concept of demodulation in theorem proving, *J. Assoc. Comput. Mach.* **14** 698–709.

EXERCISES

Section 8.1

1. Show that $(S \cup K)$ of Example 8.1 is unsatisfiable by resolution.

Section 8.2

2. Let $S \triangleq \{P(a,b), \sim P(b,a), a = b\}$. Find all E-interpretations of S. Is S E-unsatisfiable? Why?
3. Let $S \triangleq \{P(x) \vee x = b, \sim P(a), P(b)\}$. Find an E-unsatisfiable set S' of ground instances of clauses in S.

Section 8.3

4. Let
$$C_1 \triangleq P(f(x, g(x))) \vee Q(x),$$
$$C_2 \triangleq a = b \vee g(a) = a \vee f(a, g(a)) = b.$$

Find all paramodulants of C_1 and C_2.

Section 8.4

5. Consider the following set S of clauses:

(1) $R(a) \vee R(b)$

(2) $\sim D(y) \vee L(a, y)$

(3) $\sim R(x) \vee \sim Q(y) \vee \sim L(x, y)$

(4) $D(a) \vee \sim Q(a)$

(5) $Q(b) \vee \sim R(b)$

(6) $a = b.$

Let P be the ordering of the predicate symbols defined by $= > L > D > Q > R$. Find a P-hyperrefutation with resolution and paramodulation from S.

6. Repeat Exercise 5 for the ordering P defined by $R > Q > D > L > =$.

Section 8.5

7. Can you find an input refutation by R and P from the set S given in Exercise 5?
8. Can you find a unit refutation by R and P from the set S given in Exercise 5?
9. Let D be an input refutation by R and P shown in Fig. 8.7. Transform D into a unit refutation by R and P.

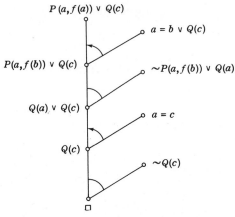

Figure 8.7

10. Let D be a unit refutation by R and P shown in Fig. 8.8. Transform D into an input refutation by R and P with top clause $a = b$.
11. Show that the set S of Example 8.1 is E-unsatisfiable by input resolution and input paramodulation.
12. Consider the following rewriting rules:

(1) $\quad x + y = y + x$ $\qquad$ (2) $\quad x + (y+z) = (x+y) + z$

(3) $\quad (x+y) - y = x$ $\qquad$ (4) $\quad x = (x+y) - y$

(5) $\quad (x-y) + z = (x+z) - y$ $\qquad$ (6) $\quad (x+y) - z = (x-z) + y.$

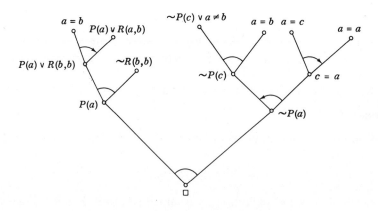

Figure 8.8

Prove each of the following equalities by input paramodulation:

(a) $(a+b) + c = a + (b+c)$ (b) $(a-b) + c = a + (c-b)$

(c) $(a-c) - (b-c) = a - b$ (d) $(a+c) - (b+c) = a - b$

(e) $a + (b-c) = (a-c) + b.$

Section 8.6

13. Consider the following set S of clauses:

(1) $P(b) \lor Q(a)$

(2) $P(a) \lor \sim Q(b)$

(3) $\sim P(a) \lor Q(b)$

(4) $\sim P(b) \lor \sim Q(a)$

(5) $a = b.$

Let clause (5) be the top clause. Find a linear refutation by R and P from S.

14. Repeat Exercise 13, using clause (1) as the top clause.

Chapter 9

Some Proof Procedures Based on Herbrand's Theorem

9.1 INTRODUCTION

In Chapter 4, we gave Herbrand's procedure (see Section 4.6) for showing that a set of clauses is unsatisfiable. Given a set S of clauses, Herbrand's procedure successively generates a sequence, say $S_1', S_2', S_3', \ldots$, in some arbitrarily defined order, and successively tests each S_i' for unsatisfiability, where S_i' is a set of ground instances of clauses in S. As discussed in Section 5.1, this procedure suffers from a major drawback. That is, in most cases, the earliest set S_k' in the sequence that is unsatisfiable turns out to be a very large set. In order to avoid such generation of ground instances, two main approaches have been proposed. One is to use Robinson's resolution principle, which has been discussed thoroughly in the previous chapters. The other is to use the idea proposed by Prawitz [1960]. Prawitz's idea is that, instead of generating the ground instances of clauses of S in some arbitrarily defined order, one should find by calculations the values that, when substituted for variables in S, give an unsatisfiable set of ground instances. Essentially, Prawitz's idea is based upon the observation that a set S of clauses is unsatisfiable if and only if there is a set M of copies (variants) of clauses in S and a ground substitution θ such that $M\theta$ is unsatisfiable. This observation is actually Herbrand's theorem in a different form. Prawitz [1960, 1969], Davis [1963], Chinlund et al. [1964], Loveland [1968, 1972], and Chang

[1971, 1972] all have proposed procedures to find such M and θ. In this chapter, we shall consider some of these procedures. We shall first briefly describe the Prawitz procedure. Then we shall discuss V-resolution and some tree procedures.

9.2   THE PRAWITZ PROCEDURE

Before giving the Prawitz procedure, we first consider an example. Suppose we have the following set of clauses:

(1) $P(x)$
(2) $\sim P(a) \vee \sim P(b)$ $\Big\}$ S.

Suppose that we want to find an unsatisfiable set S' of ground instances of clauses in S. Since x is a universal variable, we can replace x by any ground term, including a and b. If we do just that, we obtain the following set S' of clauses:

(1') $P(a)$
(2') $P(b)$ $\Big\}$ S'.
(3') $\sim P(a) \vee \sim P(b)$

S' consists of only ground clauses and is unsatisfiable. Therefore, according to Herbrand's theorem, S is unsatisfiable.

In the above case, clause (1) of S has two instances in S'. We may consider the following set M^* of clauses:

(1*) $P(x_1)$
(2*) $P(x_2)$ $\Big\}$ M^*.
(3*) $\sim P(a) \vee \sim P(b)$

By a *copy* (*variant*) of a clause C, we mean that it is C itself, or a clause obtained from C by renaming variables in C. In M^*, every clause is a copy of a clause of S, that is, M^* consists of two copies of $P(x)$ and one copy of $\sim P(a) \vee \sim P(b)$. If we replace x_1 by a and x_2 by b, we obtain S'. That is, if we let $\theta^* = \{a/x_1, b/x_2\}$, then $M^*\theta^* = S'$, which is unsatisfiable. M^* is thus of special interest to us and we shall give a formal definition for it.

Definition   Given an unsatisfiable set S of clauses, a set M^* of clauses is called a *substitutively unsatisfiable* (*S-unsatisfiable* for short) set of S if M^* satisfies the following conditions:

1. Every clause in M^* is a copy of a clause in S.
2. No two clauses in M^* have variables in common.
3. There exists a ground substitution θ^* such that $M^*\theta^*$ is unsatisfiable. (Note that a ground substitution is a substitution of ground terms for variables.)

The substitution θ^* is called a *solution* of M^*.

The reader may have already noted that variables in M^* are *not* treated as universally quantified. In fact, every variable in M^* is permitted to be replaced by only one ground term. This restriction will be followed throughout this chapter.

Given S, since S' is not known beforehand, M^* cannot be readily constructed. However, M^* is usually not a very large set. Therefore, at the beginning, we arbitrarily let M be a set consisting of a few copies of each clause of S. Rename the variables in M so that no clauses in M have variables in common. Clearly, if M is larger than M^*, we can also find a ground substitution θ such that $M\theta$ is unsatisfiable. M is called an *alleged S-unsatisfiable set* of S, and θ is called a *solution* of M, if indeed $M\theta$ is unsatisfiable. Given such M, the following procedure was originally proposed by Prawitz to find a solution of M:

Step 1   Transform M into a disjunctive normal form N.

Step 2   Seek substitutions of elements of the Herbrand universe for variables that make every conjunction of N unsatisfiable.

Step 3   If there is such a substitution θ, θ is a desired solution of M. Otherwise, M has no solution.

Example 9.1

Consider $S = \{P(x) \vee Q(y), \sim P(a), \sim Q(b)\}$. Let M be the set consisting of one copy of each clause in S. That is, $M = \{P(x) \vee Q(y), \sim P(a), \sim Q(b)\}$. Since no clauses in M have variables in common, we need not rename the variables in the clauses of M. M can also be written as $M = [P(x) \vee Q(y)] \wedge \sim P(a) \wedge \sim Q(b)$. Transforming M into a disjunctive normal form, we obtain

$$[P(x) \wedge \sim P(a) \wedge \sim Q(b)] \vee [Q(y) \wedge \sim P(a) \wedge \sim Q(b)].$$

Clearly, if x is replaced by a, the first conjunction, namely, $[P(x) \wedge \sim P(a) \wedge \sim Q(b)]$, will be unsatisfiable. Similarly, if y is replaced by b, the second conjunction, namely, $[Q(y) \wedge \sim P(a) \wedge \sim Q(b)]$, will be unsatisfiable. Therefore, there is a ground substitution, namely, $\theta = \{a/x, b/y\}$, that is a solution of M. That is, $M\theta = \{P(a) \vee Q(b), \sim P(a), \sim Q(b)\}$ is unsatisfiable.

In the above example, the substitution θ is easy to find. In general, we should develop an algorithm to find such θ. This is done by first considering the following definition.

Definition A substitution θ is called a *refuting substitution* for a conjunction C of literals if and only if there are two literals L_1 and L_2 in C such that θ is a most general unifier of L_1 and $\sim L_2$.

Consider Example 9.1. We note that $\{a/x\}$ is a refuting substitution for the conjunction $[P(x) \wedge \sim P(a) \wedge \sim Q(b)]$, and $\{b/y\}$ is a refuting substitution for $[Q(y) \wedge \sim P(a) \wedge \sim Q(b)]$. Combining $\{a/x\}$ and $\{b/y\}$, we obtain the desired substitution $\theta = \{a/x, b/y\}$, which is a solution of M.

By now, the reader may get the impression that to find a solution of M one only has to find a refuting substitution for every conjunction in a disjunctive normal form, and then combine all these refuting substitutions. However, the reader should be warned that sometimes substitutions cannot be combined. This is illustrated in the following example.

Example 9.2

Consider $S = \{P(x) \vee Q(x), \sim P(a), \sim Q(b)\}$. Let M be the set consisting of one copy of each clause in S. That is, $M = \{P(x) \vee Q(x), \sim P(a), \sim Q(b)\}$. Since no clauses in M have variables in common, we need not rename the variables in them. Let M be written as $M = [P(x) \vee Q(x)] \wedge \sim P(a) \wedge \sim Q(b)$. Transforming M into a disjunctive normal form, we obtain

$$[P(x) \wedge \sim P(a) \wedge \sim Q(b)] \vee [Q(x) \wedge \sim P(a) \wedge \sim Q(b)].$$

Now, $\{a/x\}$ is a refuting substitution for the first conjunction $[P(x) \wedge \sim P(a) \wedge \sim Q(b)]$, and $\{b/x\}$ is a refuting substitution for the second conjunction $[Q(x) \wedge \sim P(a) \wedge \sim Q(b)]$. Clearly, $\{a/x\}$ and $\{b/x\}$ cannot be combined since x cannot be simultaneously replaced by a and b. Therefore, M has no solution.

To test whether we can combine substitutions, we give the following definition.

Definition Let $\theta_1 = \{t_{11}/v_{11}, \ldots, t_{1n_1}/v_{1n_1}\}, \ldots, \theta_r = \{t_{r1}/v_{r1}, \ldots, t_{rn_r}/v_{rn_r}\}$ be substitutions, $r \geqslant 2$. From $\theta_1, \ldots, \theta_r$ we define two expressions $E_1 = (v_{11}, \ldots, v_{1n_1}, \ldots, v_{r1}, \ldots, v_{rn_r})$ and $E_2 = (t_{11}, \ldots, t_{1n_1}, \ldots, t_{r1}, \ldots, t_{rn_r})$. Then, $\theta_1, \ldots, \theta_r$ are said to be *consistent* if and only if E_1 and E_2 are unifiable. A most general unifier for $\{E_1, E_2\}$ is called a *combination* of $\theta_1, \ldots, \theta_r$. The substitutions $\theta_1, \ldots, \theta_r$ are said to be *inconsistent* if and only if they are not consistent.

Example 9.3

Consider $\theta_1 = \{a/x\}$ and $\theta_2 = \{f(a)/x\}$. For this case, we have $E_1 = (x, x)$ and $E_2 = (a, f(a))$. Since E_1 and E_2 are not unifiable, θ_1 and θ_2 are inconsistent.

Example 9.4

Let $\theta_1 = \{g(y)/x\}$ and $\theta_2 = \{f(x)/y\}$. For this case, we have $E_1 = (x, y)$ and $E_2 = (g(y), f(x))$. Since E_1 and E_2 are not unifiable, θ_1 and θ_2 are inconsistent.

Example 9.5

Let $\theta_1 = \{f(g(x_1))/x_3, f(x_2)/x_4\}$ and $\theta_2 = \{x_4/x_3, g(x_1)/x_2\}$. For this case, $E_1 = (x_3, x_4, x_3, x_2)$ and $E_2 = (f(g(x_1)), f(x_2), x_4, g(x_1))$. Since E_1 and E_2 are unifiable, θ_1 and θ_2 are consistent. A combination of θ_1 and θ_2 is $\{f(g(x_1))/x_3, f(g(x_1))/x_4, g(x_1)/x_2\}$.

The Prawitz procedure can now be described as follows:

Step 1   Transform M into a disjunctive normal form $C_1 \vee C_2 \vee \cdots \vee C_n$.

Step 2   For each conjunction C_i in the disjunctive normal form, find all refuting substitutions $\alpha_{i1}, \ldots, \alpha_{iq_i}$.

Step 3   From $\alpha_{i1}, \ldots, \alpha_{iq_i}, i = 1, \ldots, n$, find an α_{is_i} such that $\alpha_{1s_1}, \ldots, \alpha_{ns_n}$ are consistent. If such $\alpha_{is_i}, i = 1, \ldots, n$, exist, then a combination θ of $\alpha_{1s_1}, \ldots, \alpha_{ns_n}$ is a desired substitution and is a solution of M. Otherwise, M has no solution.

Example 9.6

Consider $S = \{P(x), \sim P(a) \vee \sim P(b)\}$. Let M be the set consisting of two copies of $P(x)$ and one copy of $\sim P(a) \vee \sim P(b)$. That is, $M = \{P(x), P(x), \sim P(a) \vee \sim P(b)\}$. Renaming the variable of each clause in M, we obtain $M = \{P(x_1), P(x_2), \sim P(a) \vee \sim P(b)\}$. Now, no clauses in M have variables in common. Let M be written as $M = P(x_1) \wedge P(x_2) \wedge [\sim P(a) \vee \sim P(b)]$. Transforming M into a disjunctive normal form, we obtain

$$[P(x_1) \wedge P(x_2) \wedge \sim P(a)] \vee [P(x_1) \wedge P(x_2) \wedge \sim P(b)].$$

Let

$$C_1 = P(x_1) \wedge P(x_2) \wedge \sim P(a)$$

$$C_2 = P(x_1) \wedge P(x_2) \wedge \sim P(b).$$

Clearly, $\alpha_{11} = \{a/x_1\}$ and $\alpha_{12} = \{a/x_2\}$ are the refuting substitutions for C_1. Similarly, $\alpha_{21} = \{b/x_1\}$ and $\alpha_{22} = \{b/x_2\}$ are the refuting substitutions for C_2. We know that α_{11} and α_{22} are consistent, with the combination of α_{11} and α_{22} being $\{a/x_1, b/x_2\}$. Therefore, the substitution, $\{a/x_1, b/x_2\}$, is a desired solution of M. We note that α_{12} and α_{21} are also consistent. $\{b/x_1, a/x_2\}$ is the combination of α_{12} and α_{21}. Therefore, $\{b/x_1, a/x_2\}$ is also a solution of M.

9.3   THE V-RESOLUTION PROCEDURE

The Prawitz procedure has the great virtue of generating only the substitutions actually needed for obtaining an unsatisfiable set of ground clauses. However, the use of disjunctive normal form inevitably prevents it from being efficient. As is easily seen, even for an M that consists of ten two-literal clauses, the disjunctive normal form of M will contain 2^{10} conjunctions. In order to circumvent this difficulty, Prawitz [1969] later proposed another procedure. Davis [1963], Chinlund et al. [1964], Loveland [1968], and Chang [1971, 1972] also proposed procedures that try to retain the virtues of the original Prawitz procedure but avoid the use of disjunctive normal form. In this chapter, we shall consider V-resolution proposed by Chang [1972], since it is closely related to resolution.

We shall first give a simple example of V-resolution. Let us consider the following set S of clauses

(1) $P(x)$
                                        $\left.\rule{0pt}{24pt}\right\} S.$
(2) $\sim P(a) \vee \sim P(b)$

Let M be the set that consists of one copy of clause (1) and one copy of clause (2). That is,

(1') $P(x_1)$
                                        $\left.\rule{0pt}{24pt}\right\} M.$
(2') $\sim P(a) \vee \sim P(b)$

We may resolve (1') and (2'). If we replace x_1 by a, from (1') and (2') we obtain

(3') $\sim P(b)$.

Since in V-resolution we do not allow x_1 in (1') to be replaced by some other ground term, we cannot resolve (3') and (1') again. Our algorithm could put into M another copy of clause (1), say

(4') $P(x_2)$.

We now may replace x_2 by b, and resolve (3') and (4') to obtain the empty clause.

To prevent the resolution of clauses (3′) and (1′) we shall use V-resolution, in which we keep track of substitutions for variables. In V-resolution, every clause is augmented by a substitution. We insist that every clause in M be augmented by an empty substitution ε. Thus every clause C is represented by $C|\theta$, where θ is a substitution. In the above case, since clauses (1′) and (2′) are in M, they are represented, respectively, by

(1″) $P(x_1)|\varepsilon$

(2″) $\sim P(a) \vee \sim P(b)|\varepsilon.$

Clause (3′) is represented by

(3″) $\sim P(b)|\{a/x_1\}.$

Note that $\{a/x_1\}$ in clause (3″) is the substitution obtained by resolving (1″) and (2″) upon the literals $P(x_1)$ and $\sim P(a)$. When we resolve (3″) and (1″), we need another substitution $\{b/x_1\}$. We note that this substitution is inconsistent with $\{a/x_1\}$. Therefore, this resolution will be blocked. However, since clause (4′) is put into M, clause (4′) is represented by

(4″) $P(x_2)|\varepsilon.$

Now we can resolve clauses (3″) and (4″) to obtain

(5″) $\square|\{a/x_1, b/x_2\}$

where $\{a/x_1, b/x_2\}$ indicates the substitution we have to make for variables x_1 and x_2 to resolve the clauses.

Definition   If two or more literals (with the same sign) of a clause $C|\theta$ have a most general unifier σ, and if θ and σ are consistent, with ϕ being a combination of ϕ and σ, then $C\sigma|\phi$ is called a *V-factor* of $C|\theta$.

Example 9.7

Consider a clause $C|\theta$ given as $P(a) \vee P(x) \vee Q(b)|\{b/y\}$. Since $P(a)$ and $P(x)$ have a most general unifier $\sigma = \{a/x\}$, and since $\{b/y\}$ and $\{a/x\}$ are consistent, with $\phi = \{a/x, b/y\}$ being a combination of them, $P(a) \vee Q(b)|\{a/x, b/y\}$ is a V-factor of $C|\theta$.

Definition   Let $C_1|\theta_1$ and $C_2|\theta_2$ be two clauses. If θ_1 and θ_2 are consistent, with α being a combination of θ_1 and θ_2, if there are literals L_1 and L_2 in C_1 and C_2, respectively, such that L_1 and $\sim L_2$ have a most general unifier σ, and if α and σ are consistent, with θ being a combination of α and σ, then we infer the clause

$$\{(C_1\sigma - L_1\sigma) \cup (C_2\sigma - L_2\sigma)\}|\theta.$$

The above clause is called a *binary V-resolvent* of (the *parent* clauses) $C_1|\theta_1$ and $C_2|\theta_2$. The literals L_1 and L_2 are called the *literals resolved upon.*

Example 9.8

Consider two clauses $C_1|\theta_1$ and $C_2|\theta_2$ given, respectively, as $\sim P(a) \vee Q(a,b)|\{a/x\}$ and $P(x) \vee Q(x,b)|\{b/y\}$. Let L_1 and L_2 be $\sim P(a)$ and $P(x)$, respectively. Since $\{a/x\}$ and $\{b/y\}$ are consistent, with $\alpha = \{a/x, b/y\}$ being a combination of them, since L_1 and $\sim L_2$ have a most general unifier $\sigma = \{a/x\}$, and since α and σ are consistent, with $\theta = \{a/x, b/y\}$ being a combination of α and σ, we obtain a binary V-resolvent as

$$Q(a,b)|\{a/x, b/y\}.$$

The literals $\sim P(a)$ and $P(x)$ are the literals resolved upon.

Definition A *V-resolvent* of clauses $C_1|\theta_1$ and $C_2|\theta_2$ is one of the following binary V-resolvents:

1. a binary V-resolvent of $C_1|\theta_1$ and $C_2|\theta_2$;
2. a binary V-resolvent of $C_1|\theta_1$ and a V-factor of $C_2|\theta_2$;
3. a binary V-resolvent of a V-factor of $C_1|\theta_1$ and $C_2|\theta_2$;
4. a binary V-resolvent of a V-factor of $C_1|\theta_1$ and a V-factor of $C_2|\theta_2$.

We note that V-resolution is the same as resolution except that we have to check the consistency of substitutions. V-Resolution is also called *variable-constrained resolution* because variables are treated as unknowns rather than as universal variables. Like resolution, V-resolution is complete in the sense that if an alleged S-unsatisfiable set M has a solution, then the empty clause can be derived by repeatedly applying V-resolution on M. Suppose we define

$$R^0(M) = M$$
$$R^i(M) = R^{i-1}(M) \cup \{\text{V-resolvents of all pairs of clauses in } R^{i-1}(M)\},$$
for $i = 1, 2, \ldots.$

Then, the following theorem is obvious.

Theorem 9.1 If M has a solution, then the empty clause belongs to $R^n(M)$ for some finite n.

In addition to the fact that V-resolution is complete, many refinements of resolution, such as hyperresolution, set-of-support resolution, and linear resolution introduced in Chapters 6 and 7, can be incorporated into V-resolution. The definitions of V-hyperresolution, set-of-support V-resolution, and linear V-resolution are almost the same as the definitions of their respective counterparts except that resolution (resolvent) is changed to V-resolution (V-resolvent).

The differences between resolution and V-resolution are as follows: Given a set S of clauses to prove, resolution works on the set S itself, while V-resolution works on an alleged S-unsatisfiable set M for S. In addition, variables are treated as universal variables in resolution, while they are treated as unknown parameters in V-resolution. Because of these differences, there is a marked distinction in efficiency between resolution and V-resolution. This is well illustrated in the following example. For other experimental evidence, the reader may read [Chang, 1972].

Example 9.9

Consider the following two clauses:

C_1: $\sim P(x) \vee P(f(x))$

C_2: $P(a)$.

Using resolution, from the clauses C_1 and C_2 we can generate an infinite number of clauses, $P(f(a)), P(f(f(a))), P(f(f(f(a)))), \ldots$. However, if we use a copy of each of C_1 and C_2, and augment each copy by the empty substitution ε, we have

C_1'; $\sim P(x) \vee P(f(x))|\varepsilon$

C_2': $P(a)|\varepsilon$.

Now, applying V-resolution, from C_1' and C_2' we can generate only one clause, namely, $P(f(a))|\{a/x\}$. Thus, we see that we can limit the generation of resolvents by considering a finite number of copies of clauses and by using V-resolution.

Given a set S of clauses to prove, the process that uses V-resolution may be stated as follows: *Start with an arbitrary alleged S-unsatisfiable set M for S. (Note that variables of each clause in M have to be renamed so that no clauses have any variables in common.) Apply V-resolution to M. If the empty clause is generated, we terminate the process. Otherwise, we put a few more copies of clauses of S into M, and repeat the process again. (When copies are introduced into M, variables of the new copies should be renamed so that no clauses in the enlarged set M have variables in common.) The cycle is repeated again and again until either the empty clause is generated, or some space or time bound is exceeded.* For most theorems, usually a few copies of each clause in S are sufficient.

Suppose $S = \{C_1, \ldots, C_g\}$. The initial set M can be easily specified by a set of numbers, $\{n_1, \ldots, n_g\}$, where n_i indicates the number of copies of C_i that are to be put into $M, i = 1, \ldots, g$. This notation will be used throughout this chapter. We now give a few examples to illustrate how V-resolution can be used.

Example 9.10

Let $S = \{P(x), \sim P(a) \vee \sim P(b)\}$. We want to show that S is unsatisfiable. First let an alleged S-unsatisfiable set M for S be specified by $\{2, 1\}$. Thus, $M = \{P(x), P(x), \sim P(a) \vee \sim P(b)\}$. In order that no clauses in M have variables in common, we rename the variable of each clause and obtain $M = \{P(x_1), P(x_2), \sim P(a) \vee \sim P(b)\}$. Now, augmenting every clause in M by the empty substitution ε, we have $M = \{P(x_1)|\varepsilon, P(x_2)|\varepsilon, \sim P(a) \vee \sim P(b)|\varepsilon\}$. Applying V-resolution on M, we obtain the following proof:

(1) $P(x_1)|\varepsilon$

(2) $P(x_2)|\varepsilon$ M

(3) $\sim P(a) \vee \sim P(b)|\varepsilon$

(4) $\sim P(b)|\{a/x_1\}$ a V-resolvent of (1) and (3)

(5) $\square|\{a/x_1, b/x_2\}$ a V-resolvent of (2) and (4).

Since $\{a/x_1, b/x_2\}$ is a substitution associated with $\square$, it is a solution of M.

Example 9.11

Prove the theorem: "In a group, the left identity element is also a right identity." Let $P(x, y, z)$ stand for $x \circ y = z$, where $\circ$ is the group operator. Then, the negation of this theorem can be represented by the following set S of clauses:

A: $P(i(x), x, e)$

B: $P(e, x, x)$

C: $\sim P(x, y, u) \vee \sim P(y, z, v) \vee \sim P(u, z, w) \vee P(x, v, w)$

D: $\sim P(x, y, u) \vee \sim P(y, z, v) \vee \sim P(x, v, w) \vee P(u, z, w)$

E: $\sim P(a, e, a)$.

Let an alleged S-unsatisfiable set M for S be specified by $\{2, 2, 1, 1, 1\}$. Then, using V-resolution, we have the following proof, where we use $1, 2, \ldots, 16$ to denote $x_1, x_2, \ldots, x_{16}$, respectively:

A': $P(i(1), 1, e)|\varepsilon$

A'': $P(i(2), 2, e)|\varepsilon$

B': $P(e, 3, 3)|\varepsilon$

B'': $P(e, 4, 4)|\varepsilon$

$\left.\begin{array}{l} \\ \\ \\ \\ \\ \\ \end{array}\right\} M$

C': $\sim P(5, 6, 8) \vee \sim P(6, 7, 9) \vee \sim P(8, 7, 10) \vee P(5, 9, 10)|\varepsilon$

D': $\sim P(11, 12, 14) \vee \sim P(12, 13, 15) \vee \sim P(11, 15, 16) \vee P(14, 13, 16)|\varepsilon$

E': $\sim P(a, e, a)|\varepsilon$

F: $\sim P(11, 12, a) \vee \sim P(12, e, 15) \vee \sim P(11, 15, a)|\{a/14, e/13, a/16\}$

   from E' and D'

G: $\sim P(11, e, a)|\{a/14, e/13, a/16, e/12, e/3, e/15\}$ from F and B'

H: $\sim P(5, 6, 8) \vee \sim P(6, 7, e) \vee \sim P(8, 7, a)|\{a/14, e/13, a/16, e/12, e/3, e/15,$

   $5/11, e/9, a/10\}$ from G and C'

I: $\sim P(5, 6, e) \vee \sim P(6, a, e)|\{a/14, e/13, a/16, e/12, e/3, e/15, 5/11, e/9, a/10,$

   $e/8, a/7, a/4\}$ from H and B''

J: $\sim P(1, a, e)|\{a/14, e/13, a/16, e/12, e/3, e/15, i(1)/11, e/9, a/10, e/8, a/7, a/4,$

   $i(1)/5, 1/6\}$ from I and A'

K: $\square|\{a/14, \ e/13, \ a/16, \ e/12, \ e/3, \ e/15, \ i(i(a))/11, \ e/9, \ a/10, \ e/8, \ a/7, \ a/4,$

   $i(i(a))/5, i(a)/6, i(a)/1, a/2\}$ from J and A''.

Note that A' and A'' are the two copies of A, B' and B'' are the two copies of B, and C', D', and E' are copies of C, D and E, respectively.

Example 9.12 (Graph Isomorphism Problem)

Consider the two directed graphs G_1 and G_2 in Fig. 9.1. We know that G_1 is isomorphic (topologically equivalent) to G_2 since the transformation $\{x \rightarrow a, y \rightarrow d, z \rightarrow b, u \rightarrow c\}$ can carry G_1 onto G_2. Now, suppose we do not know this fact and we want to find a transformation that can carry G_1 onto G_2. We do this by using V-resolution. First, we have to formulate this

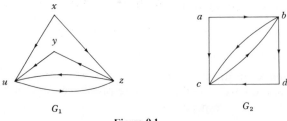

Figure 9.1

problem. Let $C(n_1, n_2)$ denote that there is an arc pointing from n_1 to n_2. Then, G_1 can be represented by the following formula:

A_1: $C(x, u) \wedge C(x, z) \wedge C(y, u) \wedge C(z, y) \wedge C(u, z) \wedge C(z, u)$.

G_2 can be represented by the formula:

A_2: $C(a, b) \wedge C(a, c) \wedge C(b, c) \wedge C(c, b) \wedge C(d, c) \wedge C(b, d)$.

Thus, showing that G_1 is isomorphic to G_2 is equivalent to showing that the formula

$$(\exists x)(\exists y)(\exists z)(\exists u)(A_2 \rightarrow A_1)$$

is valid. As usual, transforming the negation of

$$(\exists x)(\exists y)(\exists z)(\exists u)(A_2 \rightarrow A_1)$$

into a standard form, we obtain the set S of clauses as follows:

$$S = \{C(a, b), C(a, c), C(b, c), C(c, b), C(d, c), C(b, d),$$
$$\sim C(x, u) \vee \sim C(x, z) \vee \sim C(y, u) \vee \sim C(z, y) \vee \sim C(u, z) \vee \sim C(z, u)\}.$$

Let an alleged S-unsatisfiable set M for S be specified by $\{1, 1, 1, 1, 1, 1, 1\}$. Then, using V-resolution, we obtain the following proof:

(1)	$C(a, b)\|\varepsilon$	
(2)	$C(a, c)\|\varepsilon$	
(3)	$C(b, c)\|\varepsilon$	
(4)	$C(c, b)\|\varepsilon$	$\Big\} M$
(5)	$C(d, c)\|\varepsilon$	
(6)	$C(b, d)\|\varepsilon$	
(7)	$\sim C(x, u) \vee \sim C(x, z) \vee \sim C(y, u) \vee \sim C(z, y) \vee \sim C(u, z) \vee \sim C(z, u)\|\varepsilon$	

(8) $\sim C(a, z) \vee \sim C(y, c) \vee \sim C(z, y) \vee \sim C(c, z) \vee \sim C(z, c)|$
$\{a/x, c/u\}$ from (7) and (2)

(9) $\sim C(y, c) \vee \sim C(b, y) \vee \sim C(c, b) \vee \sim C(b, c)|\{a/x, c/u, b/z\}$
                                                                        from (8) and (1)

(10) $\sim C(b, d) \vee \sim C(c, b) \vee \sim C(b, c)|\{a/x, c/u, b/z, d/y\}$
                                                                        from (9) and (5)

(11) $\sim C(c, b) \vee \sim C(b, c)|\{a/x, c/u, b/z, d/y\}$ from (10) and (6)

(12) $\sim C(b, c)|\{a/x, c/u, b/z, d/y\}$ from (11) and (4)

(13) $\square|\{a/x, c/u, b/z, d/y\}$ from (12) and (3).

Since the substitution $\{a/x, c/u, b/z, d/y\}$ is associated with $\square$, we can obtain from it a transformation in which x maps to a, u to c, z to b, and y to d. Clearly, this transformation can carry G_1 onto G_2. Thus, G_1 is isomorphic to G_2.

The idea of V-resolution may also be applied to paramodulation. This is done as follows by modifying the definition of paramodulation given in Section 8.3

Definition   Let $C_1|\theta_1$ and $C_2|\theta_2$ be two clauses. If θ_1 and θ_2 are consistent, with α being a combination of θ_1 and θ_2, if C_1 is $L[t] \vee C_1'$ and C_2 is $r = s \vee C_2'$, where $L[t]$ is a literal containing the term t and C_1' and C_2' are disjunctions of literals, if t and r have a most general unifier σ, and if α and σ are consistent, with θ being a combination of α and σ, then infer

$$L\sigma[s\sigma] \cup C_1'\sigma \cup C_2'\sigma|\theta$$

where $L\sigma[s\sigma]$ denotes the result obtained by replacing one single occurrence of $t\sigma$ in $L\sigma$ by $s\sigma$. The above inferred clause is called a *binary V-paramodulant* of $C_1|\theta_1$ and $C_2|\theta_2$, and the literals L and $r = s$ are called the *literals paramodulated upon*.

The above definition of binary V-paramodulation is the same as that of binary paramodulation, except that the consistency of substitutions has to be checked.

Definition   A *V-paramodulant* of $C_1|\theta_1$ and $C_2|\theta_2$ is one of the following binary V-paramodulants:

1. a binary V-paramodulant of $C_1|\theta_1$ and $C_2|\theta_2$;
2. a binary V-paramodulant of $C_1|\theta_1$ and a V-factor of $C_2|\theta_2$;
3. a binary V-paramodulant of a V-factor of $C_1|\theta_1$ and $C_2|\theta_2$;
4. a binary V-paramodulant of a V-factor of $C_1|\theta_1$ and a V-factor of $C_2|\theta_2$.

Let S be a set of clauses including $x = x$ and the functionally reflexive axioms.

Let M be a set of copies of clauses in S. Define

$$(RP)^0(M) = M$$

$$(RP)^i(M) = (RP)^{i-1} \cup \{\text{V-resolvents and V-paramodulants of all pairs of clauses in } (RP)^{i-1}(M)\}, i = 1, 2, \ldots.$$

Then, we have the following theorem.

Theorem 9.2   If there is a ground substitution θ such that $M\theta$ is E-unsatisfiable, then the empty clause belongs to $(RP)^n(M)$ for some finite n.

Example 9.13

Let $S = \{P(x) \vee Q(x), \sim P(a), \sim Q(f(a)), x = f(x)\}$. Show that S is E-un-satisfiable. Let M be specified by $\{1, 1, 1, 1\}$. Thus, $M = \{P(x) \vee Q(x), \sim P(a), \sim Q(f(a)), x = f(x)\}$. Since we do not allow the clauses in M to have variables in common, we rename the variables and have $M = \{P(x_1) \vee Q(x_1), \sim P(a), \sim Q(f(a)), x_2 = f(x_2)\}$. Now, augmenting every clause in M by ε, we have $M = \{P(x_1) \vee Q(x_1)|\varepsilon, \sim P(a)|\varepsilon, \sim Q(f(a))|\varepsilon, x_2 = f(x_2)|\varepsilon\}$. Applying V-paramodulation and V-resolution on M, we obtain the following proof:

(1) $P(x_1) \vee Q(x_1)|\varepsilon$ $\left.\vphantom{\begin{array}{c}1\\2\\3\\4\end{array}}\right\}$

(2) $\sim P(a)|\varepsilon$

(3) $\sim Q(f(a))|\varepsilon$ $\qquad M$

(4) $x_2 = f(x_2)|\varepsilon$

(5) $Q(a)|\{a/x_1\}$ a V-resolvent of (1) and (2)

(6) $Q(f(a))|\{a/x_1, a/x_2\}$ a V-paramodulant of (4) and (5)

(7) $\square|\{a/x_1, a/x_2\}$ a V-resolvent of (3) and (6).

9.4   PSEUDOSEMANTIC TREES

In Section 4.4 we introduced the semantic tree concept. By Herbrand's theorem (Theorem 4.3), a set S of clauses is unsatisfiable if and only if there is a finite closed semantic tree of S. Therefore, to prove that S is unsatisfiable, we may try to find a finite closed semantic tree of S. In fact, we may even relax some conditions for a semantic tree. That is, we can define a pseudosemantic tree and prove the unsatisfiability of S by generating a closed pseudosemantic tree. A pseudosemantic tree satisfies fewer conditions than a semantic tree and is therefore usually easier to construct.

Definition   Given a set S of clauses, let $A = \{A_1, A_2, A_3, \ldots\}$ be the atom set of S. A (binary) *pseudosemantic* tree of S is a downward tree, T, to each link of which is attached an A_j or $\sim A_j$ in such a way that for each node N, there are only two immediate links L_1 and L_2 from N, and if Q_1 and Q_2 are the literals attached to L_1 and L_2, respectively, then $Q_1 \vee Q_2$ is a tautology.

In the sequel, if N is a node of a pseudosemantic tree of a set S of clauses, then we denote by $I(N)$ the set of all literals attached to the links of the branch from the top node to node N. In semantic trees, $I(N)$ has to be consistent, that is, it cannot contain a complementary pair. This is not the case for pseudosemantic trees.

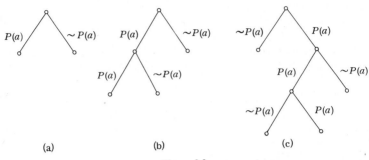

Figure 9.2

Example 9.14

Let $S = \{P(x), \sim P(a)\}$. The atom set of S is $\{P(a)\}$. Some pseudosemantic trees of S are shown in Fig. 9.2. We note that the tree shown in Fig. 9.2a is also a semantic tree.

Definition A node N of a pseudosemantic tree of a set S of clauses is called a *failure* node if and only if there is a ground instance C of a clause in S such that $I(N) \cup \{C\}$ is unsatisfiable, but there is no ground instance C' of a clause in S such that $I(N') \cup \{C'\}$ is unsatisfiable for every ancestor node N' of N.

Definition A pseudosemantic tree T is said to be *closed* if and only if every branch of T terminates at a failure node.

Example 9.15

Let $S = \{P, \sim Q, Q\}$. The atom set of S is $\{P, Q\}$. Some closed pseudosemantic trees of S are shown in Fig. 9.3. We note that the tree shown in Fig. 9.3a is also a closed semantic tree.

We note that every closed pseudosemantic tree has a closed semantic tree embedded within it. That is, we can always construct a closed semantic tree from a closed pseudosemantic tree. The following theorem is an extension of Theorem 4.3 (Herbrand's theorem, version I).

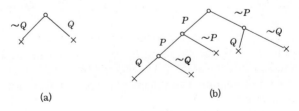

Figure 9.3

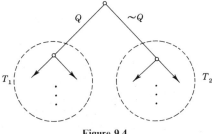

Figure 9.4

Theorem 9.3   A set S of clauses is unsatisfiable if and only if S has a closed pseudosemantic tree.

Proof ($\Rightarrow$) By Herbrand's theorem, since S is unsatisfiable, S has a closed semantic tree. However, a closed semantic tree is also a closed pseudosemantic tree. Therefore, the first half of Theorem 9.3 is obvious.

($\Leftarrow$)   Since S has a closed pseudosemantic tree T, there is a set S' of ground instances of clauses in S such that T is also a closed pseudosemantic tree for S'. We now show that S' is unsatisfiable by induction on the number n of distinct atoms in S'. If $n = 0$, clearly, S' is unsatisfiable. Suppose S' is unsatisfiable when S' consists of i distinct atoms, $0 \leqslant i \leqslant n$. To complete the induction, we consider S' that consists of exactly $n+1$ distinct atoms. Let Q and $\sim Q$ be the literals attached to the immediate links from the top node of T. Let T_1 and T_2 be the subtrees shown in Fig. 9.4. Let S_1' be the set obtained from S' by deleting any clause containing Q and deleting $\sim Q$ from the remaining clauses. Similarly, let S_2' be the set obtained from S' by deleting any clause containing $\sim Q$ and deleting Q from the remaining clauses. For any node N in T_1, if the literals attached to the immediate links from N are Q and $\sim Q$, delete these two links and attach to node N the subtree that follows the link attached by Q. Let T_1' be the pseudosemantic tree obtained from T_1 by applying the above process. Similarly, let T_2' be the pseudosemantic tree obtained from T_2 by applying the same process that is applied on T_1 except that Q and $\sim Q$ are interchanged to $\sim Q$ and Q, respectively. Clearly, T_1' and T_2' are closed pseudosemantic trees for S_1' and S_2', respectively. Since S_1' and S_2' consist of n or fewer than n atoms, by the induction hypothesis S_1' and S_2' are both unsatisfiable. Therefore, S' is unsatisfiable. Consequently, S is unsatisfiable. This completes the proof of Theorem 9.3.

9.5   A PROCEDURE FOR GENERATING CLOSED PSEUDOSEMANTIC TREES

The pseudosemantic tree concept can be used to find a solution of an alleged S-unsatisfiable set M for a set S of clauses. If M contains only *ground* clauses, a closed pseudosemantic tree, or in fact a closed semantic tree, can be easily found as follows: Start with a top node. Select an atom A occurring in M. Grow two links from the top node. Attach one of these two links with A, the other with $\sim A$. For the tree so far generated, find a tip node N, that is, a node which has no links growing from it. Test whether or not there is a clause in M that is false in $I(N)$. If yes, N is a failure node and marked by "$\times$." Otherwise, N is not a failure node. In this case, grow two links from N. Attach one of these two links with another atom A' occurring in M, and the other with $\sim A'$. This process is repeated again and again until all the tip nodes are failure nodes. When that happens, the final tree is a closed semantic tree.

The above procedure is for an M that contains only ground clauses. However, for the general case, that is, for an M which contains variables, the above procedure has to be modified. That is, we should have a method of testing whether a tip node is a failure node. In addition, we should have a mechanism of making substitutions for variables. In the sequel, we shall see that these can be easily accomplished by using *unit V-resolution*, that is, a binary V-resolution in which one of the two parent clauses is a unit clause. Furthermore, we shall use the depth-first method to generate a closed pseudosemantic tree. That is, whenever we select a tip node, we select the leftmost one in a tree. In the following procedure that we shall describe, we may come to a point where there is no substitution under which a tip node becomes a failure node. In this case, we have to back up to a preceding marked tip node and grow two links from it. To make the back-up procedure easier, we shall represent a branch of a tree by a list of literals attached to the branch. For example, consider the tree shown in Fig. 9.5, where nodes B, D, and E are tip nodes. The branch from A to B is represented by $[\sim P(x_1)]$, the branch from A to D by $[P(x_1), \sim P(x_2)]$, and the branch from A to E by $[P(x_1), P(x_2)]$.

We now describe an algorithm for generating a closed pseudosemantic tree as follows.

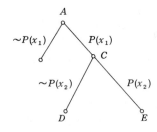

Figure 9.5

Generation of a Closed Pseudosemantic Tree

Step 0 Let M be an alleged S-unsatisfiable set for a set S of clauses. Let W be the set of all distinct positive literals occurring in M. (W may be the set of all distinct negative literals occurring in M. We use whichever is smaller.)

Step 1 Let $i = 0$, $R^0 = \{\varepsilon\}$, BRANCH $= $ (NIL), and BACKUP $=$ NIL, where NIL is a null list.

Step 2 If BRANCH is empty, terminate the algorithm; the final tree contains a closed pseudosemantic tree and every substitution in R^i is a solution of M. Otherwise, continue.

Step 3 Let B be the first element in BRANCH. Delete B from BRANCH.

Step 4 For each substitution $\theta \in R^i$ and each clause C in M, compute $K(C, \theta)$ as follows: Let $U(\theta) = \{L \mid \theta$, for all literals L in $B\}$. Test whether the empty clause can be derived from $(U(\theta) \cup \{C \mid \varepsilon\})$ by unit V-resolution. If yes, derive all the possible augmented empty clauses. $K(C, \theta)$ is the set of substitutions associated with all these augmented empty clauses. Otherwise, let $K(C, \theta)$ be the empty set. After $K(C, \theta)$ is computed for every $\theta \in R^i$ and every $C \in M$, let K be the union of all the sets $K(C, \theta)$ for all θ in R^i and all C in M, that is,

$$K = \bigcup_{C \in M} \bigcup_{\theta \in R^i} K(C, \theta).$$

Step 5 If K is empty, go to the next step. Otherwise, put the pair (B, R^i) at the beginning of BACKUP, set $R^{i+1} = K, i = i+1$, and go to Step 2.

Step 6 Choose an element L from W such that L or the negation of L does not appear in B. If this is impossible, go to Step 8. Otherwise, continue.

Step 7 Let B_1 and B_2 be the two lists obtained by appending L and $\sim L$, respectively, to the end of list B. Put B_1 and B_2 (in an arbitrary order) at the beginning of BRANCH. Go to Step 2.

Step 8 If BACKUP is empty, terminate the algorithm; there is no solution of M. (M has to be enlarged before the whole algorithm is started anew.) Otherwise, continue.

Step 9 Let (P, K) be the first pair in BACKUP. Delete (P, K) from BACKUP. Set $i = i-1, R^i = K, B = P$, and go to Step 6.

The reader may show that the above algorithm will always find a solution of M if one exists. The idea of this algorithm is that if M has a solution, then there is a ground substitution θ such that $M\theta$ is unsatisfiable. Since $M\theta$ is an unsatisfiable set of ground clauses, it is easy to find a closed pseudosemantic tree of $M\theta$. This algorithm is one that combines the search of such a substitution and the generation of a closed pseudosemantic tree. We now give an example to illustrate the algorithm.

Example 9.16

Let $S = \{P(x), \sim P(a) \vee \sim P(b)\}$. Let M be specified by $\{2, 1\}$. That is, $M = \{P(x), P(x), \sim P(a) \vee \sim P(b)\}$. After renaming variables, we have $M = \{P(x_1), P(x_2), \sim P(a) \vee \sim P(b)\}$. Since all the distinct positive literals occurring in M are $P(x_1)$ and $P(x_2)$, we obtain $W = \{P(x_1), P(x_2)\}$. We now apply the algorithm.

1. $i = 0, R^0 = \{\varepsilon\}$, BRANCH = (NIL), and BACKUP = NIL.

2. The first element of BRANCH is NIL. Therefore, $B = $ NIL and BRANCH is changed to ( ). Clearly, after Step 4, K is found to be empty. Choose $P(x_1)$ from W. Thus we obtain $B_1 = [\sim P(x_1)], B_2 = [P(x_1)]$, and BRANCH = $([\sim P(x_1)], [P(x_1)])$. The corresponding tree is shown in Fig. 9.6a.

3. Now, $B = [\sim P(x_1)]$ and BRANCH = $([P(x_1)])$. Consider each substitution θ in R^0 and each clause C in M. If $\theta = \varepsilon$ and $C = P(x_1)$, then $U(\theta) = \{\sim P(x_1)|\varepsilon\}$ and $(U(\theta) \cup \{C|\varepsilon\}) = \{\sim P(x_1)|\varepsilon, P(x_1)|\varepsilon\}$. From $(U(\theta) \cup \{C|\varepsilon\})$, we derive $\square|\varepsilon$. Therefore, $K(C, \theta) = \{\varepsilon\}$ for $C = P(x_1)$ and $\theta = \varepsilon$. Similarly, we can obtain $K(C, \theta) = \{x_1/x_2\}$ for $C = P(x_2)$ and $\theta = \varepsilon$; $K(C, \theta) = $ empty for $C = \sim P(a) \vee \sim P(b)$ and $\theta = \varepsilon$. Therefore, $K = \{\varepsilon, \{x_1/x_2\}\}$. Since K is not empty, we obtain BACKUP = $(([\sim P(x_1)], \{\varepsilon\}))$, $R^1 = \{\varepsilon, \{x_1/x_2\}\}$, and $i = 1$. The corresponding tree is shown in Fig. 9.6b, where the tip node at the end of the branch represented by $[\sim P(x_1)]$ is marked as a failure node.

4. This time $B = [P(x_1)]$ and BRANCH = ( ). We find that K is empty. Therefore, we choose an element $P(x_2)$ from W. (Note that neither $P(x_2)$ nor $\sim P(x_2)$ is in B.) Thus, $B_1 = [P(x_1), \sim P(x_2)], B_2 = [P(x_1), P(x_2)]$, and BRANCH = $([P(x_1), \sim P(x_2)], [P(x_1), P(x_2)])$. The corresponding tree is shown in Fig. 9.6c.

5. $B = [P(x_1), \sim P(x_2)]$ and BRANCH = $([P(x_1), P(x_2)])$. We find that $K = \{\varepsilon, \{x_1/x_2\}\}$. Since K is not empty, we obtain BACKUP = $(([P(x_1), \sim P(x_2)], \{\varepsilon, \{x_1/x_2\}\}), ([\sim P(x_1)], \{\varepsilon\}))$, $R^2 = \{\varepsilon, \{x_1/x_2\}\}$, and $i = 2$. The corresponding tree is shown in Fig. 9.6d.

6. $B = [P(x_1), P(x_2)]$ and BRANCH = ( ). We find that $K = \{\{a/x_1, b/x_2\}, \{b/x_1, a/x_2\}\}$. Since K is not empty, we obtain BACKUP = $(([P(x_1), P(x_2)], \{\varepsilon, \{x_1/x_2\}\}), ([P(x_1), \sim P(x_2)], \{\varepsilon, \{x_1/x_2\}\}), ([\sim P(x_1)], \{\varepsilon\}))$, $R^3 = \{\{a/x_1, b/x_2\}, \{b/x_1, a/x_2\}\}$, and $i = 3$. The corresponding tree is shown in Fig. 9.6e.

7. Since BRANCH is empty, we terminate the algorithm. Each of the substitutions in R^3 is a solution of M. Let T be the final tree shown in Fig. 9.6e. Let $T\theta$ be the tree obtained from T by replacing each literal L in T by

L0. *Tθ* is shown in Fig. 9.6f if $\theta = \{a/x_1, b/x_2\}$, or in Fig. 9.6g if $\theta = \{b/x_1, a/x_2\}$. *T$\theta$* is a closed pseudosemantic tree for *S*. Therefore, we conclude that *S* is unsatisfiable.

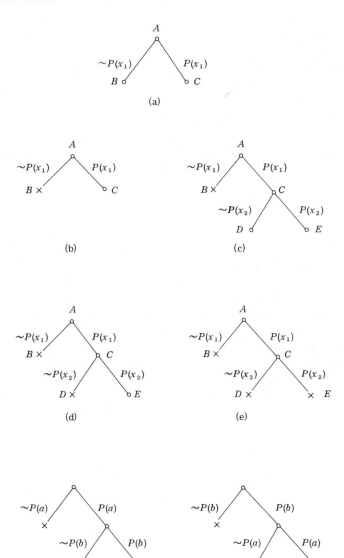

Figure 9.6

There are several ways to improve the above algorithm. Some of them are given as follows:

First, we may devise heuristic schemes for the selection of literals from the set W. (See Step 6 of the algorithm.) For example, we may prefer literals from shorter clauses to ones from longer clauses in M. It is possible that the selection of literals in a "good" order may enable us to terminate the algorithm more quickly.

Second, the set R^i generated by the algorithm at each step should be simplified as far as possible. This is done by first considering the following definition.

Definition   Let $\lambda = \{t_1/v_1, \ldots, t_n/v_n\}$ be a substitution. Let θ be another substitution. From λ, define $E = \{t_1 = v_1, \ldots, t_n = v_n\}$. Then θ is said to be *more restrictive* than λ if and only if $E\theta$ is either empty or contains only identities.

Example 9.17

$\lambda = \{a/x_1\}$ and $\theta = \{a/x_1, b/x_2\}$. From λ, we define $E = \{a = x_1\}$. Then $E\theta = \{a = a\}$, which consists of only an identity. Therefore, θ is more restrictive than λ.

The way to simplify the set R^i is this: For any θ_1 and θ_2 in R^i, if θ_1 is more restrictive than θ_2, then θ_1 should be deleted from R^i. For example, in R^1 of Example 9.16, since $\{x_1/x_2\}$ is more restrictive than ε, it should be deleted from R^1.

Third, since the efficiency of the algorithm depends heavily upon Step 4 being efficiently carried out, this step is the heart of the algorithm. In Step 4, for each substitution $\theta \in R^i$ and each clause C in M, one way to compute $K(C, \theta)$ is as follows: Let $U(\theta)$ be defined as previously. Let n be the number of literals in C. Define

$$Q^n(C, \theta) = \{C \,|\, \varepsilon\}$$

$$Q^{k-1}(C, \theta) = \{\text{Unit V-resolvents of } C_1 \text{ and } C_2, \text{ where } C_1 \in U(\theta),$$
$$C_2 \in Q^k(C, \theta), \text{ and the literal in } C_2 \text{ that is resolved upon is the}$$
$$\textit{first} \text{ literal of } C_2\}, \text{ for } k = n, n-1, \ldots, 1.$$

After $Q^0(C, \theta)$ is computed, $K(C, \theta)$ is defined as

$$K(C, \theta) = \{\alpha, \text{ for all substitutions } \alpha \text{ such that the augmented empty}$$
$$\text{clauses } \square \,|\, \alpha \text{ are in } Q^0(C, \theta)\}.$$

Finally, if the algorithm backs up through Steps 9, 6, 7, 2, 3, and 4, then K is to be recomputed again. In this case, we should make sure that substitutions previously computed are *not* computed again.

9.6   A GENERALIZATION OF THE SPLITTING RULE OF DAVIS AND PUTNAM

The principle of splitting is to break a difficult and long theorem into small and simple cases and then prove it by considering each case separately. Several splitting methods have been proposed by Prawitz [1969], Bledsoe [1971], Slagle and Koniver [1971], Ernst [1971], and Wos *et al.* [1967]. In Section 4.6 a splitting rule contained in the method of Davis and Putnam was also discussed. Davis and Putnam's method is only for ground clauses. In this section, we shall extend Davis and Putnam's method to general (nonground) clauses.

For a literal L and a set S of ground clauses, suppose we use $S(L)$ to denote the set obtained from S by deleting any clause containing L and by deleting $\sim L$ from the remaining set. In the splitting rule of Davis and Putnam, if L_1 is a literal in a set S of "ground" clauses, we compute $S(L_1)$ and $S(\sim L_1)$. We know that S is unsatisfiable if and only if both $S(L_1)$ and $S(\sim L_1)$ are unsatisfiable. The reason for applying the splitting rule is to split S, which cannot be proved by using only the one-literal rule, into two smaller sets $S(L_1)$ and $S(\sim L_2)$, which, hopefully, may be proved by the one-literal rule. Of course, $S(L_1)$ and $S(\sim L_1)$ can be split again. Let L_2 be a literal in $S(L_1)$. Clearly, we can compute $(S(L_1))(L_2)$ and $(S(L_1))(\sim L_2)$. Denote $(S(L_1))(L_2)$

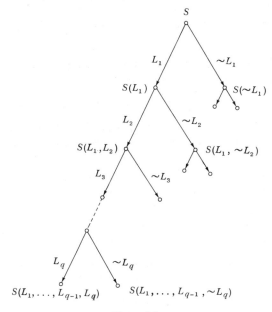

Figure 9.7

and $(S(L_1))(\sim L_2)$ by $S(L_1, L_2)$ and $S(L_1, \sim L_2)$, respectively. Thus, $S(L_1)$ is unsatisfiable if and only if both $S(L_1, L_2)$ and $S(L_1, \sim L_2)$ are unsatisfiable. The splitting rule is repeatedly applied until all the smaller sets can be proved by the one-literal rule. In general, the splitting process can be represented by the tree shown in Fig. 9.7, where $S(L_1, \ldots, L_q)$ represents $(\cdots((S(L_1))(L_2))\cdots)$ (L_q). The reader can easily see that this tree is similar to a pseudosemantic tree. We can use the depth-first method to expand this tree. That is, at every step, we try to prove the leftmost set by the one-literal rule. If we succeed, we proceed to prove the second leftmost set. Otherwise, we split the set and so on. We note that the set $S(L_1, \ldots, L_q)$ can be directly obtained from S by deleting any clause containing at least one $L_j, 1 \leqslant j \leqslant q$, and by deleting all literals $\sim L_1, \ldots, \sim L_q$ from the remaining set.

To generalize Davis and Putnam's method to nonground clauses, we shall give the following algorithm, which is similar to the one described in the last section.

A Generalized Davis and Putnam's Method

Step 0   Let M be an alleged S-unsatisfiable set for a set S of clauses. Let W be the set of all distinct positive literals occurring in M. (W may be the set of all distinct negative literals occurring in M. We use whichever is smaller.)

Step 1   Let $i = 0, R^0 = \{\varepsilon\}$, BRANCH $= $ (NIL), and BACKUP $=$ NIL, where NIL is a null list.

Step 2   If BRANCH is empty, terminate the algorithm; every substitution in R^i is a solution of M. Otherwise, continue.

Step 3   Let B be the first element in BRANCH. Delete B from BRANCH.

Step 4   Let $B = (L_1, \ldots, L_q)$. For each $\theta \in R^i$, compute $K(\theta)$ as follows: Let $M(\theta)$ be the set obtained from $M\theta$ by deleting any clause containing at least an $L_j \theta, 1 \leqslant j \leqslant q$, and by deleting all literals $\sim L_1 \theta, \ldots, \sim L_q \theta$ from the remaining set. Augment every clause in $M(\theta)$ by ε. Use *unit V-resolution* to deduce augmented empty clauses from $(\{L_1|\theta, \ldots, L_q|\theta\} \cup M(\theta))$. If augmented empty clauses can be derived, let $K(\theta)$ be the set of substitutions associated with all such derived augmented empty clauses. Otherwise, let $K(\theta)$ be empty. After $K(\theta)$ is computed for each $\theta \in R^i$, let $K = \cup_{\theta \in R^i} K(\theta)$.

Step 5   If K is empty, go to the next step. Otherwise, put the pair (B, R^i) at the beginning of BACKUP, set $R^{i+1} = K, i = i+1$, and go to Step 2.

Step 6   Choose an element L from W such that L or the negation of L does not appear in B. If this is impossible, go to Step 8. Otherwise, continue.

Step 7 Let B_1 and B_2 be the two lists obtained by appending L and $\sim L$, respectively, to the end of list B. Put B_1 and B_2 (in an arbitrary order) at the beginning of BRANCH. Go to Step 2.

Step 8 If BACKUP is empty, terminate the algorithm; there is no solution of M. (M has to be enlarged before the whole algorithm is started anew.) Otherwise, continue.

Step 9 Let (P, K) be the first pair in BACKUP. Delete (P, K) from BACKUP. Set $i = i - 1, R^i = K, B = P$, and go to Step 6.

Except for Step 4, the above algorithm is the same as the one given in the last section. It can be proved that this algorithm will also find a solution of M if one exists. In Step 4, we use "unit V-resolution," which corresponds to the one-literal rule of Davis and Putnam. The splitting process is implied by the tree-generating process of the algorithm. We note that this method is closely related to the matrix-reduction method proposed by Prawitz [1969]. For an efficient implementation of Step 4, we may use some of the features given in Appendix A.

REFERENCES

Bledsoe, W. W. (1971): Splitting and reduction heuristics in automatic theorem proving, *Artificial Intelligence* **2** 57–78.

Chang, C. L. (1971): Theorem proving by generation of pseudo-semantic trees, Division of Computer Research and Technology, National Institutes of Health, Bethesda, Maryland.

Chang, C. L. (1972): Theorem proving with variable-constrained resolution, *Information Sci.*, **4** 217–231.

Chinlund, T. J., M. Davis, P. G. Hinman, and M. D. McIlroy (1964): Theorem proving by matching, Bell Laboratory, 1964.

Davis, M. (1963): Eliminating the irrelevant from mechanical proofs, *Proc. Symp. Appl. Math.*, American Mathematical Society, vol. 15, pp. 15–30.

Ernst, G. (1971): The utility of independent subgoals in theorem proving, *Information and Control* **18** 237–252.

Loveland, D. W. (1968): Mechanical theorem proving by model elimination, *J. Assoc. Comput. Mach.* **15** 236–251.

Loveland, D. W. (1970b): Some linear Herbrand proof procedures: An analysis, Dept. Computer Science, Carnegie-Mellon University.

Loveland, D. W. (1972): A unifying view of some linear Herbrand procedures, *J. Assoc. Comput. Mach.* **19** 366–384.

Prawitz, D. (1960): An improved proof procedure, *Theoria* **33** 246–254.

Prawitz, D. (1969): Advances and problems in mechanical proof procedures, *in*: "Machine Intelligence," vol. 4 (B. Meltzer and D. Michie, eds.), American Elsevier, New York, pp. 59–71.

Slagle, J. R., and D. Koniver (1971): Finding resolution proofs and using duplicate goals in AND/OR trees, *Information Sci.* **4** 315–342.

Wos, L., G. A. Robinson, D. F. Carson, and L. Shalla (1967): The concept of demodulation in theorem proving, *J. Assoc. Comput. Mach.* **15** 698–709.

EXERCISES

Section 9.2

1. Consider $S = \{ \sim P(x), P(x) \vee Q(x, f(b)), P(x) \vee \sim Q(a, f(x)) \}$. Let M be the set consisting of two copies of $\sim P(x)$, one copy of $P(x) \vee Q(x, f(b))$, and one copy of $P(x) \vee \sim Q(a, f(x))$. Find a solution of M.

2. Consider the following set S of clauses:

$P(g(x, y), x, y)$

$P(x, h(x, y), y)$

$\sim P(x, y, u) \vee \sim P(y, z, v) \vee \sim P(x, v, w) \vee P(u, z, w)$

$\sim P(k(x), x, k(x))$.

Let M be the set consisting of one copy of each clause in S. Find a solution of M.

Section 9.3

3. Prove that the set M of Exercise 1 is S-unsatisfiable by V-resolution.
4. Prove that the set M of Exercise 2 is S-unsatisfiable by V-resolution.
5. Prove Theorem 9.1.
6. Prove that the two directed graphs in Fig. 9.8 are isomorphic.

 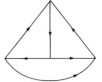

Figure 9.8

Section 9.5.

7. Let S be the set of the following clauses:

$P \vee Q$

$Q \vee R$

$R \vee W$

$\sim R \vee \sim P$

$\sim W \vee \sim Q$

$\sim Q \vee \sim R$.

Use the procedure of Section 9.5 to find a closed pseudosemantic tree for S.

8. Let S be the set of the following clauses:

$P(a)$

$\sim D(y) \lor L(a, y)$

$\sim P(x) \lor \sim Q(y) \lor \sim L(x, y)$

$D(b)$

$Q(b)$.

Use the procedure of Section 9.5 to find a closed pseudosemantic tree for S. Can you transform the tree into a resolution proof? (Consult Example 5.17.)

Section 9.6

9. Prove that the set S of Exercise 7 is unsatisfiable by the algorithm of Section 9.6.

10. Prove that the set S of Exercise 8 is unsatisfiable by the algorithm of Section 9.6.

══════ *Chapter 10*

Program Analysis

10.1 INTRODUCTION

In this and the next chapter, we shall consider the application of symbolic logic to program analysis and program synthesis. By program analysis, we mean that we are given a program and want to know the input–output relationship of the program. As for program synthesis, we are given an input–output relationship and want to synthesize a program to realize this input–output relationship. It is indeed interesting to show that symbolic logic is a very useful tool for solving these problems.

In general, given a program, we can ask the following questions:

1. *Terminating problem:* Given a certain input, will this program terminate?
2. *Response problem:* Given a certain input, if the program terminates, what is the output of the program?
3. *Correctness problem:* Given a certain input, will the output of this program satisfy the specification (input–output relationship) of the program?
4. *Equivalence problem:* Given two programs, will the programs yield the same results if the inputs are the same?
5. *Specialization problem:* Given a program P that is written to accept a set I of inputs, if we are only interested in a nonempty subset I^* of I, how can we simplify P to another P^* such that P^* runs faster on I^* than P does?

Floyd [1967] pointed out that symbolic logic can be used to analyze programs. This was further made clear in two important papers by Manna [1969a, b]. Manna [1969a] described the termination, correctness, and equivalence of programs as three different problems and gave three different algorithms for these problems. In [Manna, 1969b], it was pointed out that the termination problem is the basic problem. The correctness and equivalence problems can actually be formulated as a termination problem.

Our approach is based on the results of Floyd and Manna. We shall first point out that we can use logical formulas to describe the execution of a program. Considering these formulas as axioms, we can deduce logical consequences from these axioms. (This was pointed out in [Lee and Chang, 1971].) Some logical consequences serve as a basis to describe the input–output relationship. In particular, there is one clause, called the *halting* clause, that will be mechanically deduced (using the resolution principle) if and only if the program terminates. This clause not only tells us that the program terminates, but also tells us how it terminates. Thus using this halting clause, we can see whether the program meets the specification, or is equivalent to some other program.

We shall assume that our program can be represented by flowcharts. There are programs that cannot be represented by flowcharts, namely, programs that call themselves. But these programs can be translated into flowchart programs. The existence of LISP [McCarthy, 1962] compilers indirectly established this fact.

Before plunging into formal definitions and mathematics, we shall give an informal description of our approach in the next section.

10.2 AN INFORMAL DISCUSSION

Let us consider the flowchart of a program described in Fig. 10.1. In this program, x_1 and x_2 are the input variables, y is the program variable, and z is the output variable. To facilitate our discussion, we shall represent this program by a directed graph, as shown in Fig. 10.2. In this directed graph, every vertex represents a testing point. On every arc that leads out from vertex i and ends in vertex j, there is a predicate indicating the condition under which control is passed from vertex i to vertex j. For example, consider vertex 1. If it is found that x_1 is smaller than y, then z is assigned x_1 and the program terminates (vertex H is the halting vertex); otherwise, y is assigned x_1^2 and control is passed to vertex 2.

Let us ask ourselves the following question: How can control be passed to vertex 1? Obviously, the condition is $x_1 > 0$. After control is passed to vertex 1, y is assigned $x_2 + 1$. Therefore we can say that if control is passed to vertex 1, then $(x_1 > 0) \wedge (y = x_2 + 1)$ is true. This condition may be denoted

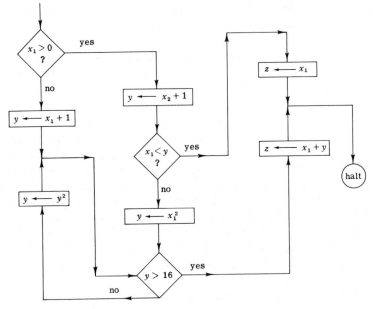

Figure 10.1

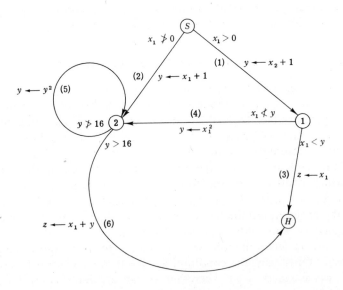

Figure 10.2

by a predicate $Q_1(x_1, x_2, y)$. Similarly, we can denote the condition under which control is passed to vertex 2 (from vertex S) by $Q_2(x_1, x_2, y)$.

From Fig. 10.2, we know that if control is at vertex 1 and x_1 is not smaller than y, then control will be passed from vertex 1 to vertex 2 through arc (4) and y will be assigned x_1^2. This fact can now be expressed by a logical formula:

$$Q_1(x_1, x_2, y) \wedge x_1 \not< y \to Q_2(x_1, x_2, x_1^2).$$

Similarly, we can also describe the passing of control to vertex 2 through arcs (2) and (5) by the following formula:

$$x_1 \not> 0 \to Q_2(x_1, x_2, x_1 + 1)$$

$$Q_2(x_1, x_2, y) \wedge y \not> 16 \to Q_2(x_1, x_2, y^2).$$

In fact, we can see that the entire program can be described by a set of logical formulas. Each formula corresponds to passing through an arc of the program. These formulas serve as a basis for analyzing the execution of the program. For example, if we know that given x_1 and x_2 there does not exist any y such that $Q_1(x_1, x_2, y)$ and $x_1 < y$ can be true simultaneously, then control can never be passed to vertex H from vertex 1. If we know further that $Q_2(x_1, x_2, y)$ cannot be true simultaneously with $y > 16$, then control cannot possibly be passed from vertex 2 to vertex H through arc (6). We can then safely say that the program will never terminate. In analyzing the correctness of programs, we follow the same kind of argument. If we can prove that whenever the program halts, the specification of the program, that is, the relationship between the input variables and the output variables, will be satisfied, then we say that the program is correct.

In the next section, we shall give formal definitions of programs.

10.3   FORMAL DEFINITIONS OF PROGRAMS

In the sequel, we shall represent programs by directed graphs [Busacker and Saaty, 1965].

Definition   A *directed graph* consists of a nonempty set V, a set A disjoint from V, and a mapping D from A into $V \times V$. The elements of V and A are called *vertices* and *arcs*, respectively, and D is called the *directed incidence mapping* associated with the directed graph. If $a \in A$ and $D(a) = (v, v')$, then a is said to have v as its *initial vertex* and v' as its *terminal vertex*. For our purpose, the number of vertices and arcs is assumed to be finite. A directed graph is usually denoted by (V, A, D), or (V, A) if D is not used explicitly. A *path* in a directed graph is a sequence of arcs $a_1, a_2, ..., a_n$, where every arc a_i has v_{i-1} as its initial vertex and v_i as its terminal vertex.

Definition   A *program* consists of an input (variable) vector $\bar{x} = (x_1, ..., \bar{x}_L)$, a program (variable) vector $\bar{y} = (y_1, ..., y_M)$, an output (variable) vector $\bar{z} = (z_1, ..., z_N)$, and a finite direct graph (V, A) such that the following conditions are satisfied:

1. In the graph (V, A), there is exactly one vertex, called the *start vertex* $S \in V$, that is not a terminal vertex of any arc; there is exactly one vertex, called the *halt vertex* $H \in V$, that is not an initial vertex of any arc; and every vertex v is on some path from S to H.

2. In (V, A), each arc a not entering H is associated with a quantifier-free formula $P_a(\bar{x}, \bar{y})$ and an assignment $\bar{y} \leftarrow f_a(\bar{x}, \bar{y})$; each arc entering the halt vertex H is associated with a quantifier-free formula $P_a(\bar{x}, \bar{y})$ and an assignment $\bar{z} \leftarrow f_a(\bar{x}, \bar{y})$. ($P_a$ is called the *testing predicate* associated with arc a and $P_a(\bar{x}, \bar{y})$ is called the *testing formula* associated with arc a.)

3. For each vertex v ($v \neq H$), let $a_1, a_2, ..., a_r$ be all the arcs leaving v and let $P_{a_1}, P_{a_2}, ..., P_{a_r}$ be the testing predicates associated with arcs $a_1, a_2, ..., a_r$, respecively. Then for all $\bar{x}$ and $\bar{y}$, one and only one of $P_{a_1}(\bar{x}, \bar{y})$, $P_{a_2}(\bar{x}, \bar{y})$, ..., $P_{a_r}(\bar{x}, \bar{y})$ is T.

Sometimes there is only one arc leading out of a certain vertex. In this case, the testing formula associated with that arc can be simply considered T and conveniently ignored.

Example 10.1

Consider a program that multiples two nonnegative integers x_1 and x_2. The only testing predicate available is the testing of whether a number is equal to zero, and the only functions used are addition and subtraction.

The program is as follows:

$$y_1 \leftarrow 0$$

$$y_2 \leftarrow x_2$$

$1:[$If $y_2 = 0$, then $[z_1 \leftarrow y_1, \text{halt}]$;

else $[y_1 \leftarrow y_1 + x_1,$

$y_2 \leftarrow y_2 - 1,$

go to $1]]$

where x_1 and x_2 are input variables, y_1 and y_2 are program variables, and z is the output variable.

This program can be represented by the directed graph shown in Fig. 10.3. We now check whether this graph satisfies all the conditions of a program. We note that there is one start vertex and one halt vertex, and every vertex

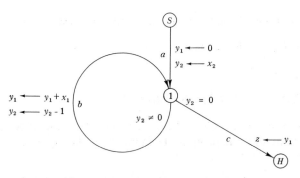

Figure 10.3

is on some path from S to H. Hence, condition 1 is satisfied. Each arc is associated with a formula and an assignment. Hence, condition 2 is satisfied. Finally, for vertex S, there is only one arc, namely arc a, leaving S; for all possible values of x_1, x_2, y_1, and y_2, the formula T is always true. As for vertex 1, there are two arcs, namely b and c, leaving vertex 1. For all values of x_1, x_2, y_1, and y_2, one and only one of $y_2 = 0$ and $y_2 \neq 0$ is true. Therefore, condition 3 is satisfied. We see that this graph satisfies all the conditions required by a program. Therefore, it indeed represents a program.

Consider the directed graph shown in Fig. 10.3 again. Given a value of x_1 and a value of x_2, the execution of the program is as follows: Assume control is initially at vertex S. It will be first passed to vertex 1. y_1 is then assigned 0 and y_2 is assigned x_2. Now, $y_2 = 0$ and $y_2 \neq 0$ are tested. If $y_2 = 0$, z is assigned y_1, control is passed to H, and the program terminates. If $y_2 \neq 0$, y_1 is assigned $y_1 + x_1$, y_2 is assigned $y_2 - 1$, and control is passed through arc b and goes back to vertex 1. Again, $y_2 = 0$ and $y_2 \neq 0$ are tested and the whole process is repeated. This process will continue forever unless at some point control is passed to vertex H.

In general, given an input $\bar{x}$ for a program, the execution of this program is carried out according to the following algorithm:

Step 1   Control is always assumed to be initially at the start vertex S.

Step 2   Let $j = 0, v^j = S$, and let $\bar{y}^j$ be the value of the program variable vector.

Step 3   If $v^j = H$, the execution is terminated; otherwise, go to the next step.

Step 4   Let a_j be the arc of which v^j is the initial vertex and $P_{a_j}(\bar{x}, \bar{y}^j)$ is true. Let v^{j+1} be the terminal vertex of a_j. Then control moves along a_j to vertex v^{j+1} and one of the following assignments is executed:

1. $\bar{y}^{j+1} \leftarrow f_{a_j}(\bar{x}, \bar{y}^j)$ if v^{j+1} is not H,

2. $\bar{z} \leftarrow f_{a_j}(\bar{x}, \bar{y}^j)$ if v^{j+1} is H.

Step 5   Let $j = j + 1$ and go to Step 3.

The above execution is said to be finite if and only if for some k, $v^k = H$. In this case, we say that the program terminates for the input $\bar{x}$.

10.4   LOGICAL FORMULAS DESCRIBING THE

EXECUTION OF A PROGRAM

As informally described in Section 10.2, the rules concerning the execution of programs can be conveniently described by logical formulas. This is done by using an ingenious concept defined below.

Definition   For each vertex v_i in a program, $Q_i(\bar{x}, \bar{y})$ (or $Q_i(\bar{x}, \bar{z})$ if $v_i = H$) is defined as the condition that control is passed (from vertex S) to vertex v_i with the input vector assigned $\bar{x}$ and the program (output) vector changed to $\bar{y}(\bar{z})$. This condition is called the *access condition* of v_i. In other words, whenever control is passed to vertex v_i, $Q_i(\bar{x}, \bar{y})$ is T. The predicate Q_i is called the *access predicate* of vertex v_i.

Since control is always assumed to be initially at S, we may simply let $Q_S(\bar{x}, \bar{y})$ be T. Because the passing of control to vertex H signifies that the program terminates, Q_H plays an important role in program analysis, as we shall see in the later sections. In the sequel, Q_H is called the *halting predicate*.

Let a be an arc of a program whose initial vertex is v_i and terminal vertex is v_j. Suppose $P_a(\bar{x}, \bar{y})$ and $f_a(\bar{x}, \bar{y})$ are the testing formula and assignment associated with arc a. If control is passed to v_i (thereby $Q_i(\bar{x}, \bar{y})$ is T) and $P_a(\bar{x}, \bar{y})$ is also T, then control will be passed through arc a to vertex v_j, with $\bar{y}$ assigned $f_a(\bar{x}, \bar{y})$ (thereby $Q_j(\bar{x}, f_a(\bar{x}, \bar{y}))$ is T). Thus, we can describe the execution of a program by a logical formula as follows.

Definition   Let a be an arc in a program. Let v_i and v_j be, respectively, the initial and terminal vertices of arc a. Suppose $P_a(\bar{x}, \bar{y})$ and $f_a(\bar{x}, \bar{y})$ are the testing formula and the assignment associated with arc a, respectively. Then we define a formula W_a associated with arc a as follows:

W_a: $Q_i(\bar{x}, \bar{y}) \wedge P_a(\bar{x}, \bar{y}) \rightarrow Q_j(\bar{x}, f_a(\bar{x}, \bar{y}))$.

W_a is called the *describing formula* for arc a. If we use a clause C_a to represent W_a, then C_a is called the *describing clause* for arc a.

Note that if $v_i = S$, then $Q_i(\bar{x}, \bar{y}) = T$, as pointed out before. Therefore, if $v_i = S$, then W_a becomes

$$P_a(\bar{x}, \bar{y}) \rightarrow Q_j(\bar{x}, f_a(\bar{x}, \bar{y})).$$

Example 10.2

Consider the program in Fig. 10.3 again. The describing formulas for arcs a, b, and c are

W_a: $Q_1(x_1, x_2, 0, x_2)$

W_b: $Q_1(x_1, x_2, y_1, y_2) \wedge y_2 \neq 0 \rightarrow Q_1(x_1, x_2, y_1 + x_1, y_2 - 1)$

W_c: $Q_1(x_1, x_2, y_1, y_2) \wedge y_2 = 0 \rightarrow Q_H(x_1, x_2, y_1)$.

Definition   Let arcs $a_1, a_2, \ldots, a_R$ be all of the arcs in a program P. Then $(\forall \bar{y})(W_{a_1} \wedge W_{a_2} \wedge \cdots \wedge W_{a_R})$ is called the *describing formula of program P*, where every W_{a_i}, $1 \leqslant i \leqslant R$, is a describing formula for arc a_i. (Note that input variables in the describing formula are treated as constants.)

In the sequel, the describing formula of a program P will be denoted as A_P. As usual, we shall represent A_P by a set of clauses. A_P may be viewed as a set of axioms. Later, it will be shown that we can deduce logical consequences from A_P. But we first have to make sure that A_P is consistent.

Theorem 10.1   Given a program P, let A_P be the set of clauses representing the describing formula of P; then A_P is satisfiable.

Proof   Note that corresponding to every arc a, the describing clause of a contains a positive literal, namely, $Q_j(\bar{x}, f_a(\bar{x}, \bar{y}))$. Therefore, in A_P every clause contains a positive literal. Let $v_1, v_2, \ldots, v_m$ be all the vertices in P except the start vertex S. Let I be an interpretation in which every $Q_i(\bar{x}, \bar{y})$, $1 \leqslant i \leqslant m$, is assigned T. Evidently, every clause in A_P is T in I, and therefore A_P is T in I. Thus A_P is satisfiable.   Q.E.D.

In the following section, we shall show how we can analyze a program by deducing logical consequences from A_P.

10.5   PROGRAM ANALYSIS BY RESOLUTION

Let us first consider a very simple example.

Example 10.3

Consider the program in Fig. 10.4, where x is an input integer, y is a program variable, and $D(x, n)$ means that n divides x. The describing clauses of the above program are:

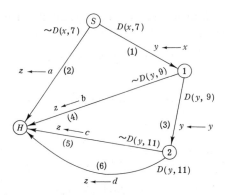

Figure 10.4

(1) C_1: $\sim D(x, 7) \vee Q_1(x, x)$

(2) C_2: $D(x, 7) \vee Q_H(x, a)$

(3) C_3: $\sim Q_1(x, y) \vee \sim D(y, 9) \vee Q_2(x, y)$

(4) C_4: $\sim Q_1(x, y) \vee D(y, 9) \vee Q_H(x, b)$

(5) C_5: $\sim Q_2(x, y) \vee D(y, 11) \vee Q_H(x, c)$

(6) C_6: $\sim Q_2(x, y) \vee \sim D(y, 11) \vee Q_H(x, d)$.

From the above clauses, we can use the resolution principle to deduce logical consequences. For instance, with the input variable x treated as a constant, we can generate the following resolvents:

(7) $\sim D(x, 7) \vee \sim D(x, 9) \vee Q_2(x, x)$ from (1) and (3)

(8) $\sim D(x, 7) \vee \sim D(x, 9) \vee \sim D(x, 11) \vee Q_H(x, d)$ from (7) and (6)

(9) $\sim D(x, 7) \vee D(x, 9) \vee Q_H(x, b)$ from (1) and (4).

In fact, we can deduce the following interesting and informative logical consequences:

(10) $D(x, 7) \vee Q_H(x, a)$

(11) $\sim D(x, 7) \vee D(x, 9) \vee Q_H(x, b)$

(12) $\sim D(x, 7) \vee \sim D(x, 9) \vee D(x, 11) \vee Q_H(x, c)$

(13) $\sim D(x, 7) \vee \sim D(x, 9) \vee \sim D(x, 11) \vee Q_H(x, d)$.

Clauses (10)–(13) can be viewed as a complete description of the input–output relationship of the program. They can be interpreted as follows:

For (10):   If 7 does not divide the input variable x, the program is terminated with output a.

For (11):   If 7 does divide x, but 9 does not divide x, the program is terminated with output b.

For (12):   If both 7 and 9 divide x, but 11 does not divide x, then the program is terminated with output c.

For (13):   If 7, 9, and 11 divide x, the program is terminated with output d.

Clauses (10)–(13) have the following common features:

a. The only access predicate that they contain is Q_H.

b. None of them contains any program variable.

c. Every clause corresponds to a path in the program starting from vertex S and ending at vertex H. For instance, clause (10) corresponds to the path that consists of arc (2) only, while clause (12) corresponds to the path consisting of arcs (1), (3), and (5).

Clearly, these clauses provide us with much information about the input–output relationship of the program. In general, if a program does not contain any loop, it is not difficult to derive all of these clauses.

Let P be a program without loops. Let $a_1, a_2, \ldots, a_R$ be a path in P from S to H. Let C_{a_i}, $1 \leqslant i \leqslant R$, be the describing clause of arc a_i. Let S and v_1 be the initial and terminal vertices of a_1, let v_R and H be the initial and terminal vertices of a_R, and let v_{i-1} and v_i be the initial and terminal vertices of a_i, $2 \leqslant i \leqslant R-1$. Obviously, C_{a_1} contains Q_1, C_{a_2} contains $\sim Q_1$, and the respective clauses are unifiable. Therefore there is one resolvent between C_{a_1} and C_{a_2} that eliminates Q_1. Let C_1 be C_{a_1}; and let C_2 be the resolvent between C_1 and C_{a_2}. C_2 contains Q_2. Thus we can resolve C_2 with C_{a_3}. This process can be repeated. In general, let C_{i+1} be the resolvent between C_i and $C_{a_{i+1}}$. The C_{R-1} is the clause describing the path $a_1, a_2, \ldots, a_R$. The reader can prove for himself that clauses (10)–(13) can all be obtained this way.

If looping is involved in a program, the input–output relationship cannot be obtained so simply as described above. In this case, some appropriate induction scheme must be used.

Example 10.4

Consider the program depicted in Fig. 10.5, where x_1 and x_2 are the input variables, y_1 and y_2 are the program variables, and z is the output variable.

The clauses describing the program are

(1) C_a: $Q_1(x_1, x_2, x_1, x_2)$

(2) C_b: $\sim Q_1(x_1, x_2, y_1, y_2) \vee y_2 = 0 \vee Q_1(x_1, x_2, y_1+1, y_2-1)$

(3) C_c: $\sim Q_1(x_1, x_2, y_1, y_2) \vee y_2 \neq 0 \vee Q_H(x_1, x_2, y_1)$.

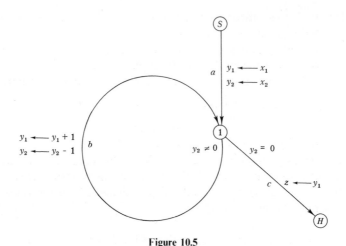

Figure 10.5

For this example, we use the following induction scheme:

$$(\exists y_2)(y_2 > 0 \wedge Q_1(x_1, x_2, x_1, y_2)) \wedge$$

$$(\forall y_3)(y_3 > 0 \wedge Q_1(x_1, x_2, x_1, y_3) \rightarrow Q_1(x_1, x_2, x_1 + 1, y_3 - 1))$$

$$\rightarrow Q_1(x_1, x_2, x_1 + x_2, 0).$$

The above axiom can be broken into three clauses, where f is a Skolem function:

(4) $y_2 \not> 0 \vee \sim Q_1(x_1, x_2, x_1, y_2) \vee f(y_2) > 0 \vee Q_1(x_1, x_2, x_1 + x_2, 0)$

(5) $y_2 \not> 0 \vee \sim Q_1(x_1, x_2, x_1, y_2) \vee Q_1(x_1, x_2, x_1, f(y_2)) \vee$

$\qquad Q_1(x_1, x_2, x_1 + x_2, 0)$

(6) $y_2 \not> 0 \vee \sim Q_1(x_1, x_2, x_1, y_2) \vee \sim Q_1(x_1, x_2, x_1 + 1, f(y_2) - 1) \vee$

$\qquad Q_1(x_1, x_2, x_1 + x_2, 0).$

Let us also assume that

(7) $x_2 > 0.$

Other axioms that we need are

(8) $0 = 0$

(9) $u \not> 0 \vee u \neq 0.$

From (1)–(9), we can generate the following resolvents:

(10) $\sim Q_1(x_1, x_2, x_1, x_2) \lor f(x_2) > 0 \lor Q_1(x_1, x_2, x_1 + x_2, 0)$

from (4) and (7)

(11) $\sim Q_1(x_1, x_2, x_1, x_2) \lor Q_1(x_1, x_2, x_1, f(x_2)) \lor Q_1(x_1, x_2, x_1 + x_2, 0)$

from (5) and (7)

(12) $\sim Q_1(x_1, x_2, x_1, x_2) \lor \sim Q_1(x_1, x_2, x_1 + 1, f(x_2) - 1)$

$\lor Q_1(x_1, x_2, x_1 + x_2, 0)$ from (6) and (7)

(13) $f(x_2) > 0 \lor Q_1(x_1, x_2, x_1 + x_2, 0)$ from (1) and (10)

(14) $Q_1(x_1, x_2, x_1, f(x_2)) \lor Q_1(x_1, x_2, x_1 + x_2, 0)$ from (1) and (11)

(15) $\sim Q_1(x_1, x_2, x_1 + 1, f(x_2) - 1) \lor Q_1(x_1, x_2, x_1 + x_2, 0)$

from (1) and (12)

(16) $f(x_2) = 0 \lor Q_1(x_1, x_2, x_1 + 1, f(x_2) - 1) \lor Q_1(x_1, x_2, x_1 + x_2, 0)$

from (2) and (14)

(17) $f(x_2) \neq 0 \lor Q_1(x_1, x_2, x_1 + x_2, 0)$ from (9) and (13)

(18) $Q_1(x_1, x_2, x_1 + 1, f(x_2) - 1) \lor Q_1(x_1, x_2, x_1 + x_2, 0)$

from (16) and (17)

(19) $Q_1(x_1, x_2, x_1 + x_2, 0)$ from (15) and (18)

(20) $0 \neq 0 \lor Q_H(x_1, x_2, x_1 + x_2)$ from (3) and (19)

(21) $Q_H(x_1, x_2, x_1 + x_2)$ from (8) and (20).

Clause (21) shows that the function of this program is to add two numbers x_1 and x_2.

Example 10.5

Consider Fig. 10.6. The reader can easily see that we can derive the following logical consequences from the formulas describing the program and other necessary axioms:

(1) $\sim (x_1 \leqslant x_2) \lor Q_H(x_1, x_2, x_1 + x_2)$

(2) $(x_1 \leqslant x_2) \lor Q_H(x_1, x_2, x_1 - x_2)$.

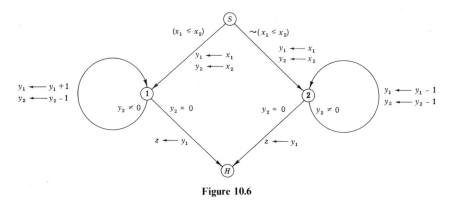

Figure 10.6

The above logical consequences indicate that if x_1 is smaller than or equal to x_2, then the output is the sum of x_1 and x_2; otherwise, it is the difference of x_1 and x_2.

10.6   THE TERMINATION AND RESPONSE OF PROGRAMS

In this section, we shall discuss the following problems:

Given a program and the specification of the input, how can we decide whether the program terminates (termination problem)?

Given a program and the specification of the input, if the program terminates, what is the output of the program (response problem)?

As we shall see, these two problems are intimately related. In fact, we shall kill these two birds with one stone.

Evidently, we need some information to answer the above questions. In general, we need the following information:

1.   Axioms describing the execution of the program. This has already been discussed in Section 10.4. As we did before, we shall denote the describing formula of the program by A_P.

2. Axioms concerning testing predicates and assignment functions. For example, we might need the axioms concerning equality or some appropriate induction schema. We denote this by A_S.

3. Axioms concerning the input. For instance, we might require the input to be positive integers, or in the LISP language we may require the input to be S-expressions. We shall denote this kind of axiom by A_I.

We assume that $A_S \wedge A_I$ is consistent. Still, we have to make sure that $A_P \wedge A_S \wedge A_I$ is consistent.

Theorem 10.2   Given a program P, let S denote the set of clauses representing $A_P \wedge A_S \wedge A_I$. Then S is satisfiable.

Proof   Let S_1, S_2, and S_3 be sets of clauses representing A_P, A_S, and A_I, respectively. According to Theorem 10.1, A_P is consistent. Therefore, there exists an interpretation I_1 (in which every $Q_i(\bar{x}, \bar{y})$ is assigned T) such that I_1 satisfies S_1. Since $A_S \wedge A_I$ is consistent, there exists an interpretation I_2 that satisfies $S_2 \wedge S_3$. Since S_1 and S_2 do not contain any access predicate Q_i's, we can define an interpretation I to be the union of I_1 and I_2. I satisfies S_1, S_2, and S_3. Thus S is satisfiable.   Q.E.D.

Since $A_P \wedge A_S \wedge A_I$ is a complete description of the program and the input, the program terminates if and only if $(\exists \bar{z})Q_H(\bar{x}, \bar{z})$ is a logical consequence of $A_P \wedge A_S \wedge A_I$. Therefore, the termination problem can be viewed as a theorem-proving problem.

But, as the reader will see, we can accomplish more than merely proving that the program terminates. In fact, we will try to deduce a logical consequence which tells us not only that the program will terminate, but also how the program will terminate.

Let us consider the program in Example 10.3 again. The reader can easily see that

$$Q_H(x, a) \vee Q_H(x, b) \vee Q_H(x, c) \vee Q_H(x, d)$$

is a logical consequence of the axioms of that program. Therefore, it will be T whenever the axioms are T. Since a clause is a disjunction of literals, at least one literal will be T. This means that, for any input x, when the program is executed, control will be passed through some path to vertex H and the program will terminate. Thus, this clause tells us that the program will terminate and its output will be a, b, c, or d when it terminates.

In Example 10.4, in addition to the axioms A_P describing the program, we also have the condition $x_2 > 0$. We can view this as an axiom A_I specifying the input. The induction schema is for the functions "+" and "−". A_S includes this induction schema, $0 = 0$, and $(u \not> 0 \vee u \neq 0)$. Since $Q_H(x_1, x_2, x_1 + x_2)$ is a logical consequence of $A_P \wedge A_S \wedge A_I$, we conclude that the program will terminate for x_2 greater than 0, and when it terminates, its output will be $x_1 + x_2$.

Definition   A clause containing only the halting predicate Q_H is called a *halting clause*.

Even if we know that a program terminates if and only if there exists a halting clause that is a logical consequence of $A_P \wedge A_S \wedge A_I$, we still want to make sure that this particular clause can be "mechanically" deduced. In the following, we shall prove that, with the help of the resolution principle, we can always deduce such a logical consequence. To prove this, we need the following lemma.

Lemma 10.1   (Program Termination Lemma)   Given a program P, let A_T be the formula obtained from the describing formula A_P by deleting any literal containing Q_H. Then P terminates if and only if $A_T \wedge A_S \wedge A_I$ is unsatisfiable.

Proof $(\Rightarrow)$ Suppose P terminates. In this case, for any value of $\bar{x}$, there exists an arc a in the program P such that control is passed through the arc a to vertex H. Let v_i be the initial vertex of a. Since the terminal vertex of a is vertex H, the describing formula for arc a is

$$Q_i(\bar{x}, \bar{y}) \wedge P_a(\bar{x}, \bar{y}) \to Q_H(\bar{x}, f_a(\bar{x}, \bar{y}))$$

where $P_a(\bar{x}, \bar{y})$ and $f_a(\bar{x}, \bar{y})$ are the testing formula and the assignment associated with arc a, respectively. Since control is passed through a from vertex v_i to vertex H, both $Q_i(\bar{x}, \bar{y})$ and $P_a(\bar{x}, \bar{y})$ must be T. Therefore, $\sim Q_i(\bar{x}, \bar{y}) \vee \sim P_a(\bar{x}, \bar{y})$ is false. However, $\sim Q_i(\bar{x}, \bar{y}) \vee \sim P_a(\bar{x}, \bar{y})$ is in A_T. Therefore $A_T \wedge A_S \wedge A_I$ is false. Hence, $A_T \wedge A_S \wedge A_I$ is unsatisfiable.

$(\Leftarrow)$ Suppose $A_T \wedge A_S \wedge A_I$ is unsatisfiable. We want to show that P terminates. By Theorem 10.2, we know that $A_P \wedge A_S \wedge A_I$ is satisfiable. Furthermore, from the proof of Theorem 10.2, we know that we can construct a model M of $A_P \wedge A_S \wedge A_I$ such that any atom containing an access predicate Q_i is assigned T in M. Since $A_T \wedge A_S \wedge A_I$ is unsatisfiable, M falsifies $A_T \wedge A_S \wedge A_I$. However, since M satisfies $A_S \wedge A_I$, M must falsify A_T. Since M contains no negation of any Q_i, M must falsify a clause C_a' in A_T such that C_a' contains no positive literal. Clearly, C_a' must be obtained from a clause C_a in A_P by deleting a literal containing Q_H. Let C_a be written as

$$\sim Q_i(\bar{x}, \bar{y}) \vee \sim P_a(\bar{x}, \bar{y}) \vee Q_H(\bar{x}, f_a(\bar{x}, \bar{y})).$$

Then C_a' is

$$\sim Q_i(\bar{x}, \bar{y}) \vee \sim P_a(\bar{x}, \bar{y}).$$

Since C_a' is false in M, both $Q_i(\bar{x}, \bar{y})$ and $P_a(\bar{x}, \bar{y})$ are T. Therefore, control will be passed from vertex v_i to vertex H and the program will terminate.

$$\text{Q.E.D.}$$

Example 10.6

Consider the program in Fig. 10.7. For this program, the formula A_T is given as follows:

W_a: $Q_1(x, x)$

W_b: $\sim Q_1(x, y) \vee \sim (y \geqslant 0)$
W_c: $\sim Q_1(x, y) \vee (y \geqslant 0).$

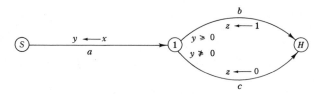

Figure 10.7

For this example, A_T itself is unsatisfiable. Therefore, the program will terminate.

Theorem 10.3   (Program Response Theorem)   Let P be a program. Let S denote the set of clauses representing $A_P \wedge A_S \wedge A_I$. Then P terminates if and only if there is a deduction of a halting clause from S (by resolution).

Proof ($\Rightarrow$)   Let S' be the set of clauses representing $A_T \wedge A_S \wedge A_I$. Since P terminates, according to Lemma 10.1 S' is unsatisfiable. Therefore, there is a deduction D' of $\square$ from S'. Let D be the deduction obtained from D' by putting literals containing Q_H back to clauses from which they were deleted. Clearly, D is a deduction of a halting clause that contains the halting predicate only. This completes the proof of the first half of Theorem 10.3.

($\Leftarrow$)   Assume there is a deduction D of a halting clause from S. Let S' be the set of clauses obtained from S by deleting all the halting predicates from S. Clearly, the same deduction D will generate an empty clause from S'. Thus S' is unsatisfiable. But S' is the set of clauses representing $A_T \wedge A_S \wedge A_I$. According to Lemma 10.1, P terminates.   Q.E.D.

If we are interested merely in whether a program terminates (not how it terminates), then we may use Lemma 10.1. Note that a program terminates if and only if $(\exists \bar{z}) Q_H(\bar{x}, \bar{z})$ is a logical consequence of $A_P \wedge A_S \wedge A_I$. Consequently, we may also try to prove that $A_P \wedge A_S \wedge A_I \wedge (\forall \bar{z})(\sim Q_H(\bar{x}, \bar{z}))$ is unsatisfiable.

10.7   THE SET-OF-SUPPORT STRATEGY AND THE DEDUCTION OF THE HALTING CLAUSE

Evidently, if we are interested only in the halting clause, we should not enumerate all possible resolvents because many of them will have no contribution to the generation of this clause. In this section, we shall discuss how to deduce the halting clause efficiently, that is, how to deduce that clause without generating too many irrelevant clauses.

Since a halting clause always contains the halting predicate Q_H, intuitively, we can limit the possible resolutions by requiring that every resolution involves at least one clause containing Q_H. Since usually the number of such clauses is small to start with, we believe that this will greatly limit the number of intermediate clauses generated. A natural question to ask is whether this kind of strategy will ensure the generation of the halting clause.

It is interesting to show that this intuitively good strategy is indeed complete for generating the halting clause. The reason is that this strategy is exactly the set-of-support stategy which we discussed in Chapter 6. Let us now give a brief review of the set-of-support strategy below.

Let S be an unsatisfiable set of clauses. Let T be a subset of S. T can be viewed as a *set of support* if $S - T$ is satisfiable. A *set-of-support resolution* is a resolution of two clauses that are not both from $S - T$. A *set-of-support deduction* is a deduction in which every resolution is a set-of-support resolution.

In our case, let S denote the set of clauses representing $A_P \wedge A_S \wedge A_I$. T is the subset of S in which every clause contains the halting predicate Q_H. Let S' and T' be the sets of clauses obtained from S and T, respectively, by deleting Q_H. Clearly, $S' - T'$ constitutes a satisfiable set of clauses, and therefore T' can be viewed as a set of support to deduce the empty clause from S'. Since a deduction of the halting clause from S corresponds to a deduction of the empty clause from S', we see that we can deduce the halting clause from S by the set-of-support strategy in which every clause in the set of support contains Q_H.

Example 10.7

Consider the case in Example 10.3. We have the following set of describing clauses:

(1) $\sim D(x, 7) \vee Q_1(x, x)$

(2) $D(x, 7) \vee Q_H(x, a)$

(3) $\sim Q_1(x, y) \vee \sim D(y, 9) \vee Q_2(x, y)$

(4) $\sim Q_1(x, y) \vee D(y, 9) \vee Q_H(x, b)$

(5) $\sim Q_2(x, y) \vee D(y, 11) \vee Q_H(x, c)$

(6) $\sim Q_2(x, y) \vee \sim D(y, 11) \vee Q_H(x, d)$.

Suppose we also have the following clauses representing the input constraint:

(7) $D(x, 7)$

(8) $D(x, 9)$.

Clauses (2), (4), (5), and (6) constitute the set of support. The following deduction is a set-of-support deduction. The reader can note that every resolution involves at least one clause containing Q_H.

(9)	$\sim Q_2(x, y) \vee Q_H(x, c) \vee Q_H(x, d)$	from (5) and (6)
(10)	$\sim Q_1(x, y) \vee \sim D(y, 9) \vee Q_H(x, c) \vee Q_H(x, d)$	from (9) and (3)
(11)	$\sim D(x, 7) \vee \sim D(x, 9) \vee Q_H(x, c) \vee Q_H(x, d)$	from (10) and (1)
(12)	$\sim D(x, 9) \vee Q_H(x, c) \vee Q_H(x, d)$	from (11) and (7)
(13)	$Q_H(x, c) \vee Q_H(x, d)$	from (12) and (8).

We by no means claim that the set-of-support strategy is the best strategy to deduce the halting clause. But, for the reader who is interested in implementing a program analysis, the set-of-support strategy is certainly easy to implement.

10.8   THE CORRECTNESS AND EQUIVALENCE OF PROGRAMS

In this section, we shall consider the correctness and equivalence of programs. A program is correct if it terminates and satisfies the specification (input–output relationship). Similarly, a program is equivalent to some other program if for the same input, both terminate and yield the same output. Since $A_P \wedge A_S \wedge A_I$ is a complete description of the program and the input, we can now formulate the above ideas as follows:

Definition   Let $R(\bar{x}, \bar{z})$ be a relation describing the specification of a program. The program is said to be *correct* with respect to $R(\bar{x}, \bar{z})$ if and only if $(\exists \bar{z})(Q_H(\bar{x}, \bar{z}) \wedge R(\bar{x}, \bar{z}))$ is a logical consequence of $A_P \wedge A_S \wedge A_I$.

Definition   Let A_P^1 and A_P^2 be the describing formulas for programs P^1 and P^2, respectively. Then the programs P^1 and P^2 are said to be *equivalent* to each other if and only if $(\exists z)(Q_H^1(\bar{x}, \bar{z}) \wedge Q_H^2(\bar{x}, \bar{z}))$ is a logical consequence of $A_P^1 \wedge A_P^2 \wedge A_S \wedge A_I$, where Q_H^1 and Q_H^2 are the halting predicates of P^1 and P^2, respectively.

Example 10.8

Consider the program in Example 10.4. Let us assume that we have the following input constraints:

(1) $x_1 > 0$

(2) $x_2 > 0.$

We also have an axiom:

(3) $\sim (x_1 > 0) \vee \sim (x_2 > 0) \vee (x_1 + x_2 > x_1).$

From Example 10.4, we know that we can derive

(4) $Q_H(x_1, x_2, x_1 + x_2)$.

Suppose the specification of the program is $z > x_1$. To show that the program is correct, we prove that

(5) $(\exists z)(Q_H(x_1, x_2, z) \wedge (z > x_1))$

is a logical consequence of the above clauses. First, using techniques of theorem proving, we negate (5) and obtain

(6) $\sim Q_H(x_1, x_2, z) \vee \sim (z > x_1)$.

Then, using resolution, we show that clauses (1)–(4) and (6) are unsatisfiable as follows:

(7)	$\sim (x_2 > 0) \vee (x_1 + x_2 > x_1)$	from (1) and (3)
(8)	$(x_1 + x_2 > x_1)$	from (7) and (2)
(9)	$\sim (x_1 + x_2 > x_1)$	from (4) and (6)
(10)	$\square$	from (9) and (8).

Thus (5) is a logical consequence of (1), (2), (3), and (4). Since clauses (1), (2), and (3) belong to A_I and clause (4) is a logical consequence of $A_P \wedge A_S \wedge A_I$, (5) is a logical consequence of $A_P \wedge A_S \wedge A_I$. Therefore, we conclude that the program is correct with respect to the specification.

10.9 THE SPECIALIZATION OF PROGRAMS

Very often we will encounter the following question: We have written a program P that is intended to accept a set I of input data. However, later on we are only interested in a restricted subset I^* of I. Of course, P can still accept I^* and produce an output. However, since I^* is smaller than I, we should be able to modify program P so that the modified program P^* can run faster on set I^*. The question now is how we are going to modify program P.

This problem has been considered by Dixon [1970, 1971], Futamura [1971], and Chang et al. [1971]. Dixon has written a program, called Specializer, to do the modification. What the specializer does is the following: Suppose P is a program that has N input variables (arguments). Suppose that specific values are assigned to M of these arguments ($M < N$). Then P^* is a program which takes only $N - M$ arguments. The other variables have been "fixed," and P^* is a specialized version of P for specific values of the fixed arguments. The advantage of specialization is the same as that of compiling: P^* usually runs faster than P. P^* may also use less storage space than P.

Dixon's specializer is written in LISP. It can only specialize LISP programs. The restriction of input is done by fixing values of some input variables. Roughly speaking, the specializer performs the following two operations: (1) deleting any instruction that can be evaluated and replacing it by a quoted value, and (2) removing a branch of a conditional instruction if it is not needed. A "pattern specializer," also defined by Dixon [1970], uses a more general approach.

In the following, we shall give a specialization procedure. Again, A_P denotes the describing formula of program P, A_S denotes axioms concerning testing predicates and assignment functions in P, and A_I* denotes axioms specifying the restricted input I*. Let S denote the set of clauses representing $A_P \wedge A_S \wedge A_I$*.

The Specialization Procedure

Step 1   Using resolution, derive a halting clause C_H from S.

Step 2   Let D be the deduction of C_H.

Step 3   Let P* be the program obtained from P by deleting any arc whose describing clause is not used in D, and by simplifying the remaining program. P* is a specialization of P for I*.

Example 10.9

Consider the program P shown in Fig. 10.8, where the input x is an integer

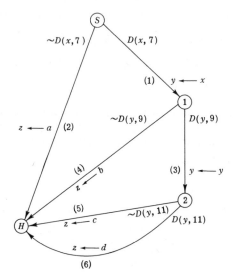

Figure 10.8

and $D(x, n)$ means that n divides x. The program P is for any integer. Suppose we know that x is less than 10. Find a specialization of P for such input. For this program, A_P is given as follows:

(1) C_1: $\sim D(x, 7) \vee Q_1(x, x)$

(2) C_2: $D(x, 7) \vee Q_H(x, a)$

(3) C_3: $\sim Q_1(x, y) \vee \sim D(y, 9) \vee Q_2(x, y)$

(4) C_4: $\sim Q_1(x, y) \vee D(y, 9) \vee Q_H(x, b)$

(5) C_5: $\sim Q_2(x, y) \vee D(y, 11) \vee Q_H(x, c)$

(6) C_6: $\sim Q_2(x, y) \vee \sim D(y, 11) \vee Q_H(x, d)$.

A_S is given as

(7) $x \not< 10 \vee \sim D(x, 11)$.

A_I^* is given as

(8) $x < 10$.

Clause (7) means that if $x < 10$, then 11 does not divide x. From $A_P \wedge A_S \wedge A_I^*$, we obtain the following deduction D of the halting clause:

(9) $\sim D(x, 11)$ from (7) and (8)

(10) $\sim Q_2(x, y) \vee Q_H(x, c)$ from (5) and (9)

(11) $\sim Q_1(x, y) \vee Q_2(x, y) \vee Q_H(x, b)$ from (3) and (4)

(12) $\sim Q_1(x, y) \vee Q_H(x, c) \vee Q_H(x, b)$ from (10) and (11)

(13) $Q_1(x, x) \vee Q_H(x, a)$ from (1) and (2)

(14) $Q_H(x, a) \vee Q_H(x, b) \vee Q_H(x, c)$ from (12) and (13).

Clause (14) is a halting clause which indicates that the output can be either a, b, or c (never d). Since C_6 is not used in the deduction, arc (6) can be deleted. We obtain P^* shown in Fig. 10.9. Clearly, P^* can be further simplified to the program shown in Fig. 10.10.

Example 10.10

Consider Fig. 10.8 again. Suppose this time we know that if 9 divides x, then 11 does not divide x. Let us see how we can simplify the program.

The describing clauses are still the same. A_I^* is given as

(7) $\sim D(x, 9) \vee \sim D(x, 11)$.

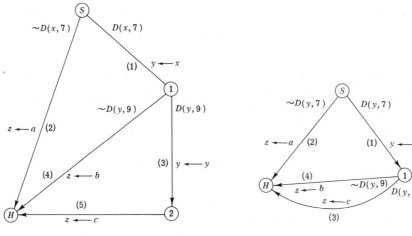

Figure 10.9   Figure 10.10

We have the following deduction.

(8) $Q_1(x,x) \vee Q_H(x,a)$ from (1) and (2)

(9) $\sim D(x,9) \vee Q_2(x,x) \vee Q_H(x,a)$ from (8) and (3)

(10) $D(x,11) \vee \sim D(x,9) \vee Q_H(x,a) \vee Q_H(x,c)$ from (9) and (5)

(11) $\sim D(x,9) \vee Q_H(x,a) \vee Q_H(x,c)$ from (10) and (7)

(12) $\sim Q_1(x,x) \vee Q_H(x,a) \vee Q_H(x,b) \vee Q_H(x,c)$ from (11) and (4)

(13) $Q_H(x,a) \vee Q_H(x,b) \vee Q_H(x,c)$ from (12) and (8).

Again, clause (6) is not used in the above deduction. Therefore the program can be finally simplified to that in Fig. 10.10.

REFERENCES

Busacker, R. G., and T. L. Saaty (1965): "Finite Graphs and Networks: an Introduction with Applications," McGraw-Hill, New York.

Chang, C. L., R. C. T. Lee, and J. Dixon (1971): Specialization of programs by theorem-proving, Division of Computer Research and Technology, National Inst. of Health, Bethesda, Maryland.

Dixon, J. (1970): "An Improved Method for Solving Deductive Problems on a Computer by Compiled Axioms," Ph.D. Thesis, University of California, Davis, California.

Dixon, J. (1971): The specializer, a method of automatically writing computer programs, Division of Computer Research and Technology, National Inst. of Health, Bethesda, Maryland.

Floyd, R. W. (1967): Assigning meaning to programs, *Proc. Symp. Appl. Math.*, American Mathematical Society, vol. 19, pp. 19–32.

Futamura, Y. (1971): Partial evaluation of computer programs: An approach to a compiler–compiler, *J. Inst. Electronics and Communication Engineers Japan.*

Lee, R. C. T., and C. L. Chang (1971): Program analysis and theorem proving, Division of Computer Research and Technology, National Inst. of Health, Bethesda, Maryland.

Manna, Z. (1969a): The correctness of programs, *J. Comput. System Sci.* **3** 119–127.

Manna, Z. (1969b): Properties of programs and the first-order predicate calculus, *J. Assoc. Comput. Math.* **16** pp. 244–255.

McCarthy, J. (1962): "LISP 1.5 Programmers Manual," The MIT Press, Cambridge, Massachusetts.

EXERCISES

Section 10.3

1. Consider the following program:

    If $x_1 > 0$, then $[y \leftarrow x_2 + 1$, go to 1$]$;

    else $[y \leftarrow x_1 + 1$, go to 2$]$;

    1: If $x_1 < y$, then $[z \leftarrow y + x_1$, halt$]$;

    else $[y \leftarrow y^2$, go to 2$]$;

    2: $z \leftarrow y + x_2$, halt.

    Represent the above program by a directed graph.

Section 10.4

2. Consider the program shown in Fig. 10.2. Find the describing clauses for all the arcs in the program.
3. Consider the program in Exercise 1. Give the describing formula.

Section 10.5

4. From the describing clauses that you have obtained in Exercise 3, derive clauses that describe the input–output relationship of the program.

Section 10.6

5. Consider the program in Fig. 10.11.

    1. Give describing clauses of arcs (a), (b), and (c).
    2. Prove that the above program will terminate if both x_1 and x_2 are positive integers.

6. From the describing clauses of the program in Exercise 1, derive a halting clause.

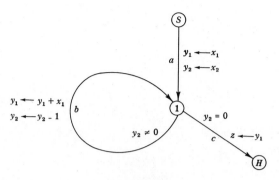

Figure 10.11

Section 10.7

7. Reconsider Example 10.4. Use the set-of-support strategy to derive the halting clause.
8. Reconsider Example 10.7. Let clause (6) be the top clause. Use linear deduction to derive the halting clause.

Section 10.8

9. In Exercise 1, assume the specification of the program is "If $x_1 < x_2 + 1$, then the output is $x_1 + x_2 + 1$." Prove that the program is correct.

Section 10.9

10. Consider the program in Fig. 10.12. Show that arc (2) can be deleted by formal theorem-proving techniques.

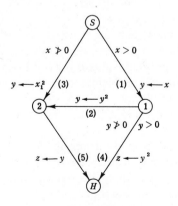

Figure 10.12

Chapter 11

Deductive Question Answering, Problem Solving, and Program Synthesis

11.1 INTRODUCTION

In this chapter, we shall discuss how theorem proving can be applied to answer questions, solve problems, and synthesize programs. Early in 1961, McCarthy [1961] first pointed out the importance of symbolic logic in programming computers to perform these tasks. This was later elaborated by many authors. For example, Black [1964], Darlington [1965], and Slagle [1965] all designed question–answering systems applying theorem-proving techniques. Slagle [1965] used a variable-matching scheme that was later used by Green and Raphael in their systems [Green and Raphael, 1968; Green, 1969a, b, c]. He also showed that a deductive question-answering system can be used as a problem-solving system. For instance, his program solved the monkey–banana problem (a modified version of a question stated by McCarthy [1961]), the Mikado problem [Safier, 1963], the end game problem [McCarthy, 1963b], and the state description compiler problem [Simon, 1963].

Green and Raphael made a breakthrough in the field of question answering by pointing out that mechanical theorem-proving techniques, such as the resolution principle, can be applied to design question-answering and problem-solving systems. The essential concept of Green and Raphael is that the set of facts necessary for question answering (or problem solving) can be viewed as axioms of a theorem, and the question (or the problem) can be viewed as the conclusion of the theorem.

Waldinger and Lee [1969] developed a program, called PROW (program writing), that writes programs automatically. They also showed that their program-writing programs can be used for question answering and problem solving.

There are two kinds of question-answering systems, the general-purpose systems and the special-purpose systems. For a special-purpose question-answering system, the data are usually concerned with a special subject and the range of questions one is allowed to ask is quite limited. Typical special-purpose question-answering systems are the Baseball system [Green et al. 1963], which answers questions about baseball in the United States, and Lindsay's system [Lindsay, 1963], which is concerned with family relationships.

The major task in designing special-purpose question-answering systems is storing the data in the memory optimally and writing efficient specialized subroutines so that the running time will be short and the memory used will be small.

Our main interest is in general-purpose question-answering systems in which the data base is not specified in advance and the range of questions that are allowed to be asked is broad. Often, the emphasis of general-purpose systems is on its deductive power. This is why symbolic logic plays such an important role in deductive question answering.

In this book, we divide questions into four classes according to the form of the answer to the question.

1. *Class A:* the kind of question that requires a "yes" or "no" answer. For instance, the answer might be "Yes, James is in Washington," or "No, James is not in Washington."

2. *Class B:* the kind of question that requires "where is," "who is," or "under what condition" as an answer. Thus the answer might be "John is in *Paris*," "John is *Mary*'s husband," or "John should run if it *rains*."

3. *Class C:* the kind of question whose answer is in the form of a sequence of actions. For example, the answer might be "Go to New York by train and then go to Paris by airplane," which involves two actions and the order of these actions is important.

4. *Class D:* the kind of question whose answer involves testing of conditions (branching). For example, the following answer is a typical one: "If bus is available, go to New York by bus; otherwise, go to New York by train."

As for problem-solving and program-writing systems, they are not much different from question-answering systems. Frequently, systems written for one purpose can also be used for other purposes. For example, the system PROW of Waldinger and Lee is for program writing. However, it can also be used to answer questions. Similarly, the system Deducom designed by Slagle [1965] is a question-answering system, yet it can also be used to solve problems. In this book, the terms "question answering," "problem solving," and "program synthesizing" will be used loosely. Roughly speaking, in practical applications, a question-answering system concerns Class A and Class B questions, a problem-solving system concerns Class C questions, and a program-synthesizing system concerns Class D questions. In the following, we shall discuss how the above four classes of questions can be handled through the use of theorem-proving techniques.

11.2   CLASS A QUESTIONS

Since the answer to a Class A question is only "yes" or "no," the question-answering problem is merely a theoreom-proving problem where the given facts are considered as axioms of a theorem and the question is presented as the conclusion of the theorem.

Example 11.1

Let us consider a very simple case here. Suppose we have the following facts:

F_1:   Every man is mortal.
F_2:   Confucius is a man.

The question to ask is

Q: Is Confucius mortal?

To answer this question, we try to prove that "Confucius is mortal" is a logical consequence of F_1 and F_2. If this is done, the answer is "yes."

Let $P(x)$ and $R(x)$ represent "x is a man" and "x is mortal," respectively. The given facts are represented by the following clauses:

(1) $\sim P(x) \lor R(x)$
(2) $P(\text{Confucius})$.

The conclusion of the theorem is represented by

(3) $R(\text{Confucius})$.

We negate (3) to obtain

(4) $\sim R(\text{Confucius})$.

The reader can easily prove that clauses (1), (2), and (4) constitute an unsatisfiable set of clauses. Therefore, (3) is a logical consequence of (1) and (2). Consequently, our answer should be "Yes, Confucius is mortal."

Sometimes we cannot prove the theorem corresponding to the question. In this case, we try to disprove it. If we succeed, the answer is "no."

Example 11.2

Let the given facts be

F_1: If one is in Paris, then one is not in Moscow.
F_2: John is in Paris.

The question is:

Q: Is John in Moscow?

Let $P(x, y)$ denote "x is in y." Then the facts are:

(1) $\sim P(x, \text{Paris}) \vee \sim P(x, \text{Moscow})$
(2) $P(\text{John}, \text{Paris})$.

The theorem corresponding to the question is

(3) $P(\text{John}, \text{Moscow})$.

It is easy to note that clause (3) is not a logical consequence of clause (1) and (2), but the negation of clause (3) is. Therefore, our answer should be "No, John is not in Moscow."

If we can neither prove nor disprove the theorem, our answer should be "Not sufficient information."

11.3 CLASS B QUESTIONS

We shall begin with a very simple example.

Example 11.3

Suppose the given fact is "John is Mary's husband" and the question is "Who is Mary's husband?" Let $P(x, y)$ denote "x is the husband of y." The given fact is

(1) $P(\text{John}, \text{Mary})$.

We then establish the following conclusion to answer our question:

(2) $(\exists x) P(x, \text{Mary})$.

If we can prove that (2) is a logical consequence of (1), at least we know that an answer exists. The reader can easily see that by tracing the variable, we can also decide what x is.

To prove this theorem, we first negate (2). We thus have

(3) $\sim P(x, \text{Mary})$.

Resolving (1) and (3), we obtain a contradiction and the theorem is proved. In the process of resolving, x is replaced by "John." If an appropriate tracing mechanism is incorporated, we can reveal this information and our answer is "*John* is Mary's husband."

A very simple way of tracing the variable is to add a predicate, called the *ANS* predicate, to clause (3). That is, in the above example, we may have

(4) $\sim P(x, \text{Mary}) \vee ANS(x)$.

Clause (4) means that "whoever Mary's husband is, he is our answer." Note that clause (4) is equivalent to the formula

(5) $(\forall x)(P(x, \text{Mary}) \rightarrow ANS(x))$.

Again, we resolve (1) and (4). This time, instead of an empty clause, we obtain

(6) $ANS(\text{John})$.

Therefore, our answer is "John is Mary's husband." We note that $ANS(\text{John})$ is a logical consequence of clauses (1) and (4). Thus we have transformed the original theorem-proving problem into a problem of deducing logical consequences.

The deduction of a clause containing only the *ANS* predicate resembles the deduction of a halting clause discussed in Chapter 10. This excellent way of tracing variables was first invented by Green [1969c]. Similar to a halting clause, a clause containing only the *ANS* predicate is called an *answering clause*.

Example 11.4

Consider the following facts:

F_1:   For all x, y, and z, if x is the father of y and z is the father of x, then z is the grandfather of y.

F_2:   Everyone has a father.

We shall ask the question: "For all x, who is the grandfather of x?"

Let $P(x, y)$ and $Q(x, y)$ denote "x is the father of y" and "x is the grand-father of y," respectively. Then the above facts are expressed as

(1) $\sim P(x, y) \vee \sim P(z, x) \vee Q(z, y)$

(2) $P(f(x), x)$

where $f(x)$ denotes the father of x. We can also have another clause corresponding to our question:

(3) $\sim Q(y, x) \vee ANS(y)$.

Clause (3) means that for all x and for all y, if y is the grandfather of x, then y is our answer.

Now, we generate the following resolvents:

(4) $\sim P(z, f(y)) \vee Q(z, y)$ from (1) and (2)

(5) $Q(f(f(y)), y)$ from (4) and (2)

(6) $ANS(f(f(x)))$ from (5) and (3).

Our answer is therefore "The father of the father of x is the grandfather of x."

In mathematics, it is common practice to find a solution through proving the existence of such a solution. For example, to find a solution of an equation, one may first prove that such a solution does exist. By extracting information from this proof, one can usually obtain a solution. In the following example, we shall show how mechanical theorem proving can be applied to solve some sophisticated mathematical problems.

Example 11.5

We are given a system that is described by the following axioms:

(1) $(\forall x)(\forall y)(\forall z)((x + y) + z = x + (y + z))$ associativity

(2) $(\forall x)(\forall y)(\exists z)(z + x = y)$ existence of a left solution

(3) $(\forall x)(\forall y)(\exists z)(x + z = y)$ existence of a right solution.

Our problem is to find a right identity of the system. In other words, we are asked to find an x such that for all $y, y + x = y$. Let us denote "$x + y = z$" by $P(x, y, z)$. Then the axioms can be represented by the following clauses:

(1) $\sim P(x, y, u) \vee \sim P(y, z, v) \vee \sim P(x, v, w) \vee P(u, z, w)$

(2) $P(g(x, y), x, y)$

(3) $P(x, h(x, y), y)$.

To find a right identity, we add the following clause:

(4) $\sim P(f(x), x, f(x)) \vee ANS(x).$

Clause (4) means that if there is an x such that $y + x = y$ for all y, then x is an answer. (Note that f in clause (4) is a Skolem function.)

(5) $\sim P(y, z, v) \vee \sim P(g(y, u), v, w) \vee P(u, z, w)$ from (1) and (2)

(6) $\sim P(v, z, v) \vee P(w, z, w)$ from (5) and (2)

(7) $P(w, h(v, v), w)$ from (3) and (6)

(8) $ANS(h(v, v))$ from (7) and (4).

Clause (8) indicates that for every $v, h(v, v)$ is a right identity.

The following table defines a specific associative system with left and right solutions.

+	a	b	c
a	a	b	c
b	b	c	a
c	c	a	b

It is easy to see that $h(a, a) = h(b, b) = h(c, c) = a$. It is also easy to see that "a" is indeed a right identity because

$$a + a = a, \qquad b + a = b, \qquad c + a = c.$$

11.4   CLASS C QUESTIONS

For this kind of question, our task is to find a sequence of actions that will achieve some goal. The most important concept involved in answering this type of question is the so called "state and state transformation" method [Green, 1969c]. At a given moment, every object under consideration is said to be in a certain state. To achieve a goal, we have to change the present state of the object to a desired state. This requires some action. It will be shown how mechanical theorem proving can be used to find such actions.

Example 11.6

Consider Fig. 11.1. Suppose the initial state of subject d is s_1, and d is initially at a. If we let $P(x, y, z)$ denote "x is located at y in state z," then the above fact can be represented by

(1) $P(d, a, s_1).$

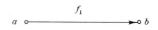

Figure 11.1

Suppose every subject x in state z can be moved from location y_1 to location y_2 by action f_1. We may consider f_1 as a function whose arguments are x, y_1, y_2, and z. The range of $f_1(x, y_1, y_2, z)$ is the new state achieved after x, which is initially in state z, is moved from y_1 to y_2 by action f_1. This is symbolized as

(2) $(\forall x)(\forall y_1)(\forall y_2)(\forall z)(P(x, y_1, z)) \rightarrow P(x, y_2, f_1(x, y_1, y_2, z))$.

We can now use (1) and (2) to answer the question: "How can we move object d from location a to location b?" We have two clauses representing the given facts:

(3) $P(d, a, s_1)$

(4) $\sim P(x, y_1, z) \vee P(x, y_2, f_1(x, y_1, y_2, z))$.

The clause representing the question is

(5) $\sim P(d, b, z) \vee ANS(z)$.

From (3), (4), and (5), we generate the following resolvents:

(6) $P(d, y_2, f_1(d, a, y_2, s_1))$ from (3) and (4)

(7) $ANS(f_1(d, a, b, s_1))$ from (5) and (6).

Thus our answer involves one action, namely, applying f_1 to move d from a to b.

In the above example, the answer to our question involves only one action. In the following example, we shall see that if the answer requires a sequence of actions, the same techniques can still be used.

Example 11.7

Consider Fig. 11.2. Subject d is initially at location a. Suppose there is no direct route between a and c. The question is "How can we move d from

Figure 11.2

a to *c*?" Let us further assume that action f_1 is required to move *d* from *a* to *b* and f_2 is required to move it from *b* to *c*. The problem can be axiomatized as follows:

(1) $P(d, a, s_1)$

(2) $\sim P(d, a, z) \vee P(d, b, f_1(d, a, b, z))$

(3) $\sim P(d, b, z) \vee P(d, c, f_2(d, b, c, z))$.

The clause corresponding to the question is

(4) $\sim P(d, c, z) \vee ANS(z)$.

From (1), (2), (3), and (4), we deduce the following resolvents:

(5) $P(d, b, f_1(d, a, b, s_1))$ from (1) and (2)

(6) $P(d, c, f_2(d, b, c, f_1(d, a, b, s_1)))$ from (5) and (3)

(7) $ANS(f_2(d, b, c, f_1(d, a, b, s_1)))$ from (6) and (4).

Clause (7) can be interpreted as the execution of two actions f_1 and f_2. That is, we first apply action f_1 to move *d* from *a* to *b*, and then apply action f_2 to move *d* from *b* to *c*.

We can now proceed to show how to solve the celebrated monkey–banana problem [McCarthy, 1963b; Slagle, 1965; Green, 1969c; Waldinger and Lee, 1969].

Example 11.8

The monkey–banana problem is concerned with a monkey who wants to eat a banana that is suspended from the ceiling of a room. The monkey is too short to reach the banana. However, he can walk around the room, carrying a chair that is in the room, and he can climb the chair to reach the banana.

The predicates are as follows:

$P(x, y, z, s)$ denotes that in state *s*, the monkey is at *x*, the banana is at *y*, and the chair is at *z*.

$R(s)$ denotes that in state *s*, the monkey can reach the banana.

The functions are as follows:

$walk(y, z, s)$: the state attained if the monkey is initially in state *s* and walks from *y* to *z*.

$carry(y, z, s)$: the state attained if the moneky is initially in state *s* and walks from *y* to *z*, carrying the chair with him.

$climb(s)$: the state attained if the money is in state *s* and climbs the chair.

We shall assume that initially, the monkey is at location a, the banana is at location b, the chair is at location c, and the monkey is in state s_1.

We thus have the following axioms:

(1) $\sim P(x, y, z, s) \vee P(z, y, z, walk(x, z, s))$

(2) $\sim P(x, y, x, s) \vee P(y, y, y, carry(x, y, s))$

(3) $\sim P(b, b, b, s) \vee R(climb(s))$

(4) $P(a, b, c, s_1)$.

The meanings of the above clauses are as follows: Clause (1) means that in any state, the monkey can walk from location x to location z. Clause (2) means that if the monkey is beside the chair which is located at x, then he can carry the chair to any place y. Clause (3) means that if the chair and the monkey are both under the banana, the monkey can climb the chair and reach the banana. Clause (4) describes the initial situation.

The clause corresponding to the question is

(5) $\sim R(s) \vee ANS(s)$.

From (1), (2), (3), (4), and (5), we derive the following resolvents:

(6) $\sim P(b, b, b, s) \vee ANS(climb(s))$ from (5) and (3)

(7) $\sim P(x, b, x, s) \vee ANS(climb(carry(x, b, s)))$ from (6) and (2)

(8) $\sim P(x, b, z, s) \vee ANS(climb(carry(z, b, walk(x, z, s))))$ from (7) and (1)

(9) $ANS(climb(carry(c, b, walk(a, c, s_1))))$ from (8) and (4).

Clause (9) gives us the answer. It can be interpreted as the execution of the following actions (start with the innermost function in clause (9) and work outward):

1. The monkey walks from a to c.
2. The monkey then walks from c to b, carrying the chair with him.
3. The monkey climbs the chair. After this action, the monkey can reach the banana.

11.5   CLASS D QUESTIONS

Let us first consider a very simple example.

Example 11.9

Assume that we have the following facts:

F_1:   If John is under five years old, he should take drug a.
F_2:   If John is not under five years old, he should take drug b.

Our question is: "What drug should John take?"

Let $P(x)$ and $R(x, y)$ denote "x is under five years old" and "x should take y," respectively. Then we have

(1) $\sim P(\text{John}) \lor R(\text{John}, a)$

(2) $P(\text{John}) \lor R(\text{John}, b)$.

The clause corresponding to our question is

(3) $\sim R(\text{John}, x) \lor ANS(x)$.

From (1), (2), and (3), we generate the following resolvents:

(4) $\sim P(\text{John}) \lor ANS(a)$ from (1) and (3)

(5) $P(\text{John}) \lor ANS(b)$ from (2) and (3)

(6) $ANS(a) \lor ANS(b)$ from (4) and (5).

Clause (6) tells us that John should either take drug a or take drug b.

Evidently, the answer extracted from clause (6) is far from satisfying. We would like to be able to tell under what condition John should take drug a and under what condition John should take drug b.

Note that clause (6) is $ANS(a) \lor ANS(b)$. We may ask the following question: "Under what condition will $ANS(a)$ be true?" or "Under what condition will $ANS(a)$ be a logical consequence of clauses (1), (2), and (3)?" A similar question can also be asked for $ANS(b)$. In the following, we shall show how information can be obtained by analyzing the deduction of clause (6), the answering clause.

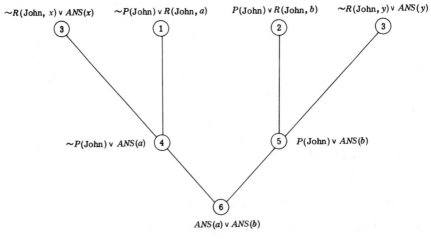

Figure 11.3. T_0.

The deduction of clause (6) is shown in Fig. 11.3, where every node denotes a clause involved in the deduction, and variables in the clauses attached to the initial nodes are renamed so that the clauses have no variables in common. Denote this deduction tree by T_0. We now analyze T_0 according to the following information-extraction algorithm.

An Information-Extraction Algorithm

Step 1   In tree T_0, if a clause C is a resolvent of clauses C_1 and C_2, with resolved literals L_1 and L_2 in C_1 and C_2, respectively, such that θ is a most general unifier of L_1 and $\sim L_2$, then on the arc from C_i to C, $i = 1, 2$, write down the negation of $L_i \theta$ and the substitution θ. (If θ is an empty substitution, we only write down the negation of $L_i \theta$). For our example, clause (4) is a resolvent of clause (3) and clause (1). The literal resolved upon in (3) is $\sim R(\text{John}, x)$. The literal resolved upon in (1) is $R(\text{John}, a)$. The most general unifier of the resolved literals is $\{a/x\}$. Therefore, we attach $R(\text{John}, a)$ and the substitution $x \leftarrow a$ to the arc from (3) to (4). Similarly, we add $\sim R(\text{John}, a)$ and the substitution $x \leftarrow a$ to the arc from (1) to (4). After doing the same for every other arc, we shall obtain a tree T_1 as shown in Fig. 11.4.

Step 2   Invert tree T_1, add arrows to the arcs, and delete all the clauses attached to the nodes. We then obtain tree T_2 as shown in Fig. 11.5, where all the tip nodes are labeled.

Step 3   For tree T_2, we now delete every node (and its associated arcs) that corresponds to a clause which does not contain the *ANS* predicate.

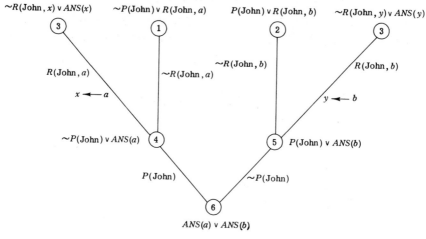

Figure 11.4. T_1.

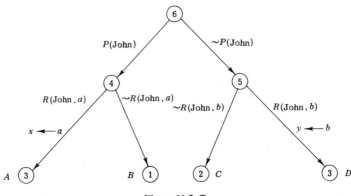

Figure 11.5. T_2.

In our example, since clauses (1) and (2) do not contain the ANS predicate, nodes B and C and their associated arcs will be deleted. The resulting tree will be T_3 as shown in Fig. 11.6.

Step 4   Let $N_1, N_2, \ldots, N_m$ be the tip nodes of T_3. For each $N_i, i = 1, \ldots, m$, let $I(N_i)$ denote the conjunction of literals attached to the path from the top node to N_i. Let $C(N_i)$ be the clause that corresponds to N_i. Find a literal $L(N_i)$ in the answering clause such that $L(N_i)$ is a logical consequence of $I(N_i) \wedge C(N_i)$. (It will be proved later in Section 11.9 that such a literal exists.) Attach $L(N_i)$ to node N_i.

In our example, since $I(A) = P(\text{John}) \wedge R(\text{John}, a), C(A) = \sim R(\text{John}, x) \vee ANS(x)$ and $L(A) = ANS(a)$. Similarly, it is very easy to see that $L(B) = ANS(b)$. The tree T_4 thus obtained is shown in Fig. 11.7.

Step 5   In T_4, let $N_1, N_2, \ldots, N_q$ be nodes where there is only one arc a_i leading out of $N_i, 1 \leq i \leq q$. Let $L(a_i)$ be the literal attached to a_i. Delete

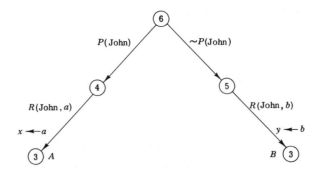

Figure 11.6. T_3.

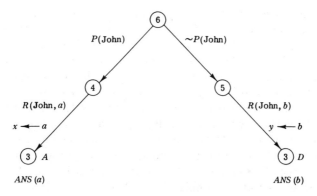

Figure 11.7. T_4.

$L(a_i), 1 \leqslant i \leqslant q$, from tree T_4 and the resulting tree will be T_5. For our example, T_5 is shown in Fig. 11.8.

Note that Fig. 11.8 can be conveniently considered as a decision tree. That is, if $P(\text{John})$ is true, $ANS(a)$ is true. Otherwise, $ANS(b)$ is true. Thus our answer is "If John is under five, he should take drug a. Otherwise, take drug b."

The reader can easily see that so far as our example is concerned, the above information-extraction algorithm is correct. That is, decision tree T_5 is a correct one. A formal proof of the correctness of this algorithm will be given in Section 11.9. Meanwhile, we shall give an informal discussion.

In Step 4, we attach $ANS(a)$ to node A after it is found that $ANS(a)$ is a logical consequence of $P(\text{John}) \wedge R(\text{John}, a) \wedge (\sim R(\text{John}, x) \vee ANS(x))$. Since $\sim R(\text{John}, x) \vee ANS(x)$ is a part of the input clauses, it is assumed to

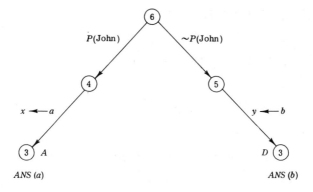

Figure 11.8. T_5.

be always true. Therefore, whenever $P(\text{John}) \wedge R(\text{John}, a)$ is true, $ANS(a)$ is true. However, consider Fig. 11.5. Node B corresponds to clause (1). Since it does not contain any ANS predicate, every literal that it contains will be resolved away to obtain the answering clause. Since the negations of all these literals are in $I(B)$, where $I(B) = \{P(\text{John}), \sim R(\text{John}, a)\}$, it is not difficult to show that $R(\text{John}, a)$ is a logical consequence of $P(\text{John})$ and clause (1). Since clause (1) is assumed true in the example, $R(\text{John}, a)$ must be true whenever $P(\text{John})$ is true. Therefore, from the fact that whenever $P(\text{John})$ and $R(\text{John}, a)$ are true, $ANS(a)$ must be true, we obtain that whenever $P(\text{John})$ is true, $ANS(a)$ is true. Similarly, we can show that whenever $\sim P(\text{John})$ is true, $ANS(b)$ is true. This proves that decision tree T_5 is correct.

Let us consider another example.

Example 11.10

Consider Fig. 11.9. Assume that d wants to go to location c from location a. He can go to c from b either through action f_2 or action f_3, depending upon whether some condition Q is satisfied. He can also reach b from a through some action f_1. Let us assume that subject d is initially at a. Our question is how to find a route for subject d to reach c, starting from a.

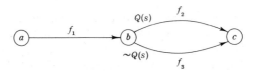

Figure 11.9

Let $P(x, y, s)$ denote "x is at y in state s." Then the facts are symbolized as

(1) $\sim P(x, a, s) \vee P(x, b, f_1(x, a, b, s))$

(2) $\sim P(x, b, s) \vee \sim Q(s) \vee P(x, c, f_2(x, b, c, s))$

(3) $\sim P(x, b, s) \vee Q(s) \vee P(x, c, f_3(x, b, c, s))$

(4) $P(d, a, s_1)$.

The clause corresponding to the question is

(5) $\sim P(d, c, s) \vee ANS(s)$.

From (1)–(5), we generate the following resolvents:

(6) $\sim P(d, b, u) \vee Q(u) \vee ANS(f_3(d, b, c, u))$ from (5) and (3)

(7) $\sim P(d, b, v) \vee \sim Q(v) \vee ANS(f_2(d, b, c, v))$ from (5) and (2)

(8) $\sim P(d, b, v) \lor ANS(f_3(d, b, c, v)) \lor ANS(f_2(d, b, c, v))$

from (6) and (7)

(9) $\sim P(d, a, w) \lor ANS(f_3(d, b, c, f_1(d, a, b, w)))$

$\lor ANS(f_2(d, b, c, f_1(d, a, b, w)))$ from (8) and (1)

(10) $ANS(f_3(d, b, c, f_1(d, a, b, s_1))) \lor$

$ANS(f_2(d, b, c, f_1(d, a, b, s_1)))$ from (9) and (4).

Clause (10) is an answering clause. Its deduction is now represented by the tree in Fig. 11.10. Applying the information-extraction algorithm to this tree, we can obtain the decision tree shown in Fig. 11.11. This tree gives us the following answer:

1. Subject d goes from a to b through action f_1.
2. If condition Q is satisfied, d goes to c through action f_2.
3. Otherwise, d goes to c through action f_3.

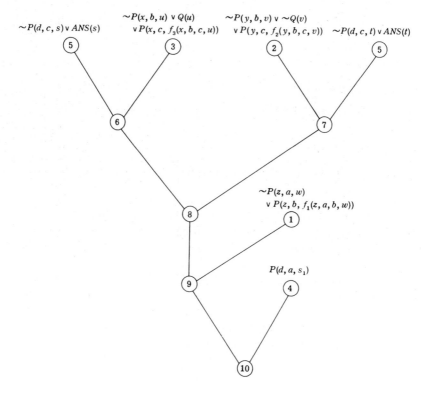

Figure 11.10

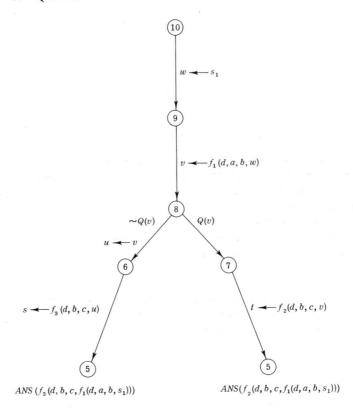

Figure 11.11

11.6 COMPLETENESS OF RESOLUTION FOR DERIVING ANSWERS

In the previous sections, we gave examples using the resolution principle to answer questions. In this section, we shall show that the resolution principle is complete in answering questions. That is, if a question is answerable, then an answering clause will be generated by repeatedly applying the resolution principle.

Let us assume that the facts from which an answer is to be derived are represented by a set S_1 of clauses. Let a question be formulated as follows: "Find values of $x_1, x_2, \ldots, x_n$ such that the relation $Q(x_1, x_2, \ldots, x_n)$ is true." We shall say that the question is *answerable* if and only if $(\exists x_1)(\exists x_2) \cdots (\exists x_n)$ $Q(x_1, x_2, \ldots, x_n)$ is a logical consequence of S_1.

Let S_2 denote the set consisting of the following clause:

$$\sim Q(x_1, x_2, \ldots, x_n) \lor ANS(x_1, x_2, \ldots, x_n).$$

Theorem 11.1   Let S be $S_1 \cup S_2$, where S_1 and S_2 are defined as above. The question is answerable if and only if there is a deduction of an answering clause from S.

Proof ($\Rightarrow$)   If the question is answerable, then $(\exists x_1) \cdots (\exists x_n) Q(x_1, \ldots, x_n)$ is a logical consequence of S_1, where S_1 is the set of clauses representing the facts. Therefore, $S_1 \wedge \sim Q(x_1, \ldots, x_n)$ is unsatisfiable. That is, there is a deduction of $\square$ from S_1. Putting $ANS(x_1, x_2, \ldots, x_n)$ onto $\sim Q(x_1, x_2, \ldots, x_n)$, we have $\sim Q(x_1, x_2, \ldots, x_n) \vee ANS(x_1, x_2, \ldots, x_n)$. Since the predicate ANS does not appear in S_1, the same deduction that deduces the empty clause from $S_1 \wedge \sim Q(x_1, x_2, \ldots, x_n)$ will now deduce an answering clause from $S_1 \wedge (\sim Q(x_1, x_2, \ldots, x_n) \vee ANS(x_1, x_2, \ldots, x_n))$. This completes the proof of the first half of Theorem 11.1.

($\Leftarrow$)   This part of the proof is easy and will be left to the reader as an exercise.

It is interesting to note that the problem of deciding whether a question is answerable is similar to that of deciding whether a program terminates. In each case, the resolution principle is complete. Besides, it not only tells us whether the question is answerable (or whether the program terminates), but also tells us what the answer is (or how the program terminates).

11.7   THE PRINCIPLES OF PROGRAM SYNTHESIS

In Chapter 10, we discussed how to apply theorem-proving techniques to program analysis. By program analysis, we mean that we are given a program and want to find its input–output relationship. In program synthesis [Waldinger and Lee, 1969; Green, 1969b; Manna and Waldinger, 1971; Nilsson, 1971; Lee *et al.*, 1972], we reverse the process; that is, we are given the specification (description) of a program, and our task is to synthesize a program to realize this specification.

The reader may have already noted that the techniques mentioned in previous sections can be used to write programs. Since we use formal theorem-proving techniques, a program written in this way will be free from logical errors.

We shall first present an example of writing a program in machine language.

Example 11.11

Let us imagine a machine with three registers a, b, and c, and one accumulator. We would like to write a program to add the contents of registers a and b and then store the result in register c.

Let ca, cb, cc, and acc represent the contents of registers a, b, c, and the accumulator, respectively. Let the initial state of the machine be d. We shall use the following functions:

$load(x, s)$: the state attained after the contents of register x are loaded into the accumulator at state s.

$add(x, s)$: the state attained after the contents of register x are added to the contents of the accumulator at state s.

$store(x, s)$: the state attained after the contents of the accumulator are stored into register x at state s.

Let $P(u, x, y, z, s)$ denote that at state s, the contents of the accumulator and registers a, b, and c are u, x, y, and z, respectively.

The problem can be axiomatized as follows:

(1) $\sim P(u, x, y, z, s) \lor P(x, x, y, z, load(a, s))$

(2) $\sim P(u, x, y, z, s) \lor P(u+y, x, y, z, add(b, s))$

(3) $\sim P(u, x, y, z, s) \lor P(u, x, y, u, store(c, s))$

(4) $P(acc, ca, cb, cc, d)$

(5) $\sim P(u, x, y, x+y, s) \lor ANS(s)$.

Using resolution, we derive the following resolvents:

(6) $\sim P(x+y, x, y, z, s) \lor ANS(store(c, s))$ from (5) and (3)

(7) $\sim P(x, x, y, z, s) \lor ANS(store(c, add(b, s)))$ from (6) and (2)

(8) $\sim P(u, x, y, z, s) \lor ANS\big(store(c, add(b, load(a, s)))\big)$ from (7) and (1)

(9) $ANS\big(store(c, add(b, load(a, d)))\big)$ from (8) and (4).

The deduction of clause (9) is shown in Fig. 11.12. Applying the information-extraction algorithm to the tree, we can obtain a tree shown in Fig. 11.13.

The substitution for s_4, s_3, s_2, and s_1 in Fig. 11.13 constitute a program and can be interpreted as follows:

1. Load the contents of register a into the accumulator.
2. Add the contents of register b to the contents of the accumulator.
3. Store the contents of the accumulator into register c.

Example 11.12

Let us write a program that tests whether 7 and/or 9 divides an integer x. If both 7 and 9 divide x, let the output be a. If 7 divides x, but 9 does not, let the output be b. If 7 does not divide x, but 9 does, let the output be c. Otherwise, let the output be d.

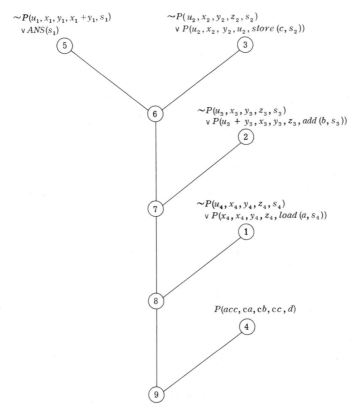

Figure 11.12

We shall use $P(x, y)$ and $R(x, z)$ to mean "y divides x" and "the output should be z when the input is x," respectively. The program is now axiomatized as follows:

(1) $\sim P(x, 7) \vee \sim P(x, 9) \vee R(x, a)$

(2) $\sim P(x, 7) \vee P(x, 9) \vee R(x, b)$

(3) $P(x, 7) \vee \sim P(x, 9) \vee R(x, c)$

(4) $P(x, 7) \vee P(x, 9) \vee R(x, d)$

(5) $\sim R(x, z) \vee ANS(z)$.

From (1), (2), (3), (4), and (5), we derive the following resolvents:

(6) $P(x, 7) \vee P(x, 9) \vee ANS(d)$ from (5) and (4)

(7) $P(x, 7) \vee \sim P(x, 9) \vee ANS(c)$ from (5) and (3)

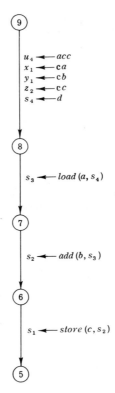

Figure 11.13

(8) $\sim P(x, 7) \lor P(x, 9) \lor ANS(b)$ from (5) and (2)

(9) $\sim P(x, 7) \lor \sim P(x, 9) \lor ANS(a)$ from (5) and (1)

(10) $P(x, 7) \lor ANS(c) \lor ANS(d)$ from (6) and (7)

(11) $\sim P(x, 7) \lor ANS(a) \lor ANS(b)$ from (8) and (9)

(12) $ANS(a) \lor ANS(b) \lor ANS(c) \lor ANS(d)$ from (10) and (11).

The deduction of clause (12) is described in Fig. 11.14. Applying the information-extraction algorithm to this tree, we can obtain a tree shown in Fig. 11.15. In this tree, since y_1, y_3, y_5, and y_7 play the role of the output variable z, we rename them z. Also, since x_1 and x_5 play the role of the input variable x, we rename them x. After this notational simplification, we can delete the substitutions $x_5 \leftarrow x_1, x_3 \leftarrow x_1$, and $x_7 \leftarrow x_5$. Thus we obtain a tree T^* shown in Fig. 11.16. From T^*, letting the top node be the starting vertex, eliminating nodes (6), (7), (8), and (9), and letting node (5) be the halting node, we now obtain a program shown in Fig. 11.17.

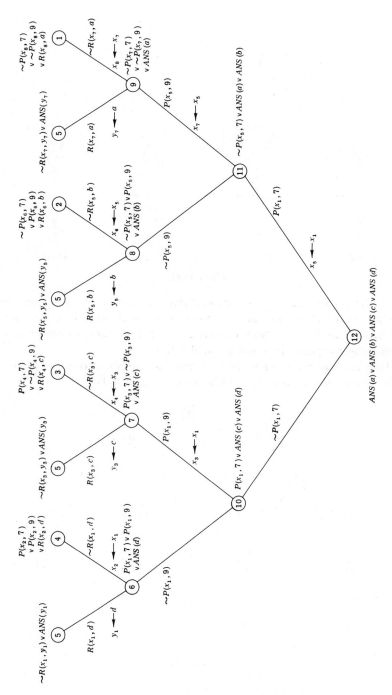

Figure 11.14

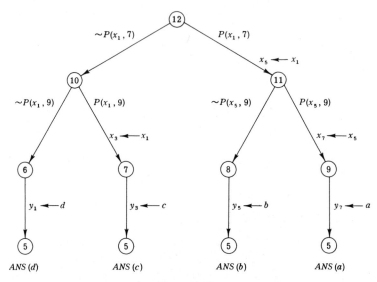

Figure 11.15

In the above example, we showed that we can obtain programs from deductions of answering clauses. However, we shall now show that not all programs obtained this way are executable. It is possible to obtain a non-executable program. This is illustrated below.

Imagine that we want to test whether an integer is positive or not. If it is, assign the output to be 1. Otherwise, assign it to be 0.

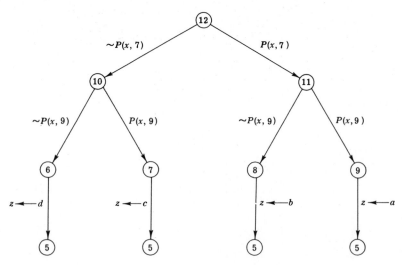

Figure 11.16

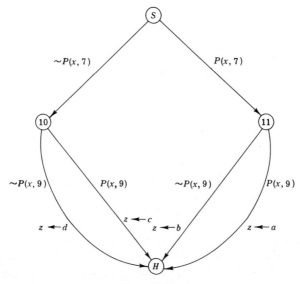

Figure 11.17

We let $P(x)$ and $R(x, z)$ denote "x is positive" and "the output is z when the input is x," respectively. Thus, our axioms are

(1) $\sim P(x) \vee R(x, 1)$

(2) $P(x) \vee R(x, 0)$

(3) $\sim R(x, z) \vee ANS(z)$

(4) $R(x, 1) \vee R(x, 0)$ from (1) and (2)

(5) $R(x, 0) \vee ANS(1)$ from (4) and (3)

(6) $ANS(1) \vee ANS(0)$ from (5) and (3).

We have derived the correct conclusion. That is, clause (6) tells us that the output is either 1 or 0. But the analysis of the deduction of clause (6) will give us a nonexecutable program as shown in Fig. 11.18.

The program corresponding to the tree in Fig. 11.18 is essentially as follows:

1. If the output should be 0, then assign the output to be 0.
2. Otherwise, assign the output to be 1.

However, since the output is not known beforehand, this program is not executable. Deductions that will produce such nonexecutable programs should be forbidden. This is achieved by asking the user to declare that

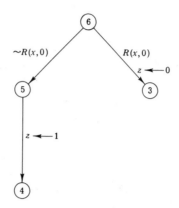

Figure 11.18

certain symbols be primitive or nonprimitive. We then use a new inference rule, called "primitive resolution." Using this special inference rule, the program that we obtain will contain only primitive symbols and will therefore be executable.

11.8   PRIMITIVE RESOLUTION AND ALGORITHM A* (A PROGRAM-SYNTHESIZING ALGORITHM)

In this section, we shall first define primitive resolution. We shall then describe an algorithm for constructing programs from primitive deductions. Again, we emphasize that our programs thus constructed do not have the ability to call themselves, as we assumed before in Chapter 10.

We shall assume that a user of our program-writing program supplies the specification of his program. Let $R(x_1, x_2, \ldots, x_r, a_{r+1}, \ldots, a_m, z_1, z_2, \ldots, z_n)$ denote "the output variables are $z_1, z_2, \ldots, z_n$ when the input variables are $x_1, \ldots, x_r$, and $a_{r+1}, \ldots, a_m$ are input constants. We then have the following formula as a part of our axioms that describe the specification:

$$(\forall x_1)(\forall x_2) \cdots (\forall x_r)(\forall z_1)(\forall z_2) \cdots (\forall z_n)(R(x_1, x_2, \ldots, x_r, a_{r+1}, \ldots, a_m, z_1, \ldots, z_n)$$
$$\rightarrow ANS(z_1, \ldots, z_n)).$$

The user then has to declare those symbols that are allowed to appear in his program. They are called *primitive symbols.*

As the reader can see, we do not have to trace all the variables involved in a deduction. Indeed, we only have to trace variables in clauses containing the *ANS* predicate.

Definition   A clause containing an *ANS* predicate is called a *vital clause.* A variable is called a *vital variable* if it appears in a vital clause.

Example 11.13

Suppose

$$C = \sim P(z, y) \lor \sim Q(y) \lor ANS(y)$$
$$D = P(x, u) \lor R(x).$$

The resolvent of C and D is

$$B = \sim Q(u) \lor R(x) \lor ANS(u).$$

B is now a vital clause, and variables u and x in clause B are now vital variables.

Definition   Let a resolvent C of clauses C_1 and C_2 be defined as

$$C = (C_1 \sigma - L_1 \sigma) \cup (C_2 \sigma - L_2 \sigma),$$

where L_1 and L_2 are literals resolved upon and σ is a unifier of L_1 and $\sim L_2$. Then C is called a *primitive resolvent* of C_1 and C_2 if one of the following conditions is satisfied:

1. C_1 and C_2 are not vital clauses.
2. C_1 is a vital clause, but C_2 is not. All the constants and function symbols occurring in terms substituted for vital variables in σ are primitive.
3. C_1 and C_2 are both vital clauses. All the constants and function and predicate symbols occurring in $L_1 \sigma$ and $L_2 \sigma$ are primitive.

Definition   A *primitive deduction* is a deduction in which every resolution is a primitive resolution.

If condition 2 is not satisfied, there might be an instruction, such as

$$y \leftarrow plus(x, z)$$

where the function *plus* is not primitive and therefore not executable. If condition 3 is not satisfied, there might be an instruction, such as

If $x < y$, then go to a.

where the predicate $<$ is not primitive and therefore not executable.

Example 11.14

Let us reconsider the case in Section 11.7, where we wanted to test whether an integer is positive. If it is, let the output be 1; otherwise, let the output be 0.

We let $P(x)$ and $R(x, z)$ denote "x is positive" and "the output should be z when the input is x," respectively. Our axioms are

(1) $\sim P(x) \lor R(x, 1)$
(2) $P(x) \lor R(x, 0)$
(3) $\sim R(x, z) \lor ANS(z).$

We further declare that symbols P, 0, and 1 are the only primitive symbols. Let us consider the following deduction:

(4) $R(x, 1) \lor R(x, 0)$ from (1) and (2)

(5) $R(x, 0) \lor ANS(1)$ from (4) and (3)

(6) $ANS(1) \lor ANS(0)$ from (5) and (3).

The reader can easily see that the resolution between clauses (5) and (3) is not primitive. Both clauses are vital clauses, but $R(x, z)$ is not primitive. Therefore, it does not satisfy condition 3 of primitive resolution. If clause (5) is allowed to resolve with clause (3), the final program would necessarily involve a testing of $R(x, z)$, which is nonexecutable.

However, the reader can easily see that the following deduction is a primitive deduction.

(4) $\sim P(x) \lor ANS(1)$ from (1) and (3)

(5) $P(x) \lor ANS(0)$ from (2) and (3)

(6) $ANS(1) \lor ANS(0)$ from (4) and (5).

The deduction of clause (6) will give the program in Fig. 11.19, which is executable.

If an answering clause can be deduced by primitive resolution, an executable program can then be constructed. In the following we shall give algorithm A*, which constructs a program from a primitive deduction of an answering clause. Algorithm A* is very similar to the information-extraction algorithm given in Section 11.5, except that it is specially designed for program synthesis. We shall use the definition of programs as given in Chapter 10. That is, we shall consider a program to be a directed graph. It should be emphasized here that this algorithm deals only with "loop-free" programs. The synthesis of iterative programs will be discussed in Section 11.10.

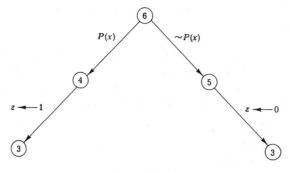

Figure 11.19

As will be shown in Section 11.11, algorithm A* will be improved to algorithm A, which is far more efficient. The advantage of algorithm A* is that a proof of the correctness of our approach can be easily obtained by analyzing algorithm A*.

Algorithm A* (A Program-Synthesizing Algorithm)

Step 1 Let T_1 be a primitive deduction tree of an answering clause from a set of clauses including the clause $\sim R(x_1, \ldots, x_r, a_{r+1}, \ldots, a_m, z_1, \ldots, z_n) \lor ANS(z_1, \ldots, z_n)$. In this deduction, the input variables $x_1, \ldots, x_r$ cannot be replaced by other terms; i.e., $x_1, \ldots, x_r$ should be treated as constants, and the variables $z_1, \ldots, z_n$ cannot appear in any term that is substituted for some variable. Also, variables in the clause (except $\sim R(x_1, \ldots, x_r, a_{r+1}, \ldots, a_m, z_1, \ldots, z_n) \lor ANS(z_1, \ldots, z_n)$) attached to initial nodes of T_1 should be renamed so that these clauses have no variables in common.

Step 2 In tree T_1, if a clause C is a resolvent of clauses C_1 and C_2 with resolved literals L_1 and L_2 in C_1 and C_2, respectively, such that σ is a unifier of L_1 and $\sim L_2$, then on the arc from C_i to $C, i = 1, 2$, write down the negation of $L_i \sigma$ and the substitution σ. (Note that if σ is an empty substitution, we need to write down only the negation of $L_i \sigma$.) The resulting tree is denoted by T_2.

Step 3 Delete every part of tree T_2 that corresponds to a nonvital clause in T_2. Delete all the clauses attached to the nodes of the remaining tree. The tree thus obtained is then inverted and arrows are added. The resulting tree is denoted by T_3.

Step 4 For every arc a that is the only arc leading out of a node N, delete the literal L_a attached to a. Denote the resulting tree by T_4.

Step 5 Label the top node of T_4 (which corresponds to the answering clause) as the starting vertex S, and all the bottom nodes (which correspond to
$$\sim R(x_1, \ldots, x_r, a_{r+1}, \ldots, a_m, z_1, \ldots, z_n) \lor ANS(z_1, \ldots, z_n))$$
as the halting vertex H. Thus the final directed graph T_5 so obtained is a program where $x_1, \ldots, x_r$ are input variables, $z_1, \ldots, z_n$ are output variables, and every other variable is a program variable.

Example 11.15

Let us consider the following specification of a program:

1. If x^2 is greater than 10, but not greater than 100, then let the output be $x + 1$.
2. If x^2 is greater than 100, let the output be $x - 1$.
3. Otherwise, let the output be x.

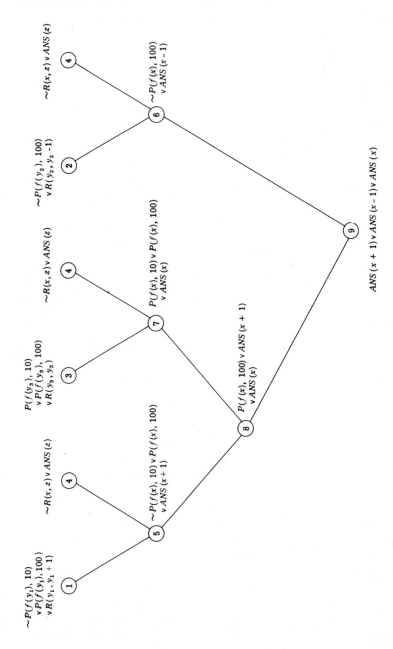

Figure 11.20. T_1

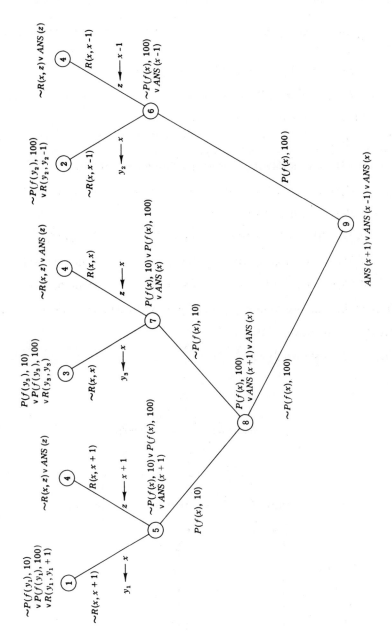

Figure 11.21. T_2

Let $f(x)$ denote x^2. Let $P(x, y)$ denote "x is greater than y." Let $R(x, z)$ denote that the output is assigned z when the input is x. We then have the following clauses (we declare that P, f, $+$, and $-$ are the only primitive symbols):

(1) $\sim P(f(y_1), 10) \vee P(f(y_1), 100) \vee R(y_1, y_1 + 1)$

(2) $\sim P(f(y_2), 100) \vee R(y_2, y_2 - 1)$

(3) $P(f(y_3), 10) \vee P(f(y_3), 100) \vee R(y_3, y_3)$

(4) $\sim R(x, z) \vee ANS(z)$.

From (1)–(4), we generate the following primitive resolvents:

(5) $\sim P(f(x), 10) \vee P(f(x), 100) \vee ANS(x+1)$ from (1) and (4)

(6) $\sim P(f(x), 100) \vee ANS(x-1)$ from (2) and (4)

(7) $P(f(x), 10) \vee P(f(x), 100) \vee ANS(x)$ from (3) and (4)

(8) $ANS(x+1) \vee P(f(x), 100) \vee ANS(x)$ from (5) and (7)

(9) $ANS(x+1) \vee ANS(x-1) \vee ANS(x)$ from (6) and (8).

In the following, we shall apply algorithm A* to synthesize the program:

Step 1 T_1 is shown in Fig. 11.20.

Step 2 T_2 is shown in Fig. 11.21.

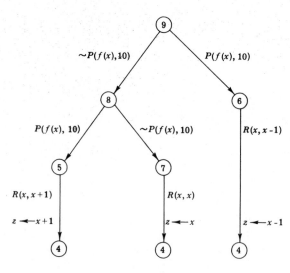

Figure 11.22. T_3

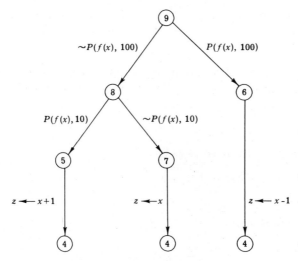

Figure 11.23. T_4

Step 3   In T_2 of Fig. 11.21, clauses (1), (2), and (3) are not vital clauses. We now delete these nodes and their associated arcs from T_2 as well as all the clauses attached to the nodes of the remaining tree, and then invert the tree and add arrows. Figure 11.22 is the resulting tree.

Step 4   For the arcs connecting (5) to (4), (7) to (4), and (6) to (4), we now delete the respective literals attached to them. The resulting tree T_4 is shown in Fig. 11.23.

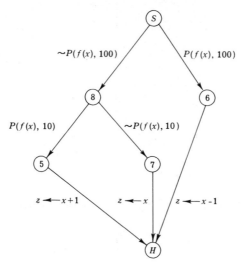

Figure 11.24. T_5

Step 5   The top node of the tree in Fig. 11.23 is 9. We denote it by S. We denote all the bottom nodes by H. The resulting tree T_5 is shown in Fig. 11.24.

In the above example, no program variable appears. We shall now give an example in which we have variables other than input variables and output variables.

Example 11.16

Let our program be specified as follows:

(1) $\sim R(x, a, y) \vee R(x, b, f_1(x, y))$

(2) $\sim R(x, b, y) \vee \sim Q(y) \vee R(x, c, f_2(x, c, y))$

(3) $\sim R(x, b, y) \vee Q(y) \vee R(x, c, f_3(x, c, y))$

(4) $R(d, a, e)$

(5) $\sim R(d, c, z) \vee ANS(z)$.

In the above program, $R(x_1, x_2, z)$ denotes "the output is z when the inputs are x_1 and x_2," and Q, a, b, c, d, e, f_1, f_2, and f_3 are the only primitive symbols.

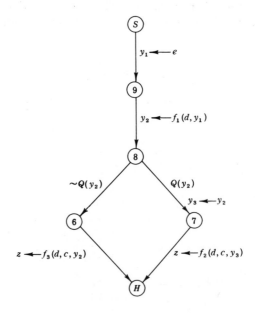

Figure 11.25

The reader should have no difficulty in proving that the following deduction is primitive:

(6) $\sim R(d,b,y_3) \vee Q(y_3) \vee ANS(f_3(d,c,y_3))$ from (5) and (3)

(7) $\sim R(d,b,y_2) \vee \sim Q(y_2) \vee ANS(f_2(d,c,y_2))$ from (5) and (2)

(8) $\sim R(d,b,y_2) \vee ANS(f_3(d,c,y_2)) \vee ANS(f_2(d,c,y_2))$
                                                                    from (6) and (7)

(9) $\sim R(d,a,y_1) \vee ANS(f_2(d,c,f_1(d,y_1))) \vee ANS(f_3(d,c,f_1(d,y_1)))$
                                                                    from (8) and (1)

(10) $ANS(f_2(d,c,f_1(d,e))) \vee ANS(f_3(d,c,f_1(d,e)))$ from (9) and (4).

Applying algorithm A* to the above deduction, we obtain the program in Fig. 11.25. In this program, y_1, y_2, and y_3 are program variables. We note that this program does not have input variables. This is because clause (5) does not contain input variables.

11.9   THE CORRECTNESS OF ALGORITHM A*

The reader can easily see that the program produced by algorithm A* satisfies all the conditions for a program as specified in Chapter 10. Besides, the program is executable. This can be seen by proving the following theorem.

Theorem 11.2   Let D be a primitive deduction of an answering clause. Let P_D be the program obtained by applying algorithm A* to D. Then P_D contains only primitive symbols.

In Step 3 of algorithm A*, since any part of D corresponding to a non-vital clause will be deleted, any substitution that contains nonprimitive constants and function symbols will be deleted. Therefore, all constants and function symbols occurring in P_D are primitive. On the other hand, there are some nonprimitive testing literals (that is, literals containing nonprimitive predicate symbols) that appear in T_3 described in algorithm A*. Let L_i be a nonprimitive symbol attached to an arc as shown in Fig. 11.26.

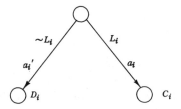

Figure 11.26

Since L_i is nonprimitive, one of C_i and D_i must not be a vital clause, according to the definition of primitive resolution. Without loss of generality, we assume that D_i is the nonvital clause. Therefore the part of the tree corresponding to D_i is deleted by Step 3 of the algorithm. In Step 4, L_i is further deleted. This means that every nonprimitive literal is deleted from T_3 and the final program P_D contains only primitive symbols.   Q.E.D.

In the following, we shall show that algorithm A* also produces a correct program; that is, the program thus produced will give a correct output. Before giving a formal proof, let us consider the program shown in Fig. 11.25, which is from Example 11.16.

In this program, after control is passed to node 8, y_2 is replaced by $f_1(d, e)$. If $\sim Q(f_1(d, e))$ is true, control is passed to node 6 and the output variable is finally assigned to be $f_3(d, c, f_1(d, e))$. We now ask, is this a correct output assignment? That is, is this result correct with respect to the specification of the program?

We shall now prove that $ANS(f_3(d, c, f_1(d, e)))$ is a logical consequence of the specification of the program and $\sim Q(f_1(d, e))$. First, we write the specification of the program as below:

(1) $\sim R(x, a, y) \vee R(x, b, f_1(x, y))$

(2) $\sim R(x, b, y) \vee \sim Q(y) \vee R(x, c, f_2(x, c, y))$

(3) $\sim R(x, b, y) \vee Q(y) \vee R(x, c, f_3(x, c, y))$

(4) $R(d, a, e)$

(5) $\sim R(d, c, z) \vee ANS(z)$.

In addition to the above clauses, we have

(6) $\sim Q(f_1(d, e))$.

From (1)–(6), we generate the following clauses:

(7) $\sim R(x, b, f_1(d, e)) \vee R(x, c, f_3(x, c, f_1(d, e)))$ from (3) and (6)

(8) $\sim R(d, a, e) \vee R(d, c, f_3(d, c, f_1(d, e)))$ from (7) and (1)

(9) $R(d, c, f_3(d, c, f_1(d, e)))$ from (8) and (4)

(10) $ANS(f_3(d, c, f_1(d, e)))$ from (9) and (5).

Since clause (10) is derived from clauses (1)–(6) by resolution, $ANS(f_3(d, c, f_1(d, e)))$ is a logical consequence of the specification of the program and $\sim Q(f_1(d, e))$. On the other hand, if $Q(f_1(d, e))$ is true, z is assigned $f_2(d, c, f_1(d, e))$. It is not difficult to see that this assignment is also correct.

In the sequel, let A denote the set of clauses representing the specification of a program. Let D be a primitive deduction of an answering clause from A. Let P_D be the program produced by algorithm A*. Let a sequence of arcs, $a_1, a_2, \ldots, a_q$ denote a path from node S to node H in program P_D. Let $\bar{y}^i$ denote the value of program variable vector after control has passed arc $a_1, a_2, \ldots, a_i, i = 1, 2, \ldots, q-1$. Let the substitution instruction attached to a_q be $\bar{z} \leftarrow f_{a_q}(\bar{x}, \bar{y}^{q-1})$. Then the value of the output vector $\bar{z}$ will be $f_{a_q}(\bar{x}, \bar{y}^{q-1})$ after control has passed arcs $a_1, a_2, \ldots, a_q$.

Along the path $a_1, a_2, \ldots, a_q$, not every arc necessarily has a testing literal attached to it. Let these arcs that have testing literals associated with them be arcs $a_{i_1}, a_{i_2}, \ldots, a_{i_k}, 1 \leqslant k \leqslant q$. Let the testing literal attached to arc a_{i_j} be $L_{i_j}(\bar{x}, \bar{y})$. Then the condition that control will pass through the path $a_1, \ldots, a_q$ is represented by $L_{i_1}(\bar{x}, \bar{y}^{i_1 - 1}) \wedge \cdots \wedge L_{i_k}(\bar{x}, \bar{y}^{i_k - 1})$. Let $I = \{L_{i_1}(\bar{x}, \bar{y}^{i_1 - 1}), \ldots, L_{i_k}(\bar{x}, \bar{y}^{i_k - 1})\}$. If program P_D is correct, then whenever every literal in I is true, $\bar{z}$ should be assigned $f_{a_q}(\bar{x}, \bar{y}^{q-1})$ according to P_D. We thus have to prove that $ANS(f_{a_q}(\bar{x}, \bar{y}^{q-1}))$ is a logical consequence of $A \cup I$.

Theorem 11.3   For every path $a_1, a_2, \ldots, a_q$ from node S to node H, $ANS(f_{a_q}(\bar{x}, \bar{y}^{q-1}))$ is a logical consequence of $A \cup I$.

Proof   Let the path be shown in tree T_5 in Fig. 11.27, where arc a_i' and subtree D_i may be present or absent, $i = 1, \ldots, q$. (T_5 is the tree obtained in Step 5 of algorithm A*.) Note that T_5 is obtained from tree T_2 shown in Fig.

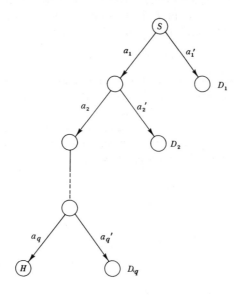

Figure 11.27. T_5

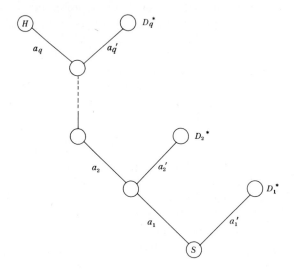

Figure 11.28. T_2

11.28, where D_i^* is a clause deduced from $A, i = 1, \ldots, q$. (T_2 is the tree obtained in Step 2 of algorithm A*.) Let $L_i(\bar{x}, \bar{y})$ be the literal attached to arc a_i of T_2, $i = 1, \ldots, q$. Let $I^* = \{L_1(\bar{x}, \bar{y}^0), \ldots, L_q(\bar{x}, \bar{y}^{q-1})\}$. Since in Step 3 and Step 4 of algorithm A*, some of the literals in I^* are deleted, I is a subset of I^*. The substitution attached to a_i in T_2 is the same as the one attached to a_i in T_5. Let C_q be attached to node H. C_q must be of the form

$$\sim R(\bar{x}, \bar{z}) \vee ANS(\bar{z}).$$

Since T_2 is a deduction of an answering clause, all of the literals in C_q except $ANS(\bar{z})$ must be resolved away. However, the negations of the instances of all these resolved literals are attached to the path, that is, they are in I^*. Therefore, $ANS(f_{a_q}(\bar{x}, \bar{y}^{q-1}))$ is a logical consequence of $(C_q \cup I^*)$. Since $C_q \in A$, $ANS(f_{a_q}(\bar{x}, \bar{y}^{q-1}))$ is also a logical consequence of $(A \cup I^*)$. We now consider the following induction process:

1. Set $j = q$ and $I_j = I^*$. Thus, $ANS(f_{a_q}(\bar{x}, \bar{y}^{q-1}))$ is a logical consequence of $(A \cup I_j)$.
2. Define

$$I_{j-1} = I_j - \{L_j(\bar{x}, \bar{y}^{j-1})\}, \qquad \text{if } L_j(\bar{x}, \bar{y}^{j-1}) \text{ is not in } I;$$
$$= I_j, \qquad\qquad\qquad \text{otherwise.}$$

We want to show that $L_j(\bar{x}, \bar{y}^{j-1})$ is a logical consequence of $(A \cup I_{j-1})$.

Clearly, if $L_j(\bar{x}, \bar{y}^{j-1})$ is in I, this is obviously true since $L_j(\bar{x}, \bar{y}^{j-1})$ is in I_{j-1}. If $L_j(\bar{x}, \bar{y}^{j-1})$ is not in I, this will be because $L_j(\bar{x}, \bar{y})$ was deleted in Step 4 of algorithm A^*. However, this is possible only when D_j^* in T_2 is a nonvital clause. (In this case, a_k' and D_k will be absent in T_5.) Since D_j^* does not contain the ANS predicate, all literals in D_j^* must be resolved away. Since $L_j(\bar{x}, \bar{y})$ is attached to a_j, $\sim L_j(\bar{x}, \bar{y})$ must be attached to a_j' of T_2. Thus, $L_j(\bar{x}, \bar{y}^{j-1})$ is an instance of a literal of D_j^*. Because of Step 2 of algorithm A^*, the negations of the instances of all other literals of D_j^* must appear in I_{j-1}. Therefore, $L_j(\bar{x}, \bar{y}^{j-1})$ is a logical consequence of $\{D_j^*\} \cup I_{j-1}$. Since D_j^* is derived from A by resolution, D_j^* is a logical consequence of A. Therefore, $L_j(\bar{x}, \bar{y}^{j-1})$ is a logical consequence of $A \cup I_{j-1}$. However, we have shown that $ANS(f_{a_q}(\bar{x}, \bar{y}^{q-1}))$ is a logical consequence of $A \cup I_j$. Therefore, $ANS(f_{a_q}(\bar{x}, \bar{y}^{q-1}))$ is a logical consequence of $A \cup I_{j-1}$.

3. Set $j = j - 1$ and go to (2).

From the above induction process, finally, we can show that $ANS(f_{a_q}(\bar{x}, \bar{y}^{q-1}))$ is a logical consequence of $A \cup I_0$. Since it is clear that $I_0 = I$, $ANS(f_{a_q}(\bar{x}, \bar{y}^{q-1}))$ is a logical consequence of $(A \cup I)$.   Q.E.D.

11.10 THE APPLICATION OF INDUCTION AXIOMS TO PROGRAM SYNTHESIS

Algorithm A^*, which was given in the above section, will only produce "loop-free" programs. In order to produce an iterative program, we have to use some induction principle. By using the induction principle, we mean that instead of deducing the main answering clause, it suffices to deduce some subordinate answering clauses. Each subordinate answering clause produces a subroutine. These subroutines are then connected to form the main program. Let us now examine an example first.

Example 11.17

We want to have a program to compute the "factorial" function that is defined as follows:

$$f(0) = 1$$

$$f(x+1) = (x+1) \cdot f(x).$$

Let $R(x, y)$ denote $y = f(x)$. Then we have the following axioms:

(1) $R(0, 1)$

(2) $\sim R(x, y) \vee R(x+1, (x+1) \cdot y)$.

We now use a famous induction axiom for natural numbers, which can be expressed as follows:

(3) $(\exists y) R(0, y) \wedge (\forall x)((\exists y_1) R(x, y_1) \rightarrow (\exists y_2) R(x+1, y_2))$

$\rightarrow (\forall x)(\exists y) R(x, y)$.

The above induction axiom suggests that, instead of asking what $f(x)$ is directly, we ask the following two questions:

1. What is the value of $f(0)$?
2. If we know the value of $f(x)$, what is the value of $f(x+1)$?

To answer the first question, we use the following axiom:

(4) $\sim R(0, z) \vee ANS(z)$.

To answer the second question, we use the following axioms:

(5) $R(y_1, y_2)$

(6) $\sim R(y_1+1, z) \vee ANS(z)$

where y_1 and y_2 in (5) and (6) are treated as constants.

If the induction axiom expressed in (3) is used, the program will be of the form illustrated in Fig. 11.29. In Fig. 11.29, Subroutine A computes the value of $f(0)$ and Subroutine B computes the value of $f(y_1+1)$ based on the assumption that $y_2 = f(y_1)$. It is not difficult to see that Subroutine A corresponds to the following instruction (obtained from clauses (1) and (4)):

$$z \leftarrow 1$$

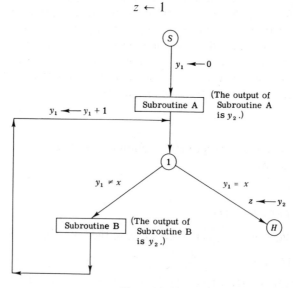

Figure 11.29

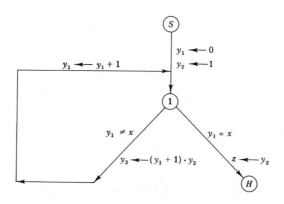

Figure 11.30

and Subroutine B corresponds to the instruction (obtained from clauses (2), (5), and (6))

$$z \leftarrow (y_1 + 1) y_2.$$

The final program that connects Subroutines A and B is shown in Fig. 11.30.

The above program consists of only one input variable. In the following example, we shall have a program with two input variables.

Example 11.18

We want to write a program that adds two integers. The only primitive function that we can use is adding 1 to a number. We first have the following definitions:

(1) $x + 0 = x$

(2) $x + (y + 1) = (x + y) + 1.$

Let $R(x_1, x_2, y)$ denote $y = x_1 + x_2$. We then have the following axioms:

(3) $R(x, 0, x)$

(4) $\sim R(x_1, x_2, z) \lor R(x_1, x_2 + 1, z + 1).$

The induction axiom that we use is

(5) $(\forall x)(\exists y) R(x, 0, y) \land$
          $(\forall x_1)(\forall x_2)((\exists y_1) R(x_1, x_2, y_1) \to (\exists y_2) R(x_1, x_2 + 1, y_2))$
          $\to (\forall x_1)(\forall x_2)(\exists y) R(x_1, x_2, y).$

We therefore have two sets of axioms. The first one is concerned with the value of $x + 0$:

(6) $\sim R(y_1, 0, z) \lor ANS(z).$

The second set of axioms is:

(7) $R(y_1, y_2, y_3)$

(8) $\sim R(y_1, y_2 + 1, z) \lor ANS(z)$.

where y_1 and y_2 in (7) and (8) are treated as constants.

Again, we let Subroutine A correspond to clause (6) and Subroutine B correspond to clauses (7) and (8). The program connecting Subroutines A and B is shown in Fig. 11.31.

To obtain Subroutine A, we resolve (3) and (6).

(9) $ANS(y_1)$ from (3) and (6).

Thus Subroutine A consists of the following instruction:

$$y_3 \leftarrow y_1.$$

To obtain Subroutine B, we use clauses (7) and (8).

(10) $R(y_1, y_2 + 1, y_3 + 1)$ from (7) and (4)

(11) $ANS(y_3 + 1)$ from (10) and (8).

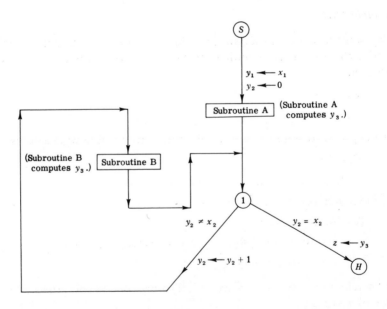

Figure 11.31

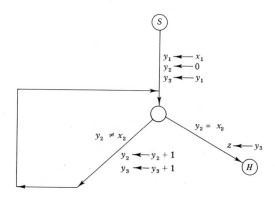

Figure 11.32

Therefore Subroutine B consists of the following instruction:

$$y_3 \leftarrow y_1 + 1.$$

The final program is shown in Fig. 11.32.

In the above two examples, the induction principle we used is called the "iterative going-up induction principle." Roughly speaking, this means that if a property holds for zero, and whenever it holds for x, it holds for $x + 1$, then this property holds for all x. We may also use the "iterative going-down induction principle." This means that if a property holds for some x, and whenever it holds for x, it holds for $x - 1$, then this property holds for $x = 0$. For a more detailed discussion of induction principles, see [Manna and Waldinger, 1971; Burstall, 1969; Park, 1970].

11.11 ALGORITHM A (AN IMPROVED PROGRAM-SYNTHESIZING ALGORITHM)

In Section 11.8, algorithm A* was given. In this section, we shall show that algorithm A* can be modified to algorithm A, which is much more efficient.

We first consider an example.

Example 11.19

Consider the following specification of a program:

(1) If $x_1 \leqslant x_2$, then assign the output to be x_1.
(2) Otherwise, assign the output to be x_2.

Let $P(x_1, x_2)$ denote "$x_1 \leqslant x_2$." Let $R(x_1, x_2, z)$ denote "the output is z if the inputs are x_1 and x_2." Then we have

(1) $\sim P(x_1, x_2) \lor R(x_1, x_2, x_1)$

(2) $P(x_1, x_2) \lor R(x_1, x_2, x_2)$

(3) $\sim R(x_1, x_2, z) \lor ANS(z)$

(4) $\sim P(x_1, x_2) \lor ANS(x_1)$ from (3) and (1)

(5) $P(x_1, x_2) \lor ANS(x_2)$ from (3) and (2).

If we resolve clauses (4) and (5), an answering clause will be obtained. Then, in the final program, a branching will occur because both clauses (4) and (5) are vital clauses. *We can make use of this information by storing the information of branching in the ANS predicate.* That is, instead of

$$ANS(x_1) \lor ANS(x_2)$$

we may have

$$ANS(P(x_1, x_2) \to ANS(x_1);$$
$$\sim P(x_1, x_2) \to ANS(x_2))$$

as the resolvent. The expression inside the *ANS* predicate is the program that we need. By recording this information, we do not have to analyze the deduction tree any more.

Since a branching occurs only when the corresponding primitive resolution involves two vital clauses, we shall now modify the definition of primitive resolution as follows:

Definition   A *modified primitive resolution* between two clauses C_1 and C_2 is defined as follows.

1. If C_1 and C_2 are not both vital clauses, then the *modified primitive resolution* is the same as primitive resolution.

2. If C_1 and C_2 are both vital clauses, let C_1 and C_2 be $C_1' \lor ANS(E_1)$ and $C_2' \lor ANS(E_2)$, respectively. Let L_1 and L_2 be the resolved literals in C_1 and C_2, respectively. Let θ be a most general unifier of L_1 and $\sim L_2$. Then the *modified primitive resolvent* of C_1 and C_2 is

$$C = (C_1'\theta - L_1\theta) \cup (C_2'\theta - L_2\theta) \cup ANS(\sim L_1\theta \to ANS(E_1\theta);$$
$$\sim L_2\theta \to ANS(E_2\theta)).$$

Definition   A *modified primitive deduction* is a deduction in which every resolution is a modified primitive resolution.

Algorithm A (An Improved Program-Synthesizing Algorithm)

Step 1   Using the modified primitive resolution, derive an answering clause C from a set of clauses including the clause $\sim R(x_1, \ldots, x_r, a_{r+1}, \ldots, a_m, z_1, \ldots, z_n) \vee ANS(z_1, \ldots, z_n)$, where in the deduction of C, the input variables $x_1, \ldots, x_r$ cannot be replaced by other terms, i.e., $x_1, \ldots, x_r$ should be treated as constants, and the output variables $z_1, \ldots, z_n$ cannot appear in any term that is substituted for some variable.

Step 2   Let S be the starting vertex. Construct the pair (S, C). Let PLIST $= ((S, C))$.

Step 3   If PLIST is empty, terminate; the final tree is a desired program. Otherwise, continue.

Step 4   Let (V, E) be the first pair in PLIST. Delete (V, E) from PLIST.

Step 5   If V does not contain $\rightarrow$ and is of the form $ANS(t_1, t_2, \ldots, t_n)$, draw an arc a from V to the halting vertex H, attach a with $z_1 \leftarrow t_1, z_2 \leftarrow t_2, \ldots$, and $z_n \leftarrow t_n$, and go to Step 3. Otherwise, continue.

Step 6 E must be of the form $ANS(L \rightarrow E_1; \sim L \rightarrow E_2)$. In this case, draw two arcs a_1 and a_2 from V. Attach a_1 and a_2 with L and $\sim L$, respectively. Let V_1 and V_2 be the vertices at the ends of a_1 and a_2, respectively. Put the pairs (V_1, E_1) and (V_2, E_2) into PLIST. Go to Step 3.

It is obvious that algorithm A is equivalent to algorithm A*; therefore, algorithm A is also correct. That is, it always produces a program that satisfies the specification of the program.

Example 11.20

In Example 11.12, we have the following set of clauses:

(1) $\sim P(x, 7) \vee \sim P(x, 9) \vee R(x, a)$

(2) $\sim P(x, 7) \vee P(x, 9) \vee R(x, b)$

(3) $P(x, 7) \vee \sim P(x, 9) \vee R(x, c)$

(4) $P(x, 7) \vee P(x, 9) \vee R(x, d)$

(5) $\sim R(x, z) \vee ANS(z)$.

The reader can see that the following is a modified primitive deduction.

(6) $\sim P(x, 7) \vee \sim P(x, 9) \vee ANS(a)$ from (1) and (5)

(7) $\sim P(x, 7) \vee P(x, 9) \vee ANS(b)$ from (2) and (5)

(8) $P(x, 7) \vee \sim P(x, 9) \vee ANS(c)$ from (3) and (5)

(9) $P(x, 7) \lor P(x, 9) \lor ANS(d)$ from (4) and (5)

(10) $\sim P(x, 7) \lor ANS(P(x, 9) \rightarrow ANS(a);$ from (6) and (7)

$$\sim P(x, 9) \rightarrow ANS(b))$$

(11) $P(x, 7) \lor ANS(P(x, 9) \rightarrow ANS(c);$ from (8) and (9)

$$\sim P(x, 9) \rightarrow ANS(d))$$

(12) $ANS(P(x, 7) \rightarrow ANS(P(x, 9) \rightarrow ANS(a);$ from (10) and (11).

$$\sim P(x, 9) \rightarrow ANS(b));$$

$$\sim P(x, 7) \rightarrow ANS(P(x, 9) \rightarrow ANS(c);$$

$$\sim P(x, 9) \rightarrow ANS(d)))$$

Applying algorithm A to clause (12), we can obtain the program shown in Fig. 11.17.

Example 11.21

Consider Example 11.16 again. We have the following set of clauses.

(1) $\sim R(x, a, y) \lor R(x, b, f_1(x, y))$

(2) $\sim R(x, b, y) \lor \sim Q(y) \lor R(x, c, f_2(x, c, y))$

(3) $\sim R(x, b, y) \lor Q(y) \lor R(x, c, f_3(x, c, y))$

(4) $R(d, a, e)$

(5) $\sim R(d, c, z) \lor ANS(z)$

(6) $\sim R(d, b, y_3) \lor Q(y_3) \lor ANS(f_3(d, c, y_3))$ from (5) and (3)

(7) $\sim R(d, b, y_2) \lor \sim Q(y_2) \lor ANS(f_2(d, c, y_2))$ from (5) and (2)

(8) $\sim R(d, b, y_2) \lor ANS(\sim Q(y_2) \rightarrow ANS(f_3(d, c, y_2));$

$$Q(y_2) \rightarrow ANS(f_2(d, c, y_2)))$$ from (6) and (7)

(9) $\sim R(d, a, y_1) \lor ANS(\sim Q(f_1(d, y_1)) \rightarrow ANS(f_3(d, c, f_1(d, y_1)));$

$$Q(f_1(d, y_1)) \rightarrow ANS(f_2(d, c, f_1(d, y_1))))$$

    from (8) and (1)

(10) $ANS(\sim Q(f_1(d, e)) \rightarrow ANS(f_3(d, c, f_1(d, e)));$

$$Q(f_1(d, e)) \rightarrow ANS(f_2(d, c, f_1(d, e))))$$ from (9) and (4).

Again, applying algorithm A to clause (10), we can obtain the program shown in Fig. 11.33. The reader can see that this program is equivalent to the one shown in Fig. 11.25.

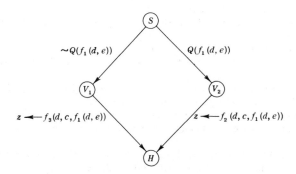

Figure 11.33

REFERENCES

Black, F. (1964): A deductive question-answering system, *in:* "Semantic Information Processing" (M. Minsky, ed.), The M.I.T. Press, Cambridge, Massachusetts, pp. 354-402.

Burstall, R. M. (1969): Proving properties of programs by structural induction, *Comput. J.* **12** 41–48.

Darlington, J. L. (1965): Machine methods for proving logical arguments expressed in English, *Machine Translation* **8** 41–67.

Green, B., Jr., A. K. Wolf, C. Chomsky, and K. Laughary (1963): Baseball: an automatic question answerer, *in:* "Computers and Thought" (E. A. Feigenbaum and J. Feldman, eds.), McGraw-Hill, New York, pp. 207–216.

Green, C. (1969a): Theorem proving by resolution as a basis for question-answering systems, *in:* "Machine Intelligence," vol. 4 (B. Meltzer and D. Michie, eds.), American Elsevier, New York, pp. 183–205.

Green, C. (1969b): "The Application of Theorem Proving to Question Answering Systems," Ph.D. Thesis, Stanford University, Stanford, California, 1969.

Green, C. (1969c): Application of theorem-proving to problem solving, *Proc. 1st Internat. Joint Conf. Artificial Intelligence*, pp. 219–239.

Green, C., and B. Raphael (1968): The use of theorem proving techniques in question-answering systems, *Proc. 23rd National Conf. ACM*, Brandon Systems Press, Princeton, New Jersey, pp. 169–181.

Lee, R. C. T., C. L. Chang, and R. J. Waldinger (1972): An improved program-synthesizing algorithm and its correctness, Division of Computer Research and Technology, National Institutes of Health, Bethesda, Maryland.

Lindsay, R. K. (1963): Inferential memory as the basis of machines which understand natural language, *in:* "Computers and Thought" (E. A. Feigenbaum and J. Feldman, eds.), McGraw-Hill, New York, pp. 217–233.

Manna, Z., and R. Waldinger (1971): Toward automatic program synthesis, *Comm. ACM.* **14** 151–165.

McCarthy, J. (1961): Programs with common sense, *Proc. Symp. Mechanization Thought Process*, vol. 1, H.M.S.O., London, pp. 75–84.

McCarthy, J. (1962): "LISP 1.5 Programmers Mannual," M.I.T. Press, Cambridge, Massachusetts.

McCarthy, J. (1963b): Situations, actions and causal laws, Memo, Stanford Artificial Intelligence Project, Stanford University, Stanford, California.

Nilsson, N. J. (1971): "Problem-solving Methods in Artificial Intelligence," McGraw-Hill, New York, Ch. 7.

Park, D. (1970): Fixpoint induction and proofs of programs properties, in: "Machine Intelligence," vol. 5 (B. Meltzer and D. Michie, eds.), American Elsevier, New York, pp. 59–78.

Safier, F. (1963): The Mikado as an advice taker problem, Memo, Stanford Artificial Intelligence Proj., Stanford University, Stanford, California.

Simon, H. A. (1963): Experiments with a heuristic compiler, J. Assoc. Comput. Mach. **10** 493–506.

Slagle, J. R. (1965): Experiments with a deductive question-answering program, Comm. ACM. **8** 792–798.

Waldinger, R. J. (1969): "Constructing Programs Automatically Using Theorem Proving," Ph.D. Thesis, Carnegie-Mellon University, Pittsburgh, Pennsylvania.

Waldinger, R. J., and R. C. T. Lee (1969): PROW: A step toward automatic program writing, Proc. 1st Internat. Joint Conf. Artificial Intelligence, pp. 241–252.

EXERCISES

Section 11.2

1. Assume we have the following facts:

   F_1:   Either it rains or it is hot.
   F_2:   If it rains, then it is hot.
   F_3:   If it does not rain, then it is not hot.

   Answer the following question:
   Is it true that if it is hot, then it must not be raining.

2. Assume that we have two premises:

   (a)   All jokes are meant to amuse.
   (b)   No act of Congress is a joke.
   Can you answer the following question:
   Is it true that every act of Congress is not meant to amuse? If you can answer this question, give your reasoning. If not, explain.

Section 11.3

3. Let us assume that we have the following facts:
   Everyone who entered this country and was not a VIP was searched by a custom official.
   William was a drug pusher.
   William entered this country.
   William was searched by drug pushers only.
   No drug pusher was a VIP.
   Solve the following problem: Find a person who was both a drug pusher and a custom official.

Section 11.4

4. Consider the following problem: There are three boxes: boxes 1, 2, and 3. Box 1 contains A, box 2 contains B, and box 3 contains nothing. We can transfer things into a box only when that box is empty. Use theorem-proving techniques to show that we can switch the contents of boxes 1 and 2.

Section 11.5

5. Consider the following problem [Waldinger and Lee, 1969]: There are two boxes, one and only one of them contains a banana. The monkey can walk freely and he does not know where the banana is. However, if he is where the box is, he can look into the box and determine whether the banana is there or not. Use the theorem-proving techniques described in Section 11.5 to determine what the monkey should do.

    Let $P(x, y, s)$ denote "The state s is attained in which the monkey is at x and the banana is at y."

    Let $find(s)$ denote "s is the state in which the monkey can find the banana."

    Let $walk(x, s)$ denote the new state achieved after the monkey walks to x in state s.

Section 11.6

6. In Section 10.7, we showed that the set-of-support strategy is complete in deducing the halting clause. Show that this is also true in the deduction of an answering clause.

7. Reconsider Examples 11.7 and 11.8. This time, use linear resolution in both cases.

Section 11.8

8. Consider Example 11.14 again. Use this example to show that the combination of linear resolution and primitive resolution is not complete.

9. Consider the following set S of clauses that represents the specification of a program:

   (1) $\sim R(a, s) \vee \sim Q(s) \vee R(b, f_1(a, b, s))$

   (2) $\sim R(a, s) \vee Q(s) \vee R(c, f_2(a, c, s))$

   (3) $\sim R(b, s) \vee R(d, f_3(b, d, s))$

   (4) $\sim R(c, s) \vee R(d, f_4(c, d, s))$

   (5) $R(a, s)$

   (6) $\sim R(d, s) \vee ANS(s).$

Derive an answering clause from S. Use algorithm A* to obtain the program. In the above example, $R, a, b, c, d, f_1, f_2, f_3$, and f_4 are the only primitive symbols.

Section 11.10

10. In Example 11.18, we used "iterative going-up induction principle" to synthesize a program adding two integers. This time, use "iterative going-down induction principle" to synthesize a program adding two nonnegative integers.

Section 11.11

11. Reconsider Exercise 9. Use algorithm A, instead of algorithm A*, to synthesize the desired program.

Chapter 12

Concluding Remarks

In the previous chapters, we have discussed mechanical theorem proving and its applications. In this closing chapter, we shall discuss some open problems concerning mechanical theorem proving.

The main problem is concerned with the efficiency of proof procedures. In this book we have introduced many resolution strategies, each of which, we believe, has its own advantages. Of course, for a specific theorem, some strategies may work well while others may perform poorly. Any study, experimental or theoretical, on the effectiveness of these strategies, will be a great contribution to the development of mechanical theorem proving.

The problem of heuristics is a very important one. This topic was only briefly discussed in Section 7.7 of this book. For more information on how to apply heuristics to theorem proving, the reader is encouraged to read [Kowalski, 1970c; Kowalski and Kuehner, 1971; Siklossy and Marinov, 1971; Norton, 1971; Brice and Derksen, 1971; Slagle and Farrell, 1971]. There are also many heuristics that are problem dependent [Bledsoe, 1971; Norton, 1966; Gelernter, 1959]. We hope that this book has provided a basic framework into which heuristics can be easily added.

Mechanization of higher-order logic is another important problem in mechanical theorem proving. For an introduction to higher-order logic, consult [Hilbert and Ackermann, 1950; Henkin, 1950]. It was pointed out that there are many problems which can be conveniently and compactly represented by higher than first-order logic [Robinson, 1968b]. Recently, there have been many papers published on the subject of mechanizing higher-order logic [Andrews, 1971; Darlington, 1971; Davis, 1968; Gould, 1966; Henschen, 1971; Robinson, 1968b, 1969, 1970; Huet, 1972; Lucchesi, 1972; Pietrzykowski, 1971; Pietrzykowski and Jensen, 1972]. However, further research along this line is still needed.

Another kind of logic that we did not discuss is modal logic. Modal logic has been shown to be important for designing problem-solving systems [McCarthy, 1961, 1963b]. For a comprehensive study of modal logic, consult [Kripke, 1963; Lewis and Langford, 1959; Snyder, 1971; von Wright, 1951].

So far, there is no computer language specifically designed for theorem proving. Most present theorem-proving programs were written in list processing languages such as LISP. However, we believe that there is a need for some language that will exploit the special data structure and manipulation rules of theorem-proving systems [Robinson, 1971a].

As for applying theorem proving to question answering, it is obvious that, given a question to answer, a question-answering system should not conduct an exhaustive search among all the facts in order to answer the question. It should only use those facts relevant to the question. Therefore, data organization is critical to the success of a question-answering system. Intuitively, we believe that, in the computer memory, similar facts should be grouped together. This leads to the subject of clustering analysis, and the reader can consult [Augustin and Minker, 1970].

Finally, in applying theorem proving to question answering, we know that we often need to translate English sentences into formulas of the first-order logic. The translation of English into the language of logic is by no means trivial. It is conceivable that the usage of "good" predicates may allow us to represent a problem simply and thus permit us to solve it easily. Although some results [Sandewall, 1972] have been obtained, more work is needed in this area. This area is also related to the general representation problem. For a comprehensive discussion of the representation problem, read [Feigenbaum, 1969; Amarel, 1968].

REFERENCES

Amarel, S. (1968): On representations of problems of reasoning about actions, *in:* "Machine Intelligence," vol. 3 (B. Meltzer and D. Michie, eds.), American Elsevier, New York, pp. 131–171.

Andrews, P. B. (1971): Resolution in type theory, *J. Symbolic Logic* **36** 414–432.

Augustin, J. G., and J. Minker (1970): An analysis of some graph theoretical cluster techniques, *J. Assoc. Comput. Mach.* **17** 571–588.

Bledsoe, W. W. (1971): Splitting and reduction heuristics in automatic theorem proving, *Artificial Intelligence* **2** 57–78.

Brice, C., and J. Derksen (1971): A heuristically guided equality rule in a resolution theorem prover, Tech. Note 45, Stanford Research Institute, Stanford, California.

Darlington, J. L. (1971): A partial mechanization of second order logic, *in:* "Machine Intelligence," vol. 6 (B. Meltzer and D. Michie, eds.), American Elsevier, New York, pp. 91–100.

Davis, M. (1968): Invited commentary on new directions in mechanical theorem proving, *Proc. IFIP Congress 1968,* vol. 1, North-Holland, Amsterdam, pp. 67–68.

Feigenbaum, E. (1969): Artificial Intelligence: Themes in the second decade, *in:* "Information Processing 68," vol. 2 (A. J. H. Morrell, ed.), North-Holland, Amsterdam, pp. 1008–1022.

Gelernter, H. (1959): Realization of a geometry theorem proving machine, *Proc. IFIP Congress 1959,* pp. 273–282.

Gould, W. E. (1966): A matching procedure for omega logic, Air Force Cambridge Research Laboratory, Report AFCRL-66-781.

Henkin, L. (1950): Completeness in the theory of types, *J. Symbolic Logic* **15** 81–91.

Henschen, J. L. (1971): "A Resolution Style Proof Procedure for Higher Order Logic," Ph.D. Thesis, University of Illinois, Urbana-Champaign, Illinois.

Hilbert, D., and W. Ackermann (1950): "Principles of Mathematical Logic," Chelsea, New York.

Huet, G. P. (1972): The undecidability of the existence of a unifying substitution between two terms in the simple theory of types, Jennings Computer Center, Case Western Reserve University, Cleveland, Ohio.

Kowalski, R. (1970c): Search strategies for theorem proving, *in:* "Machine Intelligence," vol. 5 (B. Meltzer and D. Michie, eds.), American Elsevier, New York, pp. 181–201.

Kowalski, R., and D. Kuehner (1971): Linear resolution with selection function, *Artificial Intelligence* **2** 227–260.

Kripke, S. (1963): Semantic considerations in modal logic, *Acta Philos. Fenn.* **16** 83–94.

Lewis, C. I., and C. H. Langford (1959): "Symbolic Logic," Dover Press, New York.

Lucchesi, C. L. (1972): The undecidability of the unification for third order language, CSRR 2059, Dept. of Applied Analysis and Computer Science, University of Waterloo, Waterloo, Canada.

McCarthy, J. (1961): Programs with common sense, *Proc. Symp. Mechanization Thought Process,* vol. 1, H.M.S.O., London, pp. 75–84.

McCarthy, J. (1963b): Situations, actions and causal laws, Memo, Stanford Artificial Intelligence Proj., Stanford University, Stanford, California.

Norton, L. M. (1966): "Adept-A Heuristic Program for Proving Theorems of Group Theory," Ph.D. Thesis, Massachusetts Inst. Technology, Cambridge, Massachusetts.

Norton, L. M. (1971): Experiments with a heuristic theorem-proving for the predicate calculus with equality, *Artificial Intelligence* **2** 261–284.

Pietrzykowski, T. (1971): A complete mechanization of second order logic, CSRR 2038, Dept. of Applied Analysis and Computer Science, University of Waterloo, Waterloo, Canada.

Pietrzykowski, T., and D. Jensen (1972): A complete mechanization of w-order logic, Dept. of Applied Analysis and Computer Science, University of Waterloo, Waterloo, Canada.

Robinson, J. A. (1968b): New directions in mechanical theorem proving, *Proc. IFIP Congress 1968,* vol. 1, pp. 63–68.

Robinson, J. A. (1969): Mechanizing higher order logic, *in:* "Machine Intelligence," vol. 4 (B. Meltzer and D. Michie, eds.), American Elsevier, New York, pp. 151–170.

Robinson, J. A. (1970): A note on mechanizing higher order logic, *in:* "Machine Intelligence," vol. 5 (B. Meltxer and D. Michie, eds.), American Elsevier, New York, pp. 123–133.

Robinson, J. A. (1971a): Computational logic: The unification computation, *in:* "Machine Intelligence," vol. 6 (B. Meltzer and D. Michie, eds.), American Elsevier, New York, pp. 63–72.

Sandewall, E. (1972): PCF-2, a first order calculus for expressing conceptual information, Dept. of Computer Science, Uppsala University, Uppsala, Sweden.

Siklossy, L., and V. Marinov (1971): Heuristic search and exhaustive search, *Proc. 2nd Int. Conf. Artificial Intelligence*, London, pp. 601–607.

Slagle, J. R., and C. D. Farrell (1971): Experiments in automatic learning for a multipurpose heuristic program, *Comm. ACM.* **14** 91–99.

Snyder, D. P. (1971): "Modal Logic and its Applications," Van Nostrand-Reinhold, Princeton, New Jersey.

von Wright, G. H. (1951): "An Essay in Modal Logic," North-Holland, Amsterdam.

Appendix A

A.1   A COMPUTER PROGRAM USING UNIT BINARY RESOLUTION

Let S be a set of clauses. A *unit binary resolution* is a binary resolution in which at least one of the two parent clauses is a unit clause. Our program is called TPU, which stands for *T*heorem *P*rover with *U*nit binary resolution. TPU is written in the LISP language. (For the comprehensive discussion of the LISP language, see McCarthy *et al.* [1962], Weissman [1967], and Quam [1968].) This program is adapted from the one given in [Chang, 1970a]. To call TPU to prove a given set S of clauses, we need only a statement:

$$TPU(S_1, S_2, S_3, W, N_1, N_2, N_3, N_4),$$

where

$S_1 = $ the set of all the positive unit clauses in S;
$S_2 = $ the set of all the negative unit clauses in S;
$S_3 = $ the set of all the nonunit clauses in S (it is better, but not necessary, to order S_3 according to the length of the clause: two-literal clauses first, three-literal clauses second, etc.);
$W = $ the family of sets of support for all the nonunit clauses in S_3. The set of support for each clause in S_3 is a set of unit clauses in S. They may or may not be different for different clauses in S_3;

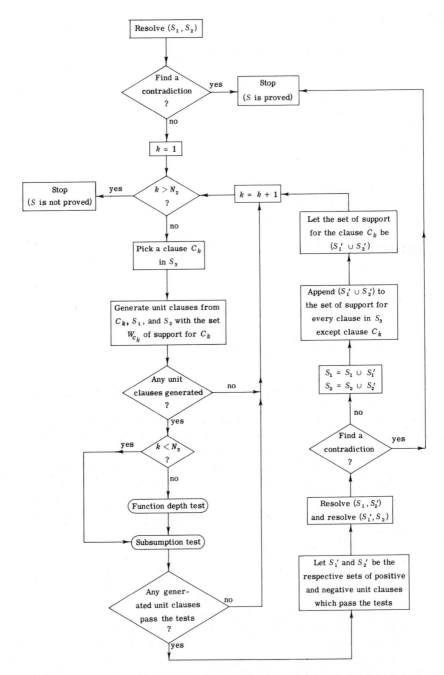

Figure A.1   The flowchart of TPU. Resolve (S_1, S_2) means that it resolves clauses in S_1 against clauses in S_2.

N_1 = total number of clauses in S;

N_2 = maximum number of times clauses in S_3 will be selected;

N_3 = maximum number of times clauses in S_3 will be selected before the function-depth test will be given to unit clauses;

N_4 = maximum function depth that a unit clause may have.

The procedure of TPU is given in the flow chart in Fig. A.1. In TPU, by the *subsumption test*, we mean that a generated unit clause is not retained if it is subsumed by some unit clause. By the *function depth test*, we mean that a generated unit clause is not retained if it contains an argument involving nesting of functions to a depth greater than N_4. For example, the function depth of $C = P(a, f(g(h(a, b))))$ is 3. C will not be retained if $N_4 = 2$. TPU works in this way: TPU first resolves the clauses in S_1 and the clauses in S_2 to test for an immediate contradiction. If no contradition is found, the clauses in S_3 are selected one at a time. At the kth step, suppose the clause selected is C_k. TPU tries to generate unit clauses from C_k, S_1, and S_2 with the set W_{C_k} of support for C_k. If no unit clauses are generated, TPU chooses another clause in S_3 and repeats the process. If some unit clauses are generated, TPU applies the subsumption test and, if applicable, the function-depth test. TPU retains only those clauses that pass these tests. TPU resolves these newly generated unit clauses against the clauses in S_1 and S_2. If a contradiction is found, it stops; otherwise it picks another clause in S_3 and repeats the process. At each step of the process, the selection of the clauses from S_3 may be crucial to the efficiency of the program. In TPU the clauses in S_3 are selected in sequence. When S_3 is exhausted, the clauses are selected again in the same order. The set-of-support strategy is incorporated into the program. In the program each clause in S_3 has its own set of support. Different clauses may have different sets of support. They are up-dated at every step.

The parameter N_3 is used to tell the program when the function-depth test will be applied. If we want to use the function-depth test throughout the entire proof, we can set $N_3 = 0$. On the other hand, if we do not need this test, we can set $N_3 = N_2 + 1$. And, of course, N_3 can be set to be any integer between 0 and $N_2 + 1$.

In addition, TPU uses the following scheme to avoid repeatedly generating the same unit clause. For example, given the clauses

(1) P

(2) Q

(3) $\sim P \vee \sim Q \vee R$

TPU will generate the following resolvents:

(4) $\sim Q \vee R$ from (1) and (3)

(5) $\sim P \vee R$ from (2) and (3)

(6) R from (1) and (5).

In TPU, to resolve a unit clause, say clause m, against a nonunit clause, say clause n'. we first find the last unit clause, say clause m', used to obtain clause n'. If m is less than m', we resolve clause m and clause n'; otherwise, they are not resolved. Thus, in the above example, clauses 2 and 4 are not resolved, since the last unit clause used to obtain clause 4 is clause 1, which is followed by clause 2, that is, $m = 2$, $m' = 1$, and m is not less than m'.

A.2 BRIEF COMMENTS ON THE PROGRAM

a. Predicates and functions are represented by Polish notation. For example, a function $f(x_1, \dots, x_n)$ is denoted by the list $(f\, x_1 \cdots x_n)$; a predicate $P(x_1, \dots, x_n)$ is denoted by the list $(P\, x_1 \cdots x_n)$; and a negative literal $\sim P(x_1, \dots, x_n)$ is denoted by the list $(\text{NOT}\, P\, x_1 \cdots x_n)$. A clause is represented by a list (N V C), where N is an integer, V is a list of universal variables occurring in C, which is a list of lists denoting literals. For example, if clause 1 is $P(f(x, g(y)), z) \vee \sim Q(a)$, it is represented by the list $\big(1\,(\text{X Y Z})\big(\big(\text{P}\,(\text{F X}\,(\text{G Y}))\text{Z}\big)$ $(\text{NOT Q A})\big)\big)$. A set of clauses is represented by a list of lists representing the clauses.

b. In our program, before a resolution is performed on two clauses C_1 and C_2, variables in C_1 and C_2 are renamed to variables in XLIST and YLIST, respectively. We assume that the number of variables in a clause is not greater than 7. Also, ZZ1, ..., ZZ7 should not be used in a clause.

c. RENAME (C XY) is defined to rename variables in the clause C to variables in the list X Y, which is either XLIST or YLIST.

d. INSIDE (A E) is defined to test whether A is inside the expression E. If yes, it returns T; otherwise NIL.

e. DISAGREE (E1 E2) is defined to find out the disagreement set of the lists E1 and E2, which are of equal length.

f. UNIFICATION (E1 E2) is defined to find out a most general unifier of the expressions E1 and E2. If E1 and E2 are not unifiable, it returns NO; otherwise, it returns a most general unifier of E1 and E2. If $\{t_1/v_1, \dots, t_n/v_n\}$ is a most general unifier, it is represented by the list $((v_1\, t_1) \cdots (v_n\, t_n))$.

g. DELETEV (X Y VAR) is defined to find $((\text{X} \cup \text{Y}) - \text{VAR})$, where X, Y, and VAR are lists representing sets of variables.

h. URESOLVE (C1 C2 N) is defined to resolve two clauses C1 and C2, where C1 is a unit clause, C2 is either a unit or nonunit clause, and N is an integer. Suppose C1 and C2 are numbered by L1 and L2, respectively. If C1 and C2 can be resolved to the empty clause, that is, if they are contradictory, it returns the list (CONTRADICTION L1 L2). Otherwise, it returns the list

(RES HIST M), where RES is the list of all resolvents of C1 and C2, say $R_1, ..., R_q$, HIST is the list indicating which literal of C2 is resolved upon C1 to obtain R_i, $i = 1, ..., q$, and M is $N + q$. The resolvents $R_1, ..., R_q$ are numbered by $N + 1$, $..., N + q$, respectively. The ordering of literals of each R_i comforms to that of literals of C2. For example, if C1 = $(1(x)((\text{NOT } Q \, x)))$, C2 = $(5(\ )((P \, A)(Q \, A)$ $(P \, B)(Q \, B)))$, and N = 10, then the program returns (RES HIST M), where

$$\text{RES} = \big((11(\ )((P \, A)(P \, B)(Q \, B)))(12(\ )((P \, A)(Q \, A)(P \, B))) \big)$$

$$\text{HIST} = ((11 \ \ 1 \ \ 5 \ \ 2)(12 \ \ 1 \ \ 5 \ \ 4))$$

$$\text{M} = 12.$$

In HIST, (11 1 5 2) indicates that resolvent 11 is obtained by resolving clause 1 upon the second literal of clause 5, while (12 1 5 4) indicates that resolvent 12 is obtained by resolving clause 1 upon the 4th literal of clause 5. Note that C1 is clause 1 and C2 is clause 5.

  i. GUNIT(S1 S2 W C N) is defined to generate unit clauses from C, S1, and S2 with W as the set of support for C, where S1 is a list of positive unit clauses, S2 is a list of negative unit clauses, W is a list of unit clauses, C is a nonunit clause, and N is an integer. The output of this function is a list of generated unit clauses.

  j. PNSORT(RES) is defined to sort a list RES of unit clauses into two lists, one of which consists of positive unit clauses and the other of negative unit clauses.

  k. FDEPTH(C) is defined to compute the function depth of a unit clause C.

  l. FTEST(RES N4) is defined to perform the function-depth test for a list RES of unit clauses.

  m. SUBSUME(C1 C2) is defined to test if a unit clause C1 subsumes another unit clause C2 or not. If yes, it returns T; otherwise, NIL.

  n. STEST(U RES) is defined to perform the subsumption test. That is, any unit clause of RES that is subsumed by another unit clause of either U or RES will be deleted.

  o. CONTRADICT(U V) is defined to find an immediate contradiction between U and V, where U is a list of positive unit clauses and V is a list of negative unit clauses.

  p. DTREE(Z HIST N1) is defined to trace a unit refutation once a contradiction (an empty clause) is found. Here, Z is of the form ("CONTRADICTION" M N), which indicates that clause M and clause N are immediately contradictory. HIST is a list that, as defined in (h), indicates the relationship among all clauses that have been generated by the program. N1 is an integer indicating the number of input clauses.

  q. TPU(S1 S2 S3 W N1 N2 N3 N4) is defined to carry out the procedure described by the flow chart given in Section A.1 of this appendix.

A.3   A LISTING OF THE PROGRAM

```
(DEFPROP XLIST (NIL X1 X2 X3 X4 X5 X6 X7) VALUE).

(DEFPROP YLIST (NIL Y1 Y2 Y3 Y4 Y5 Y6 Y7) VALUE)

(DEFPROP ZLIST (NIL ZZ1 ZZ2 ZZ3 ZZ4 ZZ5 ZZ6 ZZ7) VALUE)

(DEFPROP RENAME (LAMBDA (C XY)
        (PROG(VAR Z)
          (SETQ Z ZLIST)
          (SETQ VAR (CADR C))
        B1(COND ((NULL VAR) (GO B2)))
          (SETQ C (SUBST (CAR Z) (CAR VAR) C))
          (SETQ Z (CDR Z))
          (SETQ VAR (CDR VAR))
          (GO B1)
        B2(SETQ Z XY)
          (SETQ VAR (CADR C))
        B3(COND ((NULL VAR) (RETURN C)))
          (SETQ C (SUBST (CAR Z) (CAR VAR) C))
          (SETQ Z (CDR Z))
          (SETQ VAR (CDR VAR))
          (GO B3))) EXPR)

(DEFPROP INSIDE (LAMBDA (A E)
        (COND ((ATOM E) (EQ A E))
          ((INSIDE A (CAR E)) T) (T (INSIDE A (CDR E))))) EXPR)

(DEFPROP DISAGREE (LAMBDA (E1 E2)
        (COND ((NULL E1) NIL)
          ((OR (ATOM E1) (ATOM E2))
           (COND ((EQUAL E1 E2) NIL) (T (LIST E1 E2))))
          ((EQUAL (CAR E1) (CAR E2)) (DISAGREE (CDR E1) (CDR E2)))
          ((OR (ATOM (CAR E1)) (ATOM (CAR E2)))
           (LIST (CAR E1) (CAR E2))) (T (DISAGREE (CAR E1) (CAR E2))))) EXPR)

(DEFPROP UNIFICATION (LAMBDA  (E1 E2)
        (PROG (D U D1 D2)
          (COND ((NOT (EQUAL (LENGTH E1) (LENGTH E2)))(RETURN (QUOTE NO))))
        B1(SETQ D (DISAGREE E1 E2))
          (COND ((NULL D) (RETURN (REVERSE U))))
          (SETQ D1 (CAR D))
          (SETQ D2 (CADR D))
          (COND ((OR (MEMBER D1 XLIST) (MEMBER D1 YLIST)) (GO B3)))
          (COND ((OR (MEMBER D2 XLIST) (MEMBER D2 YLIST)) (GO B4)))
        B2(RETURN (QUOTE NO))
        B3(COND ((INSIDE D1 D2) (GO B2)))
          (SETQ U (CONS D U))
          (SETQ E1 (SUBST D2 D1 E1))
          (SETQ E2 (SUBST D2 D1 E2))
          (GO B1)
        B4(COND ((INSIDE D2 D1) (GO B2)))
          (SETQ U (CONS (REVERSE D) U))
          (SETQ E1 (SUBST D1 D2 E1))
          (SETQ E2 (SUBST D1 D2 E2))
          (GO B1))) EXPR)
```

```
(DEFPROP DELETEV (LAMBDA (X Y VAR)
          (PROG (VAR1 TX TX1 X1)
            (SETQ X (APPEND X Y))
          B1(COND ((NULL VAR) (RETURN X)))
            (SETQ VAR1 (CAR VAR))
            (SETQ TX X)
            (SETQ X1 NIL)
          B2(COND ((NULL TX) (GO B4)))
            (SETQ TX1 (CAR TX))
            (COND ((EQ TX1 VAR1) (GO B3)))
            (SETQ X1 (CONS TX1 X1))
            (SETQ TX (CDR TX))
            (GO B2)
          B3(SETQ X (APPEND X1 (CDR TX)))
          B4(SETQ VAR (CDR VAR))
            (GO B1))) EXPR)

(DEFPROP URESOLVE (LAMBDA (C1 C2 N)
          (PROG (L1 L2 VC1 VC2 X Y SIGN UNIF R RES VAR V1 V2 H HIST TC2)
            (SETQ C1 (RENAME C1 XLIST))
            (SETQ C2 (RENAME C2 YLIST))
            (SETQ L1 (CAR C1))
            (SETQ L2 (CAR C2))
            (SETQ VC1 (CADR C1))
            (SETQ VC2 (CADR C2))
            (SETQ C2 (CADDR C2))
            (SETQ X (CAR (CADDR C1)))
            (SETQ SIGN -1)
            (COND ((EQ (CAR X) (QUOTE NOT)) (GO B7)))
            (SETQ SIGN 1)
          B1(COND ((NULL C2) (RETURN (LIST (REVERSE RES) (REVERSE HIST) N))))
            (SETQ Y (CAR C2))
            (COND ((EQ (CAR Y) (QUOTE NOT)) (GO B2)))
            (GO B6)
          B2(SETQ UNIF (UNIFICATION X (CDR Y)))
          B3(COND ((EQUAL UNIF (QUOTE NO)) (GO B6)))
            (SETQ R (APPEND (REVERSE TC2) (CDR C2)))
            (COND ((NULL R )(RETURN (LIST (QUOTE CONTRADICTION) L1 L2))))
            (SETQ VAR NIL)
          B4(COND ((NULL UNIF) (GO B5)))
            (SETQ V1 (CAAR UNIF))
            (SETQ V2 (CADAR UNIF))
            (SETQ VAR (CONS V1 VAR))
            (SETQ R (SUBST V2 V1 R))
            (SETQ UNIF (CDR UNIF))
            (GO B4)
          B5(SETQ N (ADD1 N))
            (SETQ H (LIST N L1 L2 (ADD1 (LENGTH TC2))))
            (SETQ R (LIST N (DELETEV VC1 VC2 VAR) R))
            (SETQ RES (CONS R RES))
            (SETQ HIST (CONS H HIST))
          B6(SETQ TC2 (CONS Y TC2))
            (SETQ C2 (CDR C2))
            (COND ((EQUAL SIGN 1) (GO B1)))
          B7(COND ((NULL C2) (RETURN (LIST (REVERSE RES) (REVERSE HIST) N))))
            (SETQ Y (CAR C2))
            (COND ((EQ (CAR Y) (QUOTE NOT)) (GO B6)))
            (SETQ UNIF (UNIFICATION (CDR X) Y))
            (GO B3))) EXPR)
```

```
(DEFPROP GUNIT (LAMBDA (S1 S2 W C N)
        (PROG (L S3 SS3 W1 V U RES HIST M X)
          (COND ((NULL W) (RETURN (LIST RES HIST N))))
          (SETQ L (LENGTH (CADDR C)))
          (SETQ S3 (LIST (LIST 10000 C)))
          (SETQ SS3 S3)
        B1(COND ((NULL W) (GO B7)))
          (SETQ W1 (CAR W))
        B2(COND ((NULL SS3) (GO B4)))
          (SETQ V (CAR SS3))
          (COND ((GREATERP (CAR W1) (CAR V)) (GO B3)))
          (SETQ U (URESOLVE W1 (CADR V) N))
          (COND ((NULL (CAR U)) (GO B3)))
          (SETQ RES (APPEND RES (CAR U)))
          (SETQ HIST (APPEND HIST (CADR U)))
          (SETQ N (CADDR U))
        B3(SETQ SS3 (CDR SS3))
          (GO B2)
        B4(COND ((EQUAL (SUB1 L) 1) (GO B6)))
          (SETQ M (CAR W1))
        B5(COND ((NULL RES) (GO B6)))
          (SETQ X (CONS (LIST M (CAR RES)) X))
          (SETQ RES (CDR RES))
          (GO B5)
        B6(SETQ W (CDR W))
          (SETQ SS3 S3)
          (GO B1)
        B7(SETQ L (SUB1 L))
          (COND ((EQUAL L 1) (RETURN (LIST RES HIST N))))
          (SETQ S3 X)
          (SETQ SS3 S3)
          (SETQ X NIL)
          (SETQ W (APPEND S1 S2))
          (GO B1))) EXPR)

(DEFPROP PNSORT (LAMBDA (RES)
        (PROG (C POS NEG)
        B1(COND ((NULL RES) (RETURN (LIST (REVERSE POS) (REVERSE NEG)))))
          (SETQ C (CAAR (CDDAR RES)))
          (COND ((EQUAL (CAR C) (QUOTE NOT)) (GO B3)))
          (SETQ POS (CONS (CAR RES) POS))
        B2(SETQ RES (CDR RES))
          (GO B1)
        B3(SETQ NEG (CONS (CAR RES) NEG))
          (GO B2))) EXPR)

(DEFPROP FDEPTH (LAMBDA (C)
        (PROG (N U)
          (SETQ C (CAR (CADDR C)))
          (COND ((EQUAL (CAR C) (QUOTE NOT)) (GO B1)))
          (SETQ C (CDR C))
          (GO B2)
        B1(SETQ C (CDDR C))
        B2(SETQ N 0)
        B3(COND ((NULL C) (GO B5)))
          (COND ((ATOM (CAR C)) (GO B4)))
          (SETQ U (APPEND (CDAR C) U))
        B4(SETQ C (CDR C))
          (GO B3)
        B5(COND ((NULL U) (RETURN N)))
```

```
                    (SETQ N (ADD1 N))
                    (SETQ C U)
                    (SETQ U NIL)
                    (GO B3))) EXPR)

(DEFPROP FTEST (LAMBDA (RES N4)
            (PROG (C U)
            B1(COND ((NULL RES) (RETURN (REVERSE U))))
                    (SETQ C (CAR RES))
                    (COND ((GREATERP (FDEPTH C) N4) (GO B2)))
                    (SETQ U (CONS C U))
            B2(SETQ RES (CDR RES))
                    (GO B1))) EXPR)

(DEFPROP SUBSUME (LAMBDA (C1 C2)
            (PROG (Z VAR U)
                    (SETQ C1 (RENAME C1 XLIST))
                    (SETQ C1 (CAR (CADDR C1)))
                    (SETQ Z ZLIST)
                    (SETQ VAR (CADR C2))
                    (SETQ C2 (CAR (CADDR C2)))
            B1(COND ((NULL VAR) (GO B2)))
                    (SETQ C2 (SUBST (CAR Z) (CAR VAR) C2))
                    (SETQ VAR (CDR VAR))
                    (GO B1)
            B2(SETQ U (UNIFICATION C1 C2))
                    (COND ((EQUAL U (QUOTE NO)) (RETURN NIL)))
                    (RETURN T))) EXPR)

(DEFPROP STEST (LAMBDA (U RES)
            (PROG (R V W X1 Y Z)
            B1(COND ((NULL RES) (GO 35)))
                    (SETQ R (CAR RES))
                    (SETQ Z (APPEND U V))
            B2(COND ((NULL Z) (GO B3)))
                    (COND ((SUBSUME (CAR Z) R) (GO B4)))
                    (SETQ Z (CDR Z))
                    (GO B2)
            B3(SETQ V (CONS R V))
            B4(SETQ RES (CDR RES))
                    (GO B1)
            B5(COND ((NULL V) (RETURN W)))
                    (SETQ X1 (CAR V))
                    (SETQ Z (CDR V))
            B6(COND ((NULL Z) (GO B8)))
                    (COND ((SUBSUME X1 (CAR Z)) (GO B7)))
                    (SETQ Y (CONS (CAR Z) Y))
            B7(SETQ Z (CDR Z))
                    (GO B6)
            B8(SETQ W (CONS X1 W))
                    (SETQ V (REVERSE Y))
                    (SETQ Y NIL)
                    (GO B5))) EXPR)

(DEFPROP CONTRADICT (LAMBDA (U V)
            (PROG (X1 Y RES)
            B1(COND ((OR (NULL U) (NULL V)) (RETURN NIL)))
                    (SETQ X1 (CAR U))
                    (SETQ Y V)
            B2(COND ((NULL Y) (GO B3)))
```

```
                    (SETQ RES (URESOLVE X1 (CAR Y) -1))
                    (COND ((EQUAL (CAR RES) (QUOTE CONTRADICTION)) (RETURN RES)))
                    (SETQ Y (CDR Y))
                    (GO B2)
                 B3(SETQ U (CDR U))
                    (GO B1))) EXPR)

   (DEFPROP DTREE (LAMBDA (Z HIST N1)
            (PROG (X TX X1 H M1 M2 M N)
                    (SETQ HIST (REVERSE HIST))
                    (SETQ X (CDR Z))
                    (SETQ Z (LIST Z))
                    (COND ((GREATERP (CAR X) (CADR X)) (GO B0)))
                    (SETQ X (REVERSE X))
                 B0(COND ((GREATERP (CADR X) N1) (GO B1)))
                    (SETQ X (LIST (CAR X)))
                 B1(COND ((NULL X) (RETURN Z)))
                    (SETQ X1 (CAR X))
                 B2(COND ((EQUAL X1 (CAAR HIST)) (GO B3)))
                    (SETQ HIST (CDR HIST))
                    (GO B2)
                 B3(SETQ X (CDR X))
                    (SETQ H (CAR HIST))
                    (SETQ Z (CONS H Z))
                    (SETQ HIST (CDR HIST))
                    (SETQ M1 (CADR H))
                    (SETQ M2 (CADDR H))
                    (COND ((GREATERP M1 N1) (GO B5)))
                 B4(COND ((GREATERP M2 N1) (GO B6)))
                    (GO B1)
                 B5(SETQ N 1)
                    (SETQ M M1)
                    (GO B7)
                 B6(SETQ N 2)
                    (SETQ M M2)
                 B7(COND ((NULL X) (GO B8)))
                    (SETQ X1 (CAR X))
                    (COND ((EQUAL X1 M) (GO B10)))
                    (COND ((GREATERP X1 M) (GO B9)))
                 B8(SETQ X (APPEND (REVERSE TX) (CONS M X)))
                    (GO B11)
                 B9(SETQ TX (CONS X1 TX))
                    (SETQ X (CDR X))
                    (GO B7)
                 B10(SETQ X (APPEND (REVERSE TX) X))
                 B11(SETQ TX NIL)
                    (COND ((EQUAL N 2) (GO B1)))
                    (GO B4))) EXPR)

   (DEFPROP TPU (LAMBDA (S1 S2 S3 W N1 N2 N3 N4)
            (PROG (S W1 TS U1 U N K CK WCK V POS NEG HIST Y X1 X)
                    (SETQ S (APPEND S1 S2))
                    (SETQ S (REVERSE S))
                 B1(COND ((NULL W) (GO B6)))
                    (SETQ W1 (CAR W))
                 B2(SETQ TS S)
                    (COND ((NULL W1) (GO B5)))
                 B3(COND ((EQ (CAR W1) (CAAR TS)) (GO B4)))
                    (SETQ TS (CDR TS))
                    (GO B3)
```

```
B4(SETQ U1 (CONS (CAR TS) U1))
  (SETQ W1 (CDR W1))
  (GO B2)
B5(SETQ U (CONS U1 U))
  (SETQ W (CDR W))
  (SETQ U1 NIL)
  (GO B1)
B6(SETQ W (REVERSE U))
  (SETQ N N1)
  (SETQ U (CONTRADICT S1 S2))
  (COND ((NOT (NULL U)) (RETURN U)))
  (SETQ K 1)
B7(COND ((GREATERP K N2) (RETURN (QUOTE (S IS NOT PROVED)))))
  (SETQ CK (CAR S3))
  (SETQ WCK (CAR W))
  (SETQ V (GUNIT S1 S2 WCK CK N))
  (COND (( NULL (CAR V)) (GO B12)))
  (SETQ N (CADDR V))
  (SETQ HIST (APPEND HIST (CADR V)))
  (SETQ V (CAR V))
  (COND ((LESSP K N3) (GO B8)))
  (SETQ V (FTEST V N4))
B8(SETQ V (PNSORT V))
  (SETQ POS (STEST S1 (CAR V)))
  (SETQ NEG (STEST S2 (CADR V)))
  (COND ((NULL (APPEND POS NEG)) (GO B12)))
  (SETQ U (CONTRADICT S1 NEG))
  (COND ((NOT (NULL U)) (RETURN (DTREE U HIST N1))))
  (SETQ U (CONTRADICT POS S2))
  (COND ((NOT (NULL U)) (RETURN (DTREE U HIST N1))))
  (SETQ S1 (APPEND S1 POS))
  (SETQ S2 (APPEND S2 NEG))
  (SETQ W (CDR W))
  (SETQ Y (APPEND POS NEG))
B9(COND ((NULL W) (GO B10)))
  (SETQ X1 (APPEND Y (CAR W)))
  (SETQ X (CONS X1 X))
  (SETQ W (CDR W))
  (GO B9)
B10(SETQ W (APPEND (REVERSE X) (LIST Y)))
   (SETQ X NIL)
B11(SETQ S3 (APPEND (CDR S3) (LIST CK)))
   (SETQ K (ADD1 K))
   (GO B7)
B12(SETQ W (APPEND (CDR W) (LIST NIL)))
   (GO B11))) EXPR)
```

A.4   ILLUSTRATIONS

We now give examples to illustrate how the program TPU can be used. We use the PDP-10 LISP 1.6 system [Quam, 1968]. If the reader uses LISP 1.5 [McCarthy *et al.*, 1962; Weissman, 1965], he should change (DEFPROP name body EXPR) to DEFINE(((name body))), and change the expressions like (DEFPROP XLIST (NIL x1 x2 x3 x4 x5 x6 x7) VALUE) to CSET(XLIST (x1 x2 x3 x4 x5 x6 x7)).

Example 1

Prove that in an associative system with left and right solutions, there is a right identity element.

This theorem can be represented by the following set of clauses:

(1) $P(g(x, y), x, y)$

(2) $P(x, h(x, y), y)$

(3) $\sim P(k(x), x, k(x))$

(4) $\sim P(x, y, u) \lor \sim P(y, z, v) \lor \sim P(x, v, w) \lor P(u, z, w)$

(5) $\sim P(x, y, u) \lor \sim P(y, z, v) \lor \sim P(u, z, w) \lor P(x, v, w).$

To prove this theorem, the computer is called to execute the following expression

$($TPU
$\quad @\,(\,(1\,(\text{X Y})\,((\text{P}(\text{G X Y})\,\text{X Y})))$

$\qquad (2(\text{X Y})\,((\text{P X}\,(\text{H X Y})\,\text{Y})))\,)$

$\quad @\,(\,(\,(3(\text{X})\,((\text{NOT P}\,(\text{K X})\,\text{X}\,(\text{K X})))))\,)$

$\quad @\,(\,(4(\text{X Y Z U V W})\,((\text{NOT P X Y U})(\text{NOT P Y Z V})\,(\text{NOT P X V W})(\text{P U Z W})))$

$\qquad (5(\text{X Y Z U V W})\,((\text{NOT P X Y U})(\text{NOT P Y Z V})(\text{NOT P U Z W})(\text{P X V W})))\,)$

$\quad @\,(\,(3)\,\text{NIL})$

$\quad @\,5$

$\quad @\,2$

$\quad @\,3$

$\quad @\,0)$

We note that every argument of TPU is preceded by @. If E is any *S*-expression, @E means (QUOTE E) in the PDP-10 LISP system. When the above expression is executed, we obtain the following output:

$\quad ((6\ \ 3\ \ 4\ \ 4)(11\ \ 2\ \ 6\ \ 2)(15\ \ 1\ \ 11\ \ 1)(\text{CONTRADICTION}\ 1\ \ 15)).$

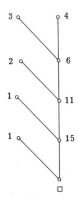

Figure A.2

From this output, we can construct a refutation. We know that

(6 3 4 4) indicates that clause 6 is obtained by resolving clause 3 with clause 4 upon the fourth literal;

(11 2 6 2) indicates that clause 11 is obtained by resolving clause 2 with clause 6 upon the second literal;

(15 1 11 1) indicates that clause 15 is obtained by resolving clause 1 with clause 11 upon the first literal;

(CONTRADICTION 1 15) indicates that clause 1 and clause 15 are resolved to the empty clause.

There, we construct a refutation shown in Fig. A.2.

In the following examples, we shall give the input and the output for each theorem.

Example 2

In an associative system with an identity element, if the square of every element is the identity, the system is commutative.

Input:

$\Big($TPU

$\quad @\big(\big(\,(1\,(\text{x})((\text{P E X X})))$

$\qquad (2\,(\text{x})\,((\text{P X E X})))$

$\qquad (3\,(\text{x})\,((\text{P X X E})))$

$\qquad (4(\ \ )\,((\text{P A B C})))\big)$

$\quad @\big(\,(5(\ \ )\,((\text{NOT P B A C})))\big)\big)$

$\quad @\big(\,(6(\text{X Y Z U V W})\,((\text{NOT P X Y U})(\text{NOT P Y Z V})(\text{NOT P X V W})(\text{P U Z W})))$

$\qquad (7(\text{X Y Z U V W})\,((\text{NOT P X Y U})(\text{NOT P Y Z V})(\text{NOT P U Z W})(\text{P X V W})))\big)\big)$

@((4) NIL)

@ 7

@ 4

@ 5

@ 0)

Output:

((8 4 6 1)(21 3 8 1)(30 2 21 1)(32 30 7 2)(42 3 32 1)(55 1 42 1)

(62 55 6 1)(112 5 62 3)(130 3 112 1)(CONTRADICTION 2 130))

Example 3

In a group the left identity element is also a right identity.

Input:

(TPU

@(((1 (X) ((P E X X)))

    (2 (X) ((P (I X) X E)))))

@(((3 ( ) ((NOT P A E A)))))

@(((4 (X Y Z U V W) ((NOT P X Y U) (NOT P Y Z V) (NOT P X V W) (P U Z W)))

    (5 (X Y Z U V W) ((NOT P X Y U) (NOT P Y Z V) (NOT P U Z W) (P X V W)))))

@(((3)(3)))

@ 5

@ 4

@ 5

@ 0)

Output:

((13 3 5 4)(16 2 13 2)(17 1 16 2)(18 17 4 4)(23 2 18 3)(24 1 23 2)

(30 24 5 4)(46 2 30 2)(56 2 46 1)(CONTRADICTION 1 56))

Example 4

In a group with left inverse and left identity every element has a right inverse.

Input:

(TPU

@((1(x)((P E X X)))

(2(x)((P(I X) X E)))))

@((3(x)((NOT P A X E)))))

@((4(X Y Z U V W)((NOT P X Y U)(NOT P Y Z V)(NOT P X V W)(P U Z W)))

(5(X Y Z U V W)((NOT P X Y U)(NOT P Y Z V)(NOT P U Z W)(P X V W)))))

@((3)(3))

@ 5

@ 4

@ 5

@ 0)

Output:

((6 3 4 4)(11 2 6 3)(12 1 11 2)(20 12 5 4)(42 2 20 2)(62 2 42 1)

(CONTRADICTION 1 62))

Example 5

If S is a nenempty subset of a group such that if x, y belong to S, then $x \cdot y^{-1}$ belongs to S, then the identity e belongs to S.

Input:

(TPU

@((1(x)((P E X X)))

(2(x)((P X E X)))

(3(x)((P X(I X) E)))

(4(x)((P(I X) X E)))

(5( )((S A)))))

@((6( )((NOT S E)))))

@((7(X Y Z)((NOT S X)(NOT S Y)(NOT P X(I Y) Z)(S Z)))

(8(X Y Z U V W)((NOT P X Y U)(NOT P Y Z V)(NOT P X V W)(P U Z W)))

(9(X Y Z U V W)((NOT P X Y U)(NOT P Y Z V)(NOT P U Z W)(P X V W)))))

@((6) NIL NIL)

@9

@4

@5

@0)

Output:

((10 6 7 4)(14 5 10 2)(18 5 14 1)(CONTRADICTION 3 18))

Example 6

If S is a nonempty subset of a group such that if x, y belong to S then $x \cdot y^{-1}$ belongs to S, then S contains x^{-1} whenever it contains x.

Input:

(TPU

@((1 (X) ((P E X X)))

(2 (X) ((P X E X)))

(3 (X) ((P X (I X) E)))

(4 (X) ((P (I X) X E)))

(5 ( ) ((S B))))

@((6 ( ) ((NOT S (I B)))))

@((7 (X Y Z) ((NOT S X) (NOT S Y) (NOT P X (I Y) Z) (S Z)))

(8 (X Y Z U V W) ((NOT P X Y U) (NOT P Y Z V) (NOT P X V W) (P U Z W)))

(9 (X Y Z U V W) ((NOT P X Y U) (NOT P Y Z V) (NOT P U Z W) (P X V W))))

@((5 6) NIL NIL)

@9

@4

@5

@0)

Output:

((11 5 7 1)(19 5 11 1)(23 3 19 1)(152 23 7 1)(169 6 152 3)

(186 5 169 1)(CONTRADICTION 1 186))

Example 7

If a is a prime and $a = b^2/c^2$ then a divides b.

Input:

$\big($ TPU

 @ $\big(\big(($ 1 ( ) $(($ P A $)))$

   $\big($ 2 ( ) $(($ M A ($ S C $)($ S B $))))\big)$

   $\big($ 3 (X) $(($ M X X ($ S X $))))\big)$ $\big)$

 @ $\big(\big(($ 4 ( ) $(($ NOT D A B $)))\big)$ $\big)$

 @ $\big(\big(($ 5 (X Y Z) $(($ NOT M X Y Z $)($ M Y X Z $)))$

   $($ 6 (X Y Z) $(($ NOT M X Y Z $)($ D X Z $)))$

   $($ 7 (X Y Z U) $(($ NOT P X $)($ NOT M Y Z U $)($ NOT D X U $)($ D X Y $)($ D X Z $)))$ $\big)$ $\big)$

 @ $\big(($ 1 2 3 4 $)($ 1 2 3 4 $)($ 1 2 3 4 $))$

 @ 7

 @ 4

 @ 5

 @ 0 $)$

Output:

$(($ 13 2 6 1 $)($ 16 13 7 3 $)($ 43 4 16 3 $)($ 66 4 43 3 $)($ 75 3 66 2 $)$

$($ CONTRADICTION 1 75 $))$

Example 8

Any number greater than 1 has a prime divisor.

Input:

$\big($ TPU

 @ $\big(\big(($ 1 ( ) $(($ L 1 A $)))$

   $\big($ 2 (X) $(($ D X X $)))\big)$ $\big)$

 @ NIL

 @ $\big(\big(($ 3 (X) $(($ P X $)($ D ($ G X $)$ X $)))$

   $\big($ 4 (X) $(($ P X $)($ L 1 ($ G X $))))\big)$

   $($ 5 (X) $(($ P X $)($ L ($ G X $)$ X $)))$

   $\big($ 6 (X) $(($ NOT P X $)($ NOT D X A $)))\big)$

$$\big(7(\text{X Y Z})\big((\text{NOT D X Y})(\text{NOT D Y Z})(\text{D X Z})\big)\big)$$

$$\big(8(\text{X})\big((\text{NOT L 1 X})(\text{NOT L X A})(\text{P}(\text{F X}))\big)\big)$$

$$\big(9(\text{X})\big((\text{NOT L 1 X})(\text{NOT L X A})(\text{D}(\text{F X})\text{X})\big)\big)\big)$$

@ ((1 2)(1 2)(1 2)(1 2)(1 2)(1 2)(1 2))

@ 9

@ 20

@ 21

@ 0)

Output:

((10 2 6 2)(15 10 3 1)(16 10 4 1)(17 10 5 1)(23 17 8 2)(25 16 23 1)

(26 17 9 2)(28 16 26 1)(32 25 6 1)(33 32 7 3)(47 28 33 1)

(CONTRADICTION 15 47))

Example 9

There exist infinitely many primes.

Input:

(TPU

@ $\big(\big(\big(1(\text{X})\big((\text{L X}(\text{F X}))\big)\big)\big)\big)$

@ $\big(\big(2(\text{X})\big((\text{NOT L X X})\big)\big)\big)$

@ $\big(\big(3(\text{X Y})\big((\text{NOT L X Y})(\text{NOT L Y X})\big)\big)$

$\big(4(\text{X Y})\big((\text{NOT D X}(\text{F Y}))(\text{L Y X})\big)\big)$

$\big(5(\text{X})\big((\text{P X})(\text{D}(\text{H X})\text{X})\big)\big)$

$\big(6(\text{X})\big((\text{P X})(\text{P}(\text{H X}))\big)\big)$

$\big(7(\text{X})\big((\text{P X})(\text{L}(\text{H X})\text{X})\big)\big)$

$\big(8(\text{X})\big((\text{NOT P X})(\text{NOT L A X})(\text{L}(\text{F A})\text{X})\big)\big)\big)$

@ ((1 2)(1 2)(1 2)(1 2)(1 2)(1 2))

@ 8

@ 20

@ 21

@ 0)

Output:

((14 2 8 3)(16 1 14 2)(17 16 5 1)(18 16 6 1)(19 16 7 1)(20 19 3 1)

(23 17 4 1)(24 23 8 2)(28 20 24 2)(CONTRADICTION 18 28))

REFERENCES

Chang, C. L. (1970a): The unit proof and the input proof in theorem proving, *J. Assoc. Comput. Mach.* **17** 698–707.

McCarthy, J., P. W. Abrahams, D. J. Edwards, T. P. Hart, and M. I. Levin (1962): "LISP 1.5 Programmer's Manual," The M.I.T. Press, M.I.T., Cambridge, Massachusetts.

Quam, L. H. (1968): "Stanford LISP 1.6 Manual," Stanford Artificial Intelligence Project, Stanford Univ., Stanford, California.

Weissman, C. (1967): "LISP 1.5 Primer," Dickenson, Belmont, California.

Appendix B

PROOF OF LEMMA 8.2

Case 1 L is derived by paramodulating L' and M

a. M is an equality denoted by $t_1 = t_2$ and L is obtained by paramodulating $t_1 = t_2$ into L'. In this case, M' is an equality that we denote by $r_1 = r_2$. If t_1 is obtained from r_1 by substituting s_2 for s_1, and if t_2 is r_2, then the deduction of Fig. 8.3a can be represented by Fig. B.1a. Clearly, L can be also obtained by the deduction shown in Fig. B.1b. If t_1 is r_1 and t_2 is obtained from r_2 by substituting s_2 for s_1, then the deduction of Fig. 8.3a can be represented by Fig. B.2a. Again, L can be also obtained by the deduction shown in Fig. B.2b. We note that Fig. B.1b and Fig. B.2b are input deductions of L with top clause L'.

$$r_1(s_1) = r_2$$
$$L'(r_1(s_2))$$
$$s_1 = s_2$$
$$r_1(s_2) = r_2$$
$$t_1 \qquad t_2$$

$$L'(r_1(s_2))$$
$$s_1 = s_2$$
$$L'(r_1(s_1))$$
$$r_1(s_1) = r_2$$
$$L'(r_2)$$

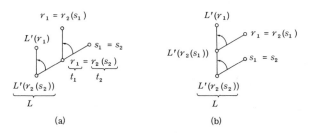

Figure B.2

b. L' is an equality denoted by $t = u$; L is obtained by paramodulating $t = u$ into M. In this case, if t and s_2 are not part of each other in M, then the deduction of Fig. 8.3a can be represented by Fig. B.3a. Clearly, L can also be obtained by the deduction shown in Fig. B.3b. If t is part of s_2, then Fig. 8.3a can be represented by Fig. B.4a. Obviously, L can also be obtained by the deduction shown in Fig. B.4b. If s_2 is part of t, Fig. 8.3a can be represented by Fig. B.5a. Again, L can also be obtained by the deduction shown in Fig. B.5b. We note that Figs. B.3b, B.4b, and B.5b are input deductions of L with top clause L'.

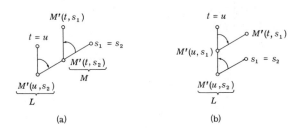

Figure B.3

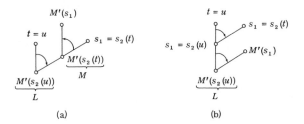

Figure B.4

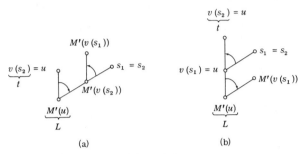

Figure B.5

Case 2 L is derived by resolving L' and M

In this case, L' is $\sim M$ and L is $\square$. The deduction of Fig. 8.3a can be represented by Fig. B.6a. Clearly, L can also be obtained by the deduction shown in Fig. B.6b. We note that Fig. B.6b is an input deduction of L with top clause L'. This completes the proof of Lemma 8.2.

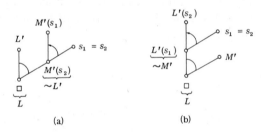

Figure B.6

Bibliography

The bibliography is divided into three parts:

Part I: the general theory of artificial intelligence and graph searching techniques,

Part II: theory and techniques of mechanical theorem proving,

Part III: application of symbolic logic and mechanical theorem proving to program analysis, program synthesis, question-answering, and problem-solving.

PART I

Banerji, R. (1969): "Theory of Problem Solving: an Approach to Artificial Intelligence," Amer. Elsevier, New York.

Banerji, R., and Mesavoric, M. (ed.) (1970): "Theoretical Approaches to Non-numerical Problem-Solving," Springer-Verlag, Berlin and New York.

Busacker, R. G., and Saaty, T. L. (1965): "Finite Graphs and Networks, an Introduction with Applications," McGraw-Hill, New York.

Chang, C. L., and Slagle, J. R. (1971): An admissible and optimal algorithm for searching AND/OR graphs, *Artif. Intelligence* **2**, No. 2, 117–128.

Collins, N. L., and Michie, D. (ed.) (1967): "Machine Intelligence," Vol. 1, American Elsevier, New York.

Dale, E., and Michie, D. (ed.) (1968): "Machine Intelligence," Vol. 2, American Elsevier, New York.

Doran, J. (1967): An approach to automatic problem-solving, "Machine Intelligence," Vol. 1 (N. L. Collins and D. Michie, eds.), American Elsevier, New York, pp. 105–123.

Doran, J. (1968): New developments of the graph traverser, "Machine Intelligence," Vol. 2 (E. Dale and D. Michie, eds.), American Elsevier, New York, pp. 119–136.

Doran, J. (1969): Planning and generalization in an automation/environment system, "Machine Intelligence," Vol. 4 (B. Meltzer and D. Michie, eds.), American Elsevier, New York, pp. 433–454.

Doran, J., and Michie, D. (1966): Experiments with the graph traverser program. *Proc. Roy. Soc., Ser. A* **294.** 235–259.

Draper, N. R., and Smith, H. (1966): "Applied Regression Analysis," Wiley, New York.

Dreyfus, H. L. (1971): "What Computers Can't Do," Harper, New York.

Dreyfus, S. (1969): An appraisal of some shortest path algorithms, *Operations Res.* **17.** No. 3, 395–412.

Ernst, G. W., and Newell, A. (1969): "GPS: a Case Study in Generality and Problem Solving," Academic Press, New York.

Feigenbaum, E. (1969): Artificial intelligence: themes in the second decade, *in:* "Information Processing 68," Vol. 2 (A. J. H. Morrell, ed.), North-Holland Publ., Amsterdam, pp. 1008–1022.

Feigenbaum, E., and Feldman, J. (eds.) (1963): "Computers and Thought," McGraw-Hill, New York.

Findler, N. V., and Meltzer, B. (eds.) (1971): "Artificial Intelligence and Heuristic Programming," American Elsevier, New York.

Fogel, L. J., Owens, A. J., and Welsh, M. J. (1966): "Artificial Intelligence through Simulated Evolution," Wiley, New York.

Hart, P. E., Nilsson, N. J., and Ralph, B. (1968): A formal basis for the heurisitic determination of minimum cost paths, *IEEE Trans. System Sci. Cybernet.* **SSC–4,** No. 2, pp. 100–107.

Lawler, E., and Wood, D. (1966): Branch and bound methods: a survey, *Operations Res.* **14,** No. 4, 699–719.

Lin, S. (1970): Heuristic techniques for solving large combinatorial problems on a computer, "Theoretical Approaches to Non-numerical Problem-Solving" (R. Banerji and M. Mesavoric, eds.), Springer-Verlag, New York, pp. 410–418.

McCarthy, J., and Hayes, P. J. (1969): Some philosophical problems from the standpoint of artificial intelligence, "Machine Intelligence," Vol. 4 (B. Meltzer and D. Michie, eds.), American Elsevier, New York, pp. 463–502.

Meltzer, B., and Michie, D. (eds.) (1969): "Machine Intelligence," Vol. 4, American Elsevier, New York.

Meltzer, B., and Michie, D. (eds.) (1970): "Machine Intelligence," Vol. 5, American Elsevier, New York.

Meltzer, B., and Michie, D. (eds.) (1971): "Machine Intelligence," Vol. 6, American Elsevier, New York.

Michie, D. (ed.) (1969): "Machine Intelligence," Vol. 3, American Elsevier, New York.

Minsky, M. (1963): Steps toward artificial intelligence, "Computers and Thought" (E. Feigenbaum and J. Feldman, eds.), McGraw-Hill, New York, pp. 406–452.

Minsky, M. (ed.) (1968): "Semantic Information Processing," M.I.T. Press, Cambridge, Massachusetts.

Newell, A., and Simon, H. (1972): "Human Problem Solving," Prentice-Hall, Englewood Cliffs, New Jersey.

Nilsson, N. J. (1969): Searching problem-solving and game-playing trees for minimal cost solutions, "Information Processing 68," Vol. 2 (A. J. H. Morrell, ed.), North-Holland Publ., Amsterdam, pp. 1556–1562.

Nilsson, N. J. (1971): "Problem Solving Methods in Artificial Intelligence," McGraw-Hill, New York.

Papert, S. (1968): "The Artificial Intelligence of Hubert L. Dreyfus. A Budget of Fallacies," M.I.T. Artificial Intelligence Memo No. 54, January.

Pohl, I. (1970a): Heuristic search viewed as path finding in a graph, *Artif. Intelligence* 1, No. 3, 193–204.

Pohl, I. (1970b): First results on the effect of error in heuristic search, "Machine Intelligence," Vol. 5 (B. Meltzer and D. Michie, eds.), American Elsevier, New York, pp. 219–236.

Pohl, I. (1971): Bi-directional search, "Machine Intelligence," Vol. 6 (B. Meltzer and D. Michie, eds.), American Elsevier, New York, pp. 127–140.

Polya, G. (1957): "How to Solve It," Doubleday, Garden City, New York.

Sandewall, E. J. (1969): A planning problem solver based on look-ahead in stochastic game trees, *J. ACM* 16, No. 3, 364–382.

Simon, R., and Lee, R. C. T. (1971): On the optimal solutions to AND/OR series-parallel graphs, *J. ACM* 18, No. 3, 354–372.

Slagle, J. R. (1963): A heuristic program that solves symbolic integration problems in freshman calculus, *J. ACM* 10, No. 4, 507–520.

Slagle, J. R. (1971): "Artificial Intelligence, the Heuristic Programming Approach," McGraw-Hill, New York.

Turing, A. M. (1950): Computing machinery and intelligence, *Mind* 59, 433–460. Reprinted in "Computers and Thought" (E. Feigenbaum and J. Feldman, eds.), McGraw-Hill, New York, pp. 11–35.

PART II

Allen, J., and Luckham, D. (1970): An interactive theorem-proving program, "Machine Intelligence," Vol. 5 (B. Meltzer and D. Michie, eds.), American Elsevier, New York, pp. 321–336.

Anderson, R. (1970): Completeness results for E-resolution, *Proc. AFIPS 1970 Spring Joint Comput. Conf.* pp. 653–656.

Anderson, R. (1970): "Some Theoretical Aspects of Automatic Theorem Proving," Ph.D. Thesis, Univ. of Texas at Austin, Texas.

Anderson, R. (1971): "Completeness of the Locking Restriction for Paramodulation," Dept. of Comput. Sci., Univ. of Houston, Houston, Texas.

Anderson, R., and Bledsoe, W. W. (1970): A linear format for resolution with merging and a new technique for establishing completeness, *J. ACM* 17, No. 3, 525–534.

Andrews, P. B. (1965): "A Transfinite Type Theory with Type Variables," North-Holland Publ., Amsterdam.

Andrews, P. B. (1968a): Resolution with merging, *J. ACM* 15, No. 3, 367–381.

Andrews, P. B. (1968b): On simplifying the matrix of a wff, *J. Symbolic Logic*, Vol. 33, No. 2, 1968, pp. 180–192.

Andrews, P. B. (1971): Resolution in type theory, *J. Symbolic Logic* 36, No. 3, 414–432.

Backer, P., and Sayre, D. (1963): The Reduced Model for Satisfiability for Two Decidable Classes of Formulae in the Predicate Calculus, IBM Res. Rep. RC 1083.

Beth, E. W. (1963): Observations concerning computing, deduction and heuristics, "Computer Programming and Formal Systems" (P. Braffort and D. Hirshberg, eds.), North-Holland Publ., Amsterdam, pp. 21–32.

Bledsoe, W. W. (1971): Splitting and reduction heuristics in automatic theorem proving, *Artif. Intelligence* 2, No. 1, 57–78.

Bledsoe, W. W., Boyer, R. S., and Henneman, W. H. (1971): Computer proofs of limit theorems, *Artif. Intelligence* **3**, No. 1, 27–60.

Bohnert, H. G., and Backer, P. (1967): Automatic English-to-Logic Translation in a Simplified Model, IBM Res. Rep. RC-1744.

Boyer, R. S. (1971): "Locking: a Restriction of Resolution," Ph.D Thesis. Univ. of Texas at Austin, Texas.

Boyer, R. S., and Moore, J. S. (1971): "The Sharing of Structure in Resolution Program," Metamathematics Unit, Univ. of Edinburgh, Edinburgh, Scotland.

Brice, C., and Derksen, J. (1971): A Heuristically Guided Equality Rule in a Resolution Theorem Prover, Tech. Note 45, Stanford Res. Inst., Artificial Intelligence Group, Menlo Park, California.

Buchi, J. R. (1958): Turing machines and the Entscheidungs-problem, *Math. Ann.* Vol. **148**, 201–213.

Bundy, A. (1971): Counterexamples and Conjectures, There is no Best Proof Procedure, Metamathematics Unit, Univ. of Edinburgh, Edinburgh, Scotland.

Burstall, R. M. (1968): A Scheme for Indexing and Retrieving Clauses for a Resolution Theorem-Prover, MIP-R-45, Univ. of Edinburgh, Edinburgh, Scotland.

Cantaralla, R. G. (1969): "Efficient Semantic Resolution Proofs Based Upon Binary Semantic Trees," Ph.D. Thesis, Syracuse Univ., Syracuse, New York.

Chang, C. L. (1970a): The unit proof and the input proof in theorem proving, *J. ACM* **17**, No. 4, 698–707.

Chang, C. L. (1970b): Renamable paramodulation for automatic theorem proving with equality, *Artif. Intelligence* **1**, No. 4, 247–256.

Chang, C. L. (1971): Theorem Proving by Generation of Pseudo-Semantic Trees, Div. of Comput. Res. and Technol., Nat. Inst. of Health, Bethesda, Maryland.

Chang, C. L. (1972): Theorem proving with variable-constrained resolution. *Information Sci.* **4**, 217–231.

Chang, C. L., and Slagle, J. R. (1971): Completeness of linear resolution for theories with equality, *J. ACM* **18**, No. 1, 126–136.

Chinlund, T. J., Davis, M., Hineman, P. G., and McIlroy, M. D. (1964): Theorem Proving by Matching, Bell Laboratory.

Church, A. (1936): An unsolvable problem of number theory, *Amer. J. Math.* **58**, 345–363.

Church, A. (1941): "The Calculi of Lambda-Conversion," Princeton Univ. Press, Princeton, New Jersey.

Collins, G. F. (1962): Computational Reduction in Tarski's Decision Method for Elementary Algebra, IBM, Yorktown Heights, New York, July.

Cook, S. A. (1965): Algebraic Techniques and the Mechanization of Number Theory, RM-4319-PR, RAND Corp., Santa Monica, California.

Cook, S. A. (1971): The complexity of theorem-proving procedures, *Proc. 3rd Ann. ACM Symp. Theory Comput.* pp. 151–158.

Cooper, D. C. (1966): Theorem proving in computers, "Advances in Programming and Non-numerical Computation" (L. Fox, ed.), pp. 155–182.

Craig, W. (1957a): Linear reasoning: a new form of the Herbrand-Gentzen theorem, *J. Symbolic Logic* **22**, 250–268.

Craig, W. (1957b): Three uses of the Herbrand-Gentzen theorem relating model theorem to proof theorem, *J. Symbolic Logic* **22**, 269–285.

Crossely, J. N., and Dummett, M. (eds.) (1967): "Formal Systems and Recursive Systems," North-Holland Publ., Amsterdam.

Darlington, J. L. (1962): A Comit Program for Davis-Putnam Algorithm, Research Laboratory, Electron. Mech. Translation Group, M.I.T., May.

Darlington, J. L. (1965): Machine methods for proving logical arguments expressed in English, *Mech. Transl.* **8**, 41–67.

Darlington, J. L. (1968a): Some theorem-proving strategies based on the resolution principle, "Machine Intelligence," Vol. 2 (E. Dale and D. Michie, eds.), American Elsevier, New York 57–71.

Darlington, J. L. (1968b): Automatic theorem proving with equality substitutions and mathematical induction, "Machine Intelligence," Vol. 3 (B. Meltzer and D. Michie, eds.), American Elsevier, New York, pp. 113–127.

Darlington, J. L. (1971): A partial mechanization of second-order logic, "Machine Intelligence," Vol. 6 (B. Meltzer and D. Michie, eds.), American Elsevier, New York, pp. 91–100.

Davis, M. (1958): "Computability and Unsolvability," McGraw-Hill, New York.

Davis, M. (1963): Eliminating the irrelevant from mechanical proofs, *Proc. Symp. Appl. Math.* **15**, 15–30.

Davis, M. (ed.) (1965): "The Undecidable," Raven Press, Hewlett, New York.

Davis, M. (1968): Invited commentary on new directions in mechanical theorem-proving, *Proc. IFIP Congress 1968,* Vol. 1, North-Holland Publ., Amsterdam, pp. 67–68.

Davis, M., Logemann, G., and Loveland, D. (1962): A machine program for theorem proving, *Comm. ACM* **5**, No. 7, 394–397.

Davis, M., and Putnam, H. (1960): A computing procedure for quantification theory, *J. ACM* **7**, No. 3, 201–215.

Dixon, J. (1970): "An Improved Method for Solving Deductive Problems on a Computer by Compiled Axioms," Ph.D. Thesis, Univ. of California, Davis, California.

Dixon, J. (1971a): The Specializer, a Method of Automatically Writing Programs, Div. of Comput. Res. and Technol., Nat. Inst. of Health, Bethesda, Maryland.

Dixon, J. (1971b): Z-resolution: theorem-proving with compiled axioms (to appear).

Dixon, J. (1971c): Experiments with a Z-Resolution Program, Div. of Comput. Res. and Technol., Nat. Inst. of Health, Bethesda, Maryland.

Dreben, B. (1952): On the completeness of quantification theory, *Proc. Nat. Acad. Sci.* **38**, 1047–1052.

Dreben, B., and Wang, H. (1964): A Refutation Procedure and its Model-Theoretic Justification, Harvard Univ., Cambridge, Massachusetts.

Dunham, B., Fridshal, R., and Sward, G. L. (1959): A nonheuristic program for proving elementary logical theorems, *Proc. IFIP Congr. 1959* pp. 282–285.

Dunham, B., and North, J. H. (1962): Theorem testing by computer, *Symp. Math. Theory Machines,* Brooklyn Poly. Inst., Brooklyn, New York, pp. 172–177.

Ernst, G. (1971): The utility of independent subgoals in theorem proving, *Information and Control* **18**, No. 3, 237–252.

Friedman, J. (1963a): A semi-decision procedure for functional calculus, *J. ACM* **10**, No. 1, 1–24.

Friedman, J. (1963b): A computer program for a solvable case of the decision problem, *J. ACM* **10**, No. 3, 348–356.

Friedman, J. (1964): "A New Decision Procedure in Logic and Its Computer Realization," Ph.D. Thesis, Harvard Univ., Cambridge, Massachusetts.

Friedman, J. (1965): Computer realization of a decision procedure in logic, *Proc. IFIP Congr. 65,* pp. 327–328.

Gelernter, H. (1959): Realization of a geometry theorem proving machine, *Proc. IFIP Congr. 1959,* pp. 273–282.

Gilmore, P. C. (1959): A procedure for the production from axioms, of proofs for theories derivable within the first order predicate calculus, *Proc. IFIP Congr. 1959,* pp. 265–273

Gilmore, P. C. (1960): A proof method for quantification theory; its justification and realization, *IBM J. Res. Develop.*, 28–35.

Gilmore, P. C. (1970): An examination of the geometry theorem machine, *Artif. Intelligence* **1**, No. 3, 171–188.

Godel, K. (1930): The completeness of the axioms of the functional calculus of logic, "From Frege to Godel: a Source Book in Mathematical Logic" (J. van Heijenoort, ed.), Harvard Univ. Press, Cambridge, Massachusetts.

Godel, K. (1931): On formally undecidable propositions on *Principia Mathematica* and related systems, "From Frege to Godel: a Source Book in Mathematical Logic" (J. van Heijenoort, ed.), Harvard Univ. Press, Cambridge, Massachusetts.

Gould, W. E. (1966): A matching procedure for omega logic, Air Force Cambridge Res. Lab., Rep. AFCRL-66-781.

Guard, J. R., Oglesby, F. C., Benneth, J. H., and Settle, L. G. (1969): Semi-automated mathematics, *J. ACM* **18**, No. 1, 49–62.

Henkins, L. (1950): Completeness in the theory of types, *J. Symbolic Logic* **15**, 81–91.

Henschen, L. J. (1971): "A Resolution Style Proof Procedure for Higher Order Logic," Ph.D. Thesis, Univ. of Illinois at Urbana-Champaign, Illinois.

Herbrand, J. (1930a): Recherches sur la theorie de la demonstration, *Travaux de la Societe des Sciences et des Lettres de Varsovie* No. 33, 128.

Herbrand, J. (1930b): Investigations in proof theory: the properties of the propositions, "From Frege to Godel: a Source Book in Mathematical Logic" (J. van Heijenoort, ed.), Harvard Univ. Press, Cambridge, Massachusetts.

Herbrand, J. (1931): On the consistency of arithmetic, "From Frege to Godel: a Source Book in Mathematical Logic" (J. van Heijenoort, ed.), Harvard Univ. Press, Cambridge, Massachusetts, pp. 618–628.

Hilbert, D. (1927): The foundations of mathematics, "From Frege to Godel: a Source Book in Mathematical Logic" (J. van Heijenoort, ed.), Harvard Univ. Press, Cambridge, Massachusetts, pp. 464–479.

Hilbert, D., and Ackerman, W. (1950): "Principle of Mathematical Logic," Chelsea, New York.

Hodes, L. (1971): Solving problems by formula manipulation in logic and linear inequality, *Proc. 2nd Internat. Joint Conf. Artif. Intelligence, London*, pp. 553–559.

Huber, H. G. M., and Morris, A. H. Jr. (1971): Primary Paramodulation, NWL Tech. Rep. TR-2552, Warfare Analysis Dept., Naval Weapons Lab., Dahlgram, Virginia.

Huet, G. P. (1972): The Undecidability of the Existence of a Unifying Substitution Between Two Terms in the Simple Theory of Types, Rep. 1120, Jennings Comput. Center, Case Western Reserve Univ., Cleveland, Ohio.

Kahr, A. S., Moore, E. F., and Wang, H. (1962): Entscheidungs-problem reduced to the AEA case, *Proc. Nat. Acad. Sci.* **48**, 365–377.

Kallick, B. (1968): A decision procedure based on the resolution method, *Proc. IFIP Congr. 1968* **1**, 269–275.

Kanger, S. (1963): A simplified proof method for elementary logic, "Computer Programming and Formal Systems" (P. Braffort and D. Hirshberg, eds.), North-Holland Publ., Amsterdam, pp. 87–94.

Kleene, S. (1967): "Mathematical Logic," Wiley, New York.

Kling, R. E. (1971): A paradigm for reasoning by analogy, *Artif. Intelligence* **2**, No. 2, 147–178.

Knuth, D. E. (1968): "The Art of Computer Programming," Addison-Wesley, Reading, Massachusetts.

Korfphage, R. R. (1966): "Logic and Algorithms," Wiley, New York.

Kowalski, R. (1970a): The case for using equality axioms in automatic demonstration, *Symp. Automatic Demonstration* Springer-Verlag, New York, pp. 112–127.

Kowalski, R. (1970b): "Studies in the Completeness and Efficiency of Theorem-proving by Resolution," Ph.D. Thesis, Univ. of Edinburgh at Edinburgh, Scotland.

Kowalski, R. (1970c): Search strategies for theorem-proving, "Machine Intelligence," Vol. 5 (B. Meltzer and D. Michie, eds.), American Elsevier, New York, pp. 181–201.

Kowalski, R., and Hayes, P. (1969): Semantic trees in automatic theorem proving, "Machine Intelligence," Vol. 4 (B. Meltzer and D. Michie, eds.), American Elsevier, New York, pp. 87–101.

Kowalski, R., and Kuehner, D. (1970): Linear Resolution with Selection Function, Meta-mathematics Unit, Edinburgh Univ., Scotland.

Kripke, S. (1963): Semantic considerations in modal logic, *Acta Philos. Fenn.* **16**, 83–94.

Krom, M. R. (1967): The decision problem for a class of first-order formulas in which all disjunctions are binary, *Z. Math. Logik Grundlagen Math.* **13**, 15–20.

Kuehner, D. G. (1971): A note on the relation between resolution and Maslov's inverse method, "Machine Intelligence," Vol. 6 (B. Meltzer and D. Michie, eds.), American Elsevier, New York, pp. 73–90.

Lee, R. C. T. (1967): "A Completeness Theorem and a Computer Program for Finding Theorems Derivable from Given Axioms," Ph.D. Thesis, Univ. of California at Berkeley, California.

Lee, R. C. T. (1972): Fuzzy logic and the resolution principle, *J. ACM* **19**, No. 1, 109–119.

Lee, R. C. T., and Chang, C. L. (1971): Some properties of fuzzy logic, *Information and Control* **19**, No. 5, 417–431.

Levine, R. E., and Maron, M. E. (1967): A computer system for inference execution and data retrieval, *Comm. ACM* **10**, No. 11, 715–721.

Lewis, C. I., and Langford, C. H. (1959): "*Symbolic Logic*," Dover, New York.

Lightstone, A. H. (1964): "The Axiomatic Method, an Introduction to Mathematical Logic," Prentice-Hall, Englewood Cliffs, New Jersey.

Loveland, D. W. (1968): Mechanical theorem proving by model elimination, *J. ACM* **15**, No. 2, 236–251.

Loveland, D. W. (1969a): A simplified format for the model elimination theorem-proving procedure, *J. ACM* **16**, No. 3, 349–363.

Loveland, D. W. (1969b): Theorem provers combining model elimination and resolution, "Machine Intelligence," Vol. 4 (B. Meltzer and D. Michie, eds.), American Elsevier, New York, pp. 73–86.

Loveland, D. W. (1970a): A linear format for resolution, *Proc. IRIA Symp. Automatic Demonstration*, Springer-Verlag, New York, pp. 147–162.

Loveland, D. W. (1970b): Some Linear Herbrand Proof Procedures: An Analysis, Dept. of Comput. Sci., Carnegie-Mellon Univ., Pittsburgh, Pennsylvania.

Loveland, D. W. (1972): A unifying view of some linear Herbrand procedures, *J. ACM* **19**, No. 2, 366–384.

Lucchesi, C. L. (1972): The Undecidability of the Unification for Third Order Language, CSRR 2059, Dept of Appl. Analysis and Comput. Sci., Univ. of Waterloo, Waterloo, Canada.

Luckham, D. (1967): The resolution principle in theorem-proving, "Machine Intelligence," Vol. 1 (N. L. Collins and D. Michie, eds.), American Elsevier, New York, pp. 47–61.

Luckham, D. (1968a): Some tree-pairing strategies for theorem-proving, "Machine Intelligence," Vol. 3 (D. Michie, ed.), American Elsevier, New York, pp. 95–112.

Luckham, D. (1968b): The Ancestry Filter Method in Automatic Demonstration, Stanford Artificial Intelligence Project Memo, Stanford, California.

Luckham, D. (1970): Refinements in resolution theory, *Proc. IRIA Symp. Automatic Demon-stration, Versailes, France, 1968*, Springer-Verlag, New York, pp. 163–190.

Maslov, S. Ju (1971): Proof-search strategies for methods of the resolution type, "Machine Intelligence," Vol. 6 (B. Meltzer and D. Michie, eds.), American Elsevier, New York, pp. 77–90.

McCarthy, J. (1961): Computer programs for checking mathematical proofs, *AMS Symp. Recursive Function Theory*, New York.

McCarthy, J. (1962): "LISP 1.5 Programmers Manual," M.I.T. Press, Cambridge, Massachusetts.

McCarthy, J. (1964): A Tough Nut for Proof Procedures, Stanford Artificial Intelligence Project, Memo 16, Stanford Univ., Stanford, California.

Meltzer, B. (1966): Theorem-proving for computers: some results on resolution and renaming, *Comput. J.* **8**, 341–343.

Meltzer, B. (1968a): A new look at mathematics and its mechanization, "Machine Intelligence," Vol. 3 (D. Michie, ed.), American Elsevier, New York, pp. 63–70.

Meltzer, B. (1968b): Some notes on resolution strategies, "Machine Intelligence," Vol. 3 (D. Michie, ed.), American Elsevier, New York, pp. 71–76.

Meltzer, B. (1970a): Generation of hypothesis and theories, *Nature (London)* **225**, 972.

Meltzer, B. (1970b): The semantics of induction and the possibility of complete systems of inductive inference, *Artif. Intelligence* **1**, No. 3, 189–192.

Meltzer, B. (1970c): Power amplification for theorem-provers, "Machine Intelligence," Vol. 5 (B. Meltzer and D. Michie, eds.), American Elsevier, New York, pp. 165–179.

Meltzer, B. (1971): Prolegomener to a theory of efficiency of proof procedures, "Artificial Intelligence and Heuristic Programming" (N. V. Findler and B. Meltzer, eds.), American Elsevier, New York, pp. 15–33.

Mendelson, E. (1964): "Introduction to Mathematical Logic," Van Nostrand–Reinhold, Princeton, New Jersey.

Morris, J. B. (1969): E-resolution: extension of resolution to include the equality, *Proc. Internat. Joint Conf. Artif. Intelligence, Washinton. D.C.. 1969* pp. 287–294.

Nagel, E., and Newman, J. R. (1968): "Gödel's Proof," New York Univ. Press, New York.

Nevins, A. J. (1971): A Human Oriented Logic for Automatic Theorem Proving, Inst. for Management Sci. and Eng., George Washington Univ., Washington, D.C.

Newell, A., Shaw, J. C., and Simon, H. A. (1956): The logic theory machine, *IRE Trans. Information Theory*, **IT-2**, No. 3, 61–79.

Newell, A., Shaw, J., and Simon, H. (1957): Empirical explorations of the logic theory machine, *Proc. West. Joint Comput. Conf.* **15**, 218–239.

Norton, L. (1966): "Adept-A Heuristic Program for Proving Theorems of Group Theory," Ph.D. Thesis, M.I.T., Cambridge, Massachusetts.

Norton, L. M. (1971): Experiments with a heuristic theorem-proving for the predicate calculus with equality, *Artif. Intelligence* **2**, No. 3/4, 261–284.

Pfeiffer, P. E. (1964): "Sets, Events and Switching," McGraw-Hill, New York.

Pietrzykowski, T. (1971): A Complete Mechanization of Second Order Logic, Dept. of Appl. Analysis and Comput. Sci., Univ. of Waterloo at Waterloo, Canada.

Pietrzykowski, T., and Jensen, D. (1972): A Complete Mechanization of W-Order Logic Dept. of Appl. Analysis and Comput. Sci., Univ. of Waterloo at Waterloo, Canada.

Pitrat, J. (1965): Realization of a program which chooses the theorems it proves, *Proc. IFIP Congr. 1965* pp. 324–325.

Pitrat, J. (1966): "Realization de Programmes de Demonstraction de Theorem Utilicant des Methodes Heuristiques," Ph.D. Thesis, Univ. of Paris at Paris, France.

Plotkin, G. D. (1970): A note on inductive generalization, "Machine Intelligence," Vol. 5 (B. Meltzer and D. Michie, eds.), American Elsevier, New York.

Popplestone, R. J. (1967): Beth tree methods in theorem proving, "Machine Intelligence," Vol. 1 (N. L. Collins and D. Michie, eds.), American Elsevier, New York, pp. 31–46.

Pratt, T. W. (1971): Kernel equivalence of programs and proving kernel equivalence and correctness by test cases, *Proc. 2nd Inter. Joint Conf. Artif. Intelligence*, pp. 474–480.

Prawitz, D. (1960): An improved proof procedure, *Theoria* **26**, 102–139.

Prawitz, D. (1967): Completeness and Hauptsatz for second order logic, *Theoria* **33**, 246–254.

Prawitz, D. (1968): Hauptsatz for higher order logic, *J. Symbolic Logic* **33**, 452–457.

Prawitz, D. (1969): Advances and problems, in mechanical proof procedures, "Machine Intelligence," Vol. 4 (B. Meltzer and D. Michie, eds.), American Elsevier, New York, pp. 59–71.

Prawitz, D., Prawitz, H., and Voghera, N. (1960): A mechanical proof procedure and its realization in an electronic computer, *J. ACM* **7**, No. 1–2, 102–128.

Quam. L. H. (1968): "Stanford LISP 1.6 Manual." Stanford Artificial Intelligence Project, Stanford Univ., Stanford, California.

Quine, W. V. (1955): A proof procedure for quantification theory, *J. Symbolic Logic* **20**, No. 2, 141–149.

Quine, W. V. (1959): "Methods of Logic," Holt, New York.

Quinland, J. R., and Hunt, E. B. (1968): A formal deductive problem-solving system, *J. ACM* **15**, No. 4, 625–646.

Reiter, R. (1971): Two results on ordering for resolution with merging and linear format, *J. ACM* **18**, No. 4, 630–646.

Reynolds, J. C. (1968): A generalized resolution principle based upon context-free grammers, *Proc. IFIP Congr. 1968*, pp. 1405–1411.

Reynolds, J. C. (1970): Transformational systems and the algebraic structure of atomic formulas, "Machine Intelligence," Vol. 5 (B. Meltzer and D. Michie, eds.), American Elsevier, New York, pp. 135–151.

Robinson, A. (1957): Proving theorems, as done by man, machine and logician, Summaries of Talks Presented at the Summer Institute for Symbolic Logic, Communications Res. Div., Inst. for Defense Analysis, Princeton, New Jersey, 2nd ed. 1960.

Robinson, A. (1960): On the mechanization of the theory of equations, *Bull. Res. Council Israel*, **9F**, 47–70.

Robinson, A. (1967): A basis for the mechanization of the theory of equations, "Computer Programming and Formal Systems" (P. Braffort and D. Hirshberg, eds.), North-Holland Publ., Amsterdam, pp. 95–99.

Robinson, G. A., and Wos, L. (1969): Paramodulation and theorem proving in first order theories with equality, "Machine Intelligence," Vol. 4 (B. Meltzer and D. Michie, eds.), American Elsevier, New York, pp. 135–150.

Robinson, G. A., and Wos, L. (1970): Axiom systems in automatic theorem proving, *Symp. Automatic Demonstration*, Springer-Verlag, New York.

Robinson, J. A. (1963): Theorem proving on the computer, *J. ACM* **10**, No. 2, 163–174.

Robinson, J. A. (1965a): A machine oriented logic based on the resolution principle, *J. ACM* **12**, No. 1, 23–41.

Robinson, J. A. (1965b): Automatic deduction with hyper-resolution, *Internat. J. Comput. Math.* **1**, 227–234.

Robinson, J. A. (1967): A review of automatic theorem proving, *Proc. Symp. Appl. Math. Amer. Math. Soc.* **19**, 1–18.

Robinson, J. A. (1968a): The generalized resolution principle, "Machine Intelligence," Vol. 3 (D. Michie, ed.), American Elsevier, New York, pp. 77–94.

Robinson, J. A. (1968b): New directions in mechanical theorem proving, *Proc. IFIP Congr. 1968* **1**, 63–68.

Robinson, J. A. (1969): Mechanizing higher order logic, "Machine Intelligence," Vol. 4 (B. Meltzer and D. Michie. eds.). American Elsevier. New York. pp. 151–170.

Robinson, J. A. (1970): A note on mechanizing higher order logic, "Machine Intelligence," Vol. 5 (B. Meltzer and D. Michie, eds.), Amercian Elsevier, New York, pp. 123–133.

Robinson, J. A. (1971a): Computational logical: the unification computation, "Machine Intelligence," Vol. 6 (B. Meltzer and D. Michie, eds.), American Elsevier, New York, pp. 63–72.

Robinson, J. A. (1971b): Building deduction machines, "Artificial Intelligence and Heuristic Programming" (N. V. Findler and B. Meltzer, eds.), American Elsevier. New York, pp. 3–13.

Russell, B. (1908): Mathematical logic as based on the theory of types, "From Frege to Godel: a Source Book to Mathematical Logic" (J. van Heijenoort, ed.), Harvard Univ. Press, Cambridge, Massachusetts, pp. 150–182.

Schutte, K. (1960): Syntactical and semantic properties of simple type theory, *J. Symbolic Logic* **25**, 305–326.

Schoenfield, J. R. (1967): "Mathematical Logic," Addison-Wesley, Reading, Massachusetts.

Sibert, E. E. (1969): A machine oriented logic incorporating the equality relation, "Machine Intelligence," Vol. **4** (B. Meltzer and D. Michie, eds.), American Elsevier, New York, pp. 103–134.

Siklossy, L., and Marinov, V. (1971): Heuristic search vs. exhaustive search, *Proc. 2nd Internat. Conf. Artif. Intelligence London,* pp. 601–607.

Siklossy, L., and Rich, A. (1971): The Logic Theorist Revisited or a Defense of the British Museum Algorithm, Dept. of Comput. Sci., Univ. of Texas at Austin, Texas.

Skolem, T. (1920): Logisch-kombinaterische Untersuchungen über die Erfüllbarkeit oder Beweibarkeit mathematischer Sätze, *Skrifter utgit av Videnskapsselskapet i Kristiania,* No. 4, 4–36.

Slagle, J. R. (1965a): A Proposed Preference Strategy Using Sufficiency Resolution for Answering Question, UCRL-14361, Lawrence Radiation Lab., Livermore, California.

Slagle, J. R. (1965b): A multipurpose, theorem proving, heuristic program that learns, *Proc. IFIP Cong. 1965,* **2**, 323–328.

Slagle, J. R. (1967): Automatic theorem proving with renamable and semantic resolution, *J. ACM* **14**, No. 4, 687–697.

Slagle, J. R. (1970a): Heuristic search programs, "Theoretical Approaches to Non-numerical Problem Solving" (R. Banergi and M. Mesavoric, eds.), Springer-Verlag, New York, pp. 246–273.

Slagle, J. R. (1970b): Interpolation theorem for resolution in lower predicate calculus, *J. ACM* **17**, No. 3, 535–542.

Slagle, J. R. (1971a): An approach for finding C-linear complete inference systems, *J. ACM* (to appear).

Slagle, J. R. (1971b): Automatic Theorem-Proving for the Theories of Partial and Total Ordering, Div. of Comput. Res. and Technol., Nat. Inst. of Health, Bethesda, Maryland.

Slagle, J. R. (1972): Automatic theorem proving with built-in theories including equality, partial ordering and sets, *J. ACM* **19**, No. 1, 120–135.

Slagle, J. R., and Bursky, P. (1968): Experiments with a multipurpose, theorem-proving, heuristic program, *J. ACM* **15**, No. 1, 85–99.

Slagle, J. R., Chang, C. L., and Lee, R. C. T. (1969): Completeness theorems for semantic resolution in consequence finding, *Proc. 1st Internat. Joint Conf. Artif. Intelligence,* pp. 281–285.

Slagle, J. R., and Farrell, C. D. (1971): Experiments in automatic learning for a multipurpose heuristic program, *Comm. ACM* **14**, No. 2, 91–99.

Slagle, J. R., and Koniver, D. (1971): Finding resolution proofs and using duplicate goals in AND/OR trees, *Information Sci.* **4**, No. 4 315–342.

Slagle, J. R., and Norton, L. (1971): Experiments with an Automated Theorem Prover Having Partial Ordering Rules, Div. of Comput. Res. and Tech., Nat. Inst. of Health, Bethesda, Maryland.

Smullyan, R. M. (1963): A unifying principle in quantification theory, *Proc. Nat. Acad. Sci.* **49**, 828–832.

Snyder, D. P. (1971): "Modal Logic and Its Applications," Van Nostrand–Reinhold, Princeton, New Jersey.

Stoll, R. R. (1961): "Sets, Logic and Axiomatic Theories," Freeman, San Francisco, California.

Stoll, R. R. (1963): "Set Theory and Logic," Freeman, San Francisco, California.

Turing, A. M. (1936): On computable numbers, with an application to the entscheindungs-problem, *Proc. London Math. Soc.* **42**, 230–265.

Van Veijenoort, J. (ed.) (1967): "From Frege to Gödel: a Source Book in Mathematical Logic," Harvard Univ. Press, Cambridge, Massachusetts.

von Wright, G. H. (1951): "An Essay in Modal Logic," North Holland Publ., Amsterdam.

Wang, H. (1960a): Proving theorems by pattern recognition I, *Comm. ACM* **3**, No. 3, 220–234.

Wang, H. (1960b): Towards mechanical mathematics, *IBM J. Res. Develop.* **4**, 224–268.

Wang, H. (1961): Proving theorems by pattern recognition II, *Bell Syst. Tech. J.* **40**, 1–41.

Wang, H. (1962): Dominoes and the AEA case of the decision problem, *Symp. Math. Theory Machines*, Brooklyn Polytechnic Inst., pp. 23–56.

Wang, H. (1963): The mechanization of mathematical arguments, *Proc. Symp. Appl. Math.* **15**, 31–40.

Wang, H. (1965a): Games, logic and computers, Scientific American, November, pp. 98–107.

Wang, H. (1965b): Formalization and automatic theorem-proving, *Proc. IFIP Congr., 1965* pp. 51–58.

Wang, H. (1967a): Mechanical mathematics and inferential analysis, "Computer Programming and Formal Systems" (P. Braffort and D. Hirschberg, eds.), North-Holland Publ., Amsterdam, pp. 1–20.

Wang, H. (1967b): Remarks on machine, sets, and the decision problem, "Formal Systems and Recursive Functions" (J. N. Crossley and M. Dummett, eds.), North-Holland Publ., Amsterdam, pp. 304–320.

Weissman, C. (1967): "LISP 1.5 Primer." Dickenson, Belmont, California.

Westrhenen, S. C. (1972): Statistical studies of threshold in classical propositional and first order predicate calculus, *J. ACM* **19**, No. 2, 347–365.

Whitehead, A. N., and Russell, B. (1927): "Principia Mathematica," Cambridge Univ. Press, London and New York.

Wos, L., Carson, D., and Robinson, G. A. (1964): The unit preference stategy in theorem proving, *Proc. AFIPS 1964 Fall Joint Comput. Conf.* **26**, 616–621.

Wos, L., Robinson, G. A., and Carson, D. F. (1965): Efficiency and completeness of the set of support strategy in theorem proving, *J. ACM* **12**, No. 4, 536–541.

Wos, L., Robinson, G. A., and Carson, D. F. (1965): Automatic generation of proofs in the language of mathematics, *Proc. IFIP Congr. 1965* **2**, 325–326.

Wos, L., and Robinson, G. A. (1970): Paramodulation and set of support, *Proc. Symp. Automatic Demonstration, Versailles, France, 1968* Springer-Verlag, New York, pp. 276–310.

Wos, L., Robinson, G. A., Carson, D. F., and Shalla, L. (1967): The concept of demodulation in theorem proving, *J. ACM* **14**, No. 4, 698–709.

Yasuhara, A. (1971): "Recursive Function Theory and Logic," Academic Press, New York.

Yates, R., Raphael, B., and Hart, T. (1970): Resolution graphs, *Artif. Intelligence* **1**, No. 4, 257–290.

PART III

Allen, D. (1972): Derivation of axiomatic definitions of programming languages from algebraic definitions, *Proc. ACM Conf. Proving Assertions about Programs*, pp. 15–26.

Amarel, S. (1968): On representations of problems of reasoning about actions, "Machine Intelligence," Vol. 3 (B. Meltzer and D. Michie, eds.), American Elsevier, New York, pp. 131–171.

Ashcroft, E. A. (1969): Functional Programs as Axiomatic Theories, Center for Computing and Automation, Rep. No. 9, Imperial College, London.

Ashcroft, E. A. (1970): "Mathematical Logic Applied to the Semantics of Computer Programs," Ph.D. Thesis, Imperial College, London.

Ashcroft, E. A. (1972): Program correctness methods and language definitions, *Proc. ACM Conf. Proving Assertions about Programs*, pp. 51–57.

Ashcroft, E. A., and Manna, Z. (1971): Formalization of properties of parallel programs, "Machine Intelligence," Vol. 6 (B. Meltzer and D. Michie, eds.), American Elsevier, New York, pp. 17–41.

Black, F. (1964): A deductive question-answering system, "Semantic Information Processing" (M. Minsky, ed.), M.I.T. Press, Cambridge, Massachusetts, pp. 354–402.

Bliss, K., Chien, R., and Stohl, F. (1971): R2 a natural language question-answering system, *Proc. AFIPS, 1971*, pp. 303–308.

Braffort, F., and Hirshberg, D. (eds.) (1963): "Computer Programming and Formal Systems," North-Holland, Publ., Amsterdam.

Bruce, B. C. (1972): A model for temporal references and its applications in a question answering program, *Artif. Intelligence* 3, No. 1, 1 26.

Burstall, R. M. (1968): Semantics of assignment, "Machine Intelligence," Vol. 2 (E. Dale and D. Michie, eds.), American Elsevier, New York, pp. 3 20.

Burstall, R. M. (1969): Proving properties of programs by structural induction, *Comput. J.* 12, 41 48.

Burstall, R. M. (1970): Formal description of program structure and semantics in first order logic, "Machine Intelligence," Vol. 5 (B. Meltzer and D. Michie, eds.), American Elsevier, New York, pp. 79–98.

Burstall, R. M. (1972): An algebraic description of programs with assertions, verification and simulation, *Proc. ACM Conf. Proving Assertions about Programs*, pp. 7 14.

Burstall, R. M., and London, R. J. (1969): Programs and their proofs: an algebraic approach, "Machine Intelligence," Vol. 4 (B. Meltzer and D. Michie, eds.), American Elsevier, New York, pp. 17 43.

Cadiou, J. M. (1972): "Recursive Definitions of Partial Functions and their Computations," Ph.D. Thesis, Stanford Univ., Stanford, California.

Chang, C. L., Lee, R. C. T., and Dixon, J. (1971): Specialization of Programs by Theorem-Proving, Div. of Comput. Res. and Technol., Nat. Inst. of Health, Bethesda, Maryland.

Cooper, D. C. (1967): Mathematical proofs about computer programs, "Machine Intelligence," Vol. 1 (N. L. Collins and D. Michie, eds.), American Elsevier, New York, pp. 17–30.

Cooper, D. C. (1968): Some transformations and standard forms of graphs with applications to computer programs, "Machine Intelligence," Vol. 2 (E. Dale and D. Michie, eds.), American Elsevier, New York, pp. 21 32.

Cooper, D. C. (1969): Program scheme and second order logic, "Machine Intelligence," Vol. 4 (B. Meltzer and D. Michie, eds.), American Elsevier, New York, pp. 3–15.

Cooper, D. C. (1971): Programs for mechanical program verification, "Machine Intelligence," Vol. 6 (B. Meltzer and D. Michie, eds.), American Elsevier, New York, pp. 43 59.

Cooper, W. S. (1964): Fact retrieval and deductive question answering information retrieval systems, *J. ACM* **11**, No. 2, 117–137.

Darlington, J. L. (1969): Theorem proving and information retrieval, "Machine Intelligence," Vol. 4 (B. Meltzer and D. Michie, eds.), American Elsevier, New York, 173–181.

Davidson, D. (1967): The logical form of action sentences, "The Logic of Decision and Action" (N. Rescher, ed.), Univ. Pittsburgh Press, Pittsburgh, Pennsylvania.

Dijkstra, E. W. (1968): A constructive approach to the problem of program correctness, B.I.T. **8**, No. 3, 174–186.

Evans, C. O. (1967): States, activities and performances, *Australian J. Philos.* **45**, 293–308.

Fikes, R. E., and Nilsson, N. J. (1971): STRIPS: a new approach to the application of theorem proving to problem solving, *Proc. 2nd Internat. Joint Conf. Artif. Intelligence, London*, pp. 608–620.

Florentin, J. J. (1968): Language definition and compiler validation, "Machine Intelligence," Vol. 3 (D. Michie, ed.), American Elsevier, New York, pp. 33–41.

Floyd, R. W. (1967): Assigning meaning to programs, *Proc. Symp. Appl. Math.* **19**, 19–32.

Floyd, R. W. (1971): Toward interactive design of correct programs, Stanford Artificial Intelligence Proj., Memo AIM-150, Stanford Univ., Stanford, California.

Futamura, Y. (1971): Partial evaluation of computer programs: an approach to a compiler-compiler, *J. Inst. of Electron. Commun. Eng. Japan.*

Good, D. I. (1970): "Toward a Man-machine System for Proving Program Correctness," Ph.D. Thesis, Univ. of Wisconsin at Madison, Wisconsin.

Green, B. Jr., Wolf, A. K., Chomsky, C., and Laughary, K. (1963): Baseball: an automatic question answerer, "Computers and Thought" (E. A. Feigenbaum and J. Feldman, eds.), McGraw-Hill, New York, pp. 207–216.

Green, C. (1969a): Theorem proving by resolution as a basis for question-answering systems, "Machine Intelligence," Vol. 4 (B. Meltzer and D. Michie, eds.), American Elsevier, New York, pp. 183–205.

Green, C. (1969b): "The application of Theorem Proving to Question Answering Systems," Ph.D. Thesis, Stanford Univ. at Stanford, California.

Green, C. (1969c): Application of theorem proving to problem solving, *Proc. 1st Internat. Joint Conf. Artif. Intelligence* pp. 219–239.

Green, C., and Raphael, B. (1968): The use of theorem proving techniques in question answering systems, *Proc. 23rd Nat. Conf. ACM*, Brandon Systems Press, Princeton, New Jersey, pp. 169–181.

Hayes, P. (1969): A Machine-Oriented Formulation of the Extended Functional Calculus, Stanford Artif. Intelligence Proj. Memo 86, Stanford Univ., Stanford, California.

Hayes, P. (1970): Robotologic, "Machine Intelligence," Vol. 5 (B. Meltzer and D. Michie, eds.), American Elsevier, New York, pp. 533–554.

Hayes, P. (1971): A logic of actions, "Machine Intelligence," Vol. 6 (B. Meltzer and D. Michie, eds.), American Elsevier, New York, pp. 495–520.

Hewitt, C. (1969): Planner: a language for proving theorems in robots, *Proc. 1st Internat. Joint Conf. Artif. Intelligence* pp. 295–302.

Hoare, C. A. R. (1969): An axiomatic basis for computer programs, *Comm. ACM* **12**, No. 10, 576–580.

Hull, T. E., Enright, W. H., and Sedgwick, A. E. (1972): The correctness of numerical algorithms, *Proc. ACM Conf. Proving Assertions About Programs*, pp. 66–73.

Ianov, In (1960): The logical schemes of algorithms, "Problems of Cybernetis," Vol. 1, Pergamon, Oxford, pp. 82–140 (English translation).

Kaplan, D. M. (1968): Some completeness results in the mathematical theory of computation, *J. ACM* **15**, No. 1, 124–134.

King, J. C. (1969): "Program Verifier," Ph.D. Thesis, Carnegie-Mellon Univ., Pittsburgh, Pennsylvania.

King, J. C. (1971): Proving programs to be correct, *IEEE Trans. Comput.* **C-20**, No. 11, 1331–1336.

King, J. C., and Floyd, R. W. (1970): Interpretation oriented theorem prover over integers, *2nd Ann. ACM Symp. Theory Comput. Northampton, Mass.*, pp. 169–170.

Kuhns, J. L. (1967): Answering Questions by Computer: A Logical Study, Memo. RM-5428-PR, Dec. 1967, The Rand Corp., Sanata Monica, California.

Landin, P. J. (1964): The mechanical evaluation of expressions, *Comput. J.* **6**, 1964, pp. 308–320.

Lee, R. C. T., and Chang, C. L. (1971): Program Analysis and Theorem Proving, Div. of Comput. Res. and Technol., Nat. Inst. of Health, Bethesda, Maryland.

Lee, R. C. T., Chang, C. L., and Waldinger, R. J. (1972): An Improved Program-Synthesizing Algorithm and its Correctness, Div. of Comput. Res. and Technol., Nat. Inst. of Health, Bethesda, Maryland.

Lindsay, R. K. (1963): Inferential memory as the basis of machines which understand natural language, "Computers and Thought" (E. Feigenbaum and J. Feldman, eds.), McGraw-Hill, New York, pp. 217–236.

London, R. (1970): Bibliography on proving the correctness of computer programs, "Machine Intelligence," Vol. 5 (B. Meltzer and D. Michie, eds.), American Elsevier, New York, pp. 569–580.

Luckham, D., and Nilsson, N. J. (1971): Extracting information from resolution proof trees, *Artif. Intelligence* **2**, No. 1, 27–54.

Luckham, D., Park, D. M. R., and Paterson, M. S. (1967): Formalized computer programs, *J. Comput. and Syst. Sci.* **4**, No. 3, 220–249.

Manna, Z. (1969a): The correctness of programs, *J. Comput. and Syst. Sci.* **3**, No. 2, 119–127.

Manna, Z. (1969b): Properties of programs and the first order predicate calculus, *J. ACM* **16**, No. 2, 244–255.

Manna, Z. (1970a): The correctness of non-deterministic programs, *Artif. Intelligence* **1**, No. 1, 1–26.

Manna, Z. (1970b): Mathematical theory of partial correctness, *Symp. Semantics Algorithmic Languages* (E. Engler, ed.), Springer-Verlag, New York.

Manna, Z., and McCarthy, J. (1970): Properties of programs and partial function logic, "Machine Intelligence," Vol. 5 (B. Meltzer and D. Michie, eds.), American Elsevier, New York, pp. 27–38.

Manna, Z., Ness, S., and Vuillemin, J. (1972): Inductive methods for proving properties of programs, *Proc. ACM Conf. Proving Assertions About Programs*, pp. 27–50.

Manna, Z., and Pnueli, A. (1968): The Validity Problem of the 91-Function, Stanford Artif. Intelligence Project, Memo 68, Stanford Univ. at Stanford, California.

Manna, Z., and Pnueli, A. (1970): Formalization of properties of functional programs, *J. ACM* **17**, No. 3, 555–569.

Manna, Z., and Waldinger, R. (1971): Toward automatic program synthesis, *Comm. ACM* **14**, No. 3, 151–165.

McCarthy, J. (1961): Programs with common sense, *Proc. Symp. Mechanization Thought Process* **1**, H.M.S.O., London, pp. 75–84.

McCarthy, J. (1962): Towards a mathematical science of computation, *Proc. IFIP Congr. 62*, pp. 21–28.

McCarthy, J. (1963a): Predicate Calculus with "Undefined" as a Truth-Value, Stanford Artif. Intelligence Project, Memo 1, Stanford Univ. at Stanford, California.

McCarthy, J. (1963b): Situations, Actions and Causal Laws, Memo, Stanford Artif. Intelligence Proj. Stanford Univ., Stanford, California.

McCarthy, J. (1965): Problems in the theory of computation, *Proc. IFIP Congress 65*, pp. 219–222.

McCarthy, J. (1967): A basis for a mathematical theory of computation, "Computer Programming and Formal Systems" (P. Braffort and D. Michie, eds.), North-Holland Publ., Amsterdam, pp. 33–70.

Milner, R. (1972): Implementation and applications of Scott's logic for computable functions, *Proc. ACM Conf. Proving Assertions About Programs*, pp. 1–6.

Minker, J., and Sable, J. D. (1971): Relational Data Systems Study, Dept. of Comput. Sci., Univ. of Maryland, College Park, Maryland.

Morgan, C. G. (1971): Hypothesis generating by machine, *Artif. Intelligence* **2**, No. 2, 179–187.

Naur, P. (1966): Proofs of algorithms by general snapshots, *B.I.T.* **6**, 310–316.

Painter, J. A. (1967): Semantic Correctness of a Compiler for an Algol-like Language, Stanford Artif. Intelligence Project, Memo 44, Stanford Univ., Stanford, California.

Park, D. (1970): Fixpoint induction and proofs of programs properties, "Machine Intelligence," Vol. 5 (B. Meltzer and D. Michie, eds.), American Elsevier, New York, pp. 59–78.

Paterson, M. (1967): "Equivalence Problems in a Model of Computation," Ph.D. Thesis, Cambridge University, 1967.

Paterson, M. S. (1968): Program schemata, *Machine Intelligence 3*. (D. Michie, ed.), American Elsevier, New York, pp. 19–31.

Quinlan, J. R., and Hunt. E. B. (1968): A formal deductive problem-solving system, *J. ACM* **25**, No. 4, 625–646.

Raphael, B. (1964): A computer program which understands, *Proc. AFIPS Fall Joint Comput. Conf. 1964* pp. 577–589.

Rescher, N. (1964): "Hypothetical Reasoning," North-Holland Publ., Amsterdam.

Rescher, N. (1967): "The Logic of Decision and Actions," Univ. of Pittsburgh Press, Pittsburgh, Pennsylvania.

Safier, F. (1963): The Mikado as an Advice Taker Problem, Memo, Stanford Artif. Intelligence Project, Stanford Univ. at Stanford, California.

Sandewall, E. (1971a): Representing language information in predicate calculus, "Machine Intelligence," Vol. 6 (B. Meltzer and D. Michie, eds.), American Elsevier, New York, pp. 255–280.

Sandewall, E. (1971b): Formal methods in the design of question-answering systems, *Artif. Intelligence* **2**, No. 2, 129–146.

Sandewall, E. (1972): PCF-2, a First Order Calculus for Expressing Conceptual Information, Dept. of Comput. Sci., Uppsala Univ. at Uppsala, Sweden.

Scott, D. (1970): Outline of a mathematical theory of computation, *Proc. 4th Princeton Conf. Inform. Sci. Syst.* pp. 169–178.

Simmons, R. (1965): Answering English questions by computer, a survey, *Comm. ACM* **8**, No. 1, pp. 53–70.

Simon, H. A. (1963): Experiments with a heuristic compiler, *J. ACM* **10**, No. 4, 493–506.

Simon, H. A. (1967): The logic of heuristic decision making, "The Logic of Decision and Action" (N. Rescher, ed.), Univ. of Pittsburgh Press, Pittsburgh, Pennsylvania.

Slagle, J. R. (1965): Experiments with a deductive question-answering program, *Comm. ACM* **8**, 792–798.

Snyder, D. P. (1971): "Modal Logic and Its Applications," Van Nostrand–Reinhold, Princeton, New Jersey.

Von Wright, G. H. (1968): "An Essay in Deontic Logic and the General Theory of Actions," North-Holland Publ., Amsterdam.

Vuillemin, J. (1972): "Proof Techniques for Recursive Programs," Ph.D. Thesis, Stanford Univ. at Stanford, California.

Waldinger, R. (1969): "Constructing Programs Automatically Using Theorem Proving," Ph.D. Thesis, Carnegie-Mellon Univ., Pittsburgh, Pennsylvania.

Waldinger, R., and Lee, R. C. T. (1969): PROW: a step toward automatic program writing, *Proc. 1st Internat. Joint Conf. Artif. Intelligence*, pp. 241–252.

Index

Computer Science and Applied Mathematics

A SERIES OF MONOGRAPHS AND TEXTBOOKS

Editor
Werner Rheinboldt
University of Pittsburgh

HANS P. KÜNZI, H. G. TZSCHACH, and C. A. ZEHNDER. Numerical Methods of Mathematical Optimization: With ALGOL and FORTRAN Programs, Corrected and Augmented Edition

AZRIEL ROSENFELD. Picture Processing by Computer

JAMES ORTEGA AND WERNER RHEINBOLDT. Iterative Solution of Nonlinear Equations in Several Variables

AZARIA PAZ. Introduction to Probabilistic Automata

DAVID YOUNG. Iterative Solution of Large Linear Systems

ANN YASUHARA. Recursive Function Theory and Logic

JAMES M. ORTEGA. Numerical Analysis: A Second Course

G. W. STEWART. Introduction to Matrix Computations

CHIN-LIANG CHANG AND RICHARD CHAR-TUNG LEE. Symbolic Logic and Mechanical Theorem Proving

C. C. GOTLIEB AND A. BORODIN. Social Issues in Computing

ERWIN ENGELER. Introduction to the Theory of Computation

F. W. J. OLVER. Asymptotics and Special Functions

DIONYSIOS C. TSICHRITZIS AND PHILIP A. BERNSTEIN. Operating Systems

ROBERT R. KORFHAGE. Discrete Computational Structures

PHILIP J. DAVIS AND PHILIP RABINOWITZ. Methods of Numerical Integration

A. T. BERZTISS. Data Structures: Theory and Practice, Second Edition

N. CHRISTOPHIDES. Graph Theory: An Algorithmic Approach

ALBERT NIJENHUIS AND HERBERT S. WILF. Combinatorial Algorithms

AZRIEL ROSENFELD AND AVINASH C. KAK. Digital Picture Processing

SAKTI P. GHOSH. Data Base Organization for Data Management

DIONYSIOS C. TSICHRITZIS AND FREDERICK H. LOCHOVSKY. Data Base Management Systems

JAMES L. PETERSON. Computer Organization and Assembly Language Programming

WILLIAM F. AMES. Numerical Methods for Partial Differential Equations, Second Edition